The Food Processing Industry's 200th Anniversary: Bioactive Compounds, Antioxidants, and Potential Health Benefits

The Food Processing Industry's 200th Anniversary: Bioactive Compounds, Antioxidants, and Potential Health Benefits

Guest Editors

Vassilis Athanasiadis
Theodoros G. Chatzimitakos

MDPI

Basel • Beijing • Wuhan • Barcelona • Belgrade • Novi Sad • Cluj • Manchester

Guest Editors
Vassilis Athanasiadis
University of Thessaly
Karditsa
Greece

Theodoros G. Chatzimitakos
University of Thessaly
Karditsa
Greece

Editorial Office
MDPI AG
Grosspeteranlage 5
4052 Basel, Switzerland

This is a reprint of the Special Issue, published open access by the journal *International Journal of Molecular Sciences* (ISSN 1422-0067), freely accessible at: https://www.mdpi.com/journal/ijms/special_issues/0YLSQP87GC.

For citation purposes, cite each article independently as indicated on the article page online and as indicated below:

Lastname, A.A.; Lastname, B.B. Article Title. *Journal Name* **Year**, *Volume Number*, Page Range.

ISBN 978-3-7258-2767-1 (Hbk)
ISBN 978-3-7258-2768-8 (PDF)
https://doi.org/10.3390/books978-3-7258-2768-8

Contents

International Journal of *Molecular Sciences*

MDPI

Article

Oilseed Cakes: A Promising Source of Antioxidant, and Anti-Inflammatory Agents—Insights from *Lactuca sativa*

Mayye Majed [1], Amal A. Galala [1,2], Mohamed M. Amer [1], Dirk Selmar [3,*] and Sara Abouzeid [1,3,*]

1 Pharmacognosy Department, Faculty of Pharmacy, Mansoura University, Mansoura 35516, Egypt; mayye.majed@mans.edu.eg (M.M.); amal_galala@yahoo.com (A.A.G.); pharmamer_47@yahoo.com (M.M.A.)
2 Pharmacognosy Department, Faculty of Pharmacy, Horus University in Egypt (HUE), New Damietta 34517, Egypt
3 Institute for Plant Biology, Technical University of Braunschweig, Mendelssohnsstr. 4, 38106 Braunschweig, Germany
* Correspondence: d.selmar@tu-bs.de (D.S.); sara86@mans.edu.eg (S.A.)

Abstract: This study evaluated the antioxidant and antibacterial properties of methanolic extracts derived from oilseed cakes of *Lactuca sativa* (lettuce), *Nigella sativa* (black seed), *Eruca sativa* (rocket), and *Linum usitatissimum* (linseed). Lettuce methanolic extract showed the highest potential, so it was selected for further investigation. High-performance liquid chromatography (HPLC-DAD) analysis and bioassay-guided fractionation of lettuce seed cake extract led to the isolation of five compounds: 1,3-propanediol-2-amino-1-(3′,4′-methylenedioxyphenyl) (**1**), luteolin (**2**), luteolin-7-O-β-D-glucoside (**3**), apigenin-7-O-β-D-glucoside (**4**), and β-sitosterol 3-O-β-D-glucoside (**5**). Compound (**1**) was identified from *Lactuca* species for the first time, with high yield. The cytotoxic effects of the isolated compounds were tested on liver (HepG2) and breast (MCF-7) cancer cell lines, compared to normal cells (WI-38). Compounds (**2**), (**3**), and (**4**) exhibited strong activity in all assays, while compound (**1**) showed weak antioxidant, antimicrobial, and cytotoxic effects. The anti-inflammatory activity of lettuce seed cake extract and compound (**1**) was evaluated in vivo using a carrageenan-induced paw oedema model. Compound (**1**) and its combination with ibuprofen significantly reduced paw oedema, lowered inflammatory mediators (IL-1β, TNF-α, PGE2), and restored antioxidant enzyme activity. Additionally, compound (**1**) showed promising COX-1 and COX-2 inhibition in an in vitro enzymatic anti-inflammatory assay, with IC_{50} values of 17.31 ± 0.65 and 4.814 ± 0.24, respectively. Molecular docking revealed unique interactions of compound (**1**) with COX-1 and COX-2, suggesting the potential for targeted inhibition. These findings underscore the value of oilseed cakes as a source of bioactive compounds that merit further investigation.

Keywords: oilseed cake; lettuce; antioxidant; antimicrobial assay; cytotoxicity; anti-inflammatory

Citation: Majed, M.; Galala, A.A.; Amer, M.M.; Selmar, D.; Abouzeid, S. Oilseed Cakes: A Promising Source of Antioxidant, and Anti-Inflammatory Agents—Insights from *Lactuca sativa*. *Int. J. Mol. Sci.* **2024**, *25*, 11077. https://doi.org/10.3390/ijms252011077

Academic Editors: Vassilis Athanasiadis and Theodoros G. Chatzimitakos

Received: 13 September 2024
Revised: 12 October 2024
Accepted: 12 October 2024
Published: 15 October 2024

1. Introduction

Oilseed crops are rich and well-known sources of nutrients and diverse biologically active compounds [1]. Cold-pressed seed oils are mostly utilized as dietary supplements and as ingredients in different skin care products. They are reliable sources of compounds that are medicinally and biologically active [2]. Worldwide oil plant production is expanding as oils extracted from seeds have become increasingly popular as food additives in recent years [3]. Moreover, oilseed meals/cakes, which are the remaining after the oil pressing process, are considered agro-industrial waste. Recently, they have been used in the production of biodiesel as well as animal feed [4]. Flaxseed cakes have undergone engine testing as a substitute feedstock for biodiesel, indicating that the future output of this oil cake might be anticipated to rise [5]. There are limited phytochemical and biological studies of seed cakes, although they are rich sources of carbohydrates, proteins, minerals, fibers, and certain lipids, as well as phytochemicals like phenolics (e.g., *p*-hydroxybenzoic, syringic, and *p*-coumaric acids) and flavonoids [6]. In this context, oilseed cakes are a very good example of agro-industrial waste that needs more attention in order to provide secondary metabolites with implementation in the medical and business sectors. Accordingly, it encouraged us to do further phytochemical and biological investigations of various medicinally important oilseed cakes (e.g., lettuce, flaxseed, black seed, and rocket seed meals).

Lettuce, black seed, rocket, and flaxseed seeds contain different compounds under the classes of flavonoids and phenolic acids, which are known and reported for their antioxidant activity [7–9]. Moreover, lettuce and rocket seeds were reported to have analgesic and anti-inflammatory activities [9,10]. Additionally, black seed was stated for various biological activities such as antioxidant, antibacterial, and neuroprotective effects [11]. In vitro screening of flaxseed demonstrated potentials for antibacterial, antioxidant, anti-diabetic, and anti-inflammatory properties [12,13]. Concerns about health and safety have led to the use of natural antioxidants in the food sector as a substitute for synthetic ones (e.g., phenolic compounds). Natural phenolic compounds are anticipated to have a steady increase in interest since many of these molecules have been shown to exhibit a wide range of biological significance, including antimutagenic, antitumor, having protective properties against inflammations, platelet aggregation inhibitory, neuroprotective, antimicrobial, and many more [14]. Consequently, adding these substances to everyday products would stabilize food as well as cosmetic products and provide significant health benefits to customers.

In this study, the antibacterial and antioxidant activities of four seed cake extracts (e.g., lettuce, flaxseed, black seed, and rocket) were examined. Moreover, further chromatographic and anti-inflammatory investigations were performed on the seed cake with the highest bioactivity.

2. Results and Discussion

Agro-industrial wastes are a hidden treasure of valuable bioactive compounds that require more investigation. Oilseed cakes, which are leftovers from the oil extraction of many seeds and fruits, are a very good example of plant waste that needs more attention as a source of bioactive compounds. They occur as by-products at the end of the pressing process, and they are rich in some major and minor components. In addition, oilseed cakes contain a high percentage of nutrients besides antioxidant and antimicrobial bioactivity and thus could be used as nutraceuticals and functional foods [15,16]. In our study, we focused on antioxidant and antimicrobial screening of the different seed cakes/meals. Furthermore, the highest bioactive seed cake was subjected to further bio-guided isolation of active constituents.

2.1. Anti-Oxidant and Antimicrobial Screening of Total Methanolic Extracts from Different Oilseed Cakes

Lettuce, black seed, rocket, and flaxseed seed contain different compounds under the class flavonoids and phenolic acids, which are known and reported for their antioxidant and antimicrobial activity [7–9]. The antimicrobial and antioxidant properties of the seed cakes' methanolic extract were assessed at 1 mg/mL. Lettuce seed cake extract showed the highest anti-oxidant activity with 36.1% inhibition, followed by black seed 17.3%, rocket 14%, and flaxseed 10.5%, respectively, compared to ascorbic acid standard 88.1% (Figure S1). Whilst the only lettuce seed cake extracts showed antimicrobial activities against *S. aureus*, *E. coli*, and *C. albicans* (Table S1). The highest inhibitory activity of lettuce was observed against *C. albicans*, followed by *S. aureus* and *E. coli*, with a zone of inhibitions at 6.83 ± 0.29, 3.75 ± 0.25, and 2.8 ± 0.29 mm, respectively. In addition, flaxseed and rocket seed cakes showed low antimicrobial activities only against *S. aureus* (1.75 ± 0.25, 1.8 ± 0.25 mm) and *C. albicans* (4.75 ± 0.25, 5.85 ± 0.13 mm), respectively. Black seed cake could not inhibit the growth of all tested microorganisms (Table S1). It was noticeable that lettuce extract was the only extract among the tested ones to give broad-spectrum activity against *S. aureus*, *E. coli*, and *C. albicans*, while other extracts gave activity against only one organism or two compared to ampicillin and clotrimazole standards. Accordingly, oilseed cakes represent a promising agro-industrial waste with considerable biological activities. It has been reported that oilseed cakes from various plants, such as canola, sunflower, and mustard, exhibit antioxidant and antimicrobial properties, likely due to their phenolic and flavonoid content [17]. Furthermore, *Jatropha curcas* seed cake could be combined with commercial antibiotics to enhance their effectiveness, offering a potential solution for treating clinical infections, including those caused by multidrug-resistant bacteria [18].

2.2. Antioxidant and Antimicrobial Activities of Different Fractions of Lettuce Seed Cake Extract

Based on the previous activity results, the most promising cake was selected for further biological and phytochemical investigation. As lettuce methanolic extract showed dual activity, and due to the scarcity of previous work on this seed and its waste, it was a valuable candidate for further phytochemical and biological investigations. Lettuce seed cake methanolic extract was subjected to fractionation, which afforded petroleum ether, methylene chloride, and ethyl acetate fractions. Ethyl acetate fraction showed the highest antioxidant activity, followed by the methylene chloride fraction with IC_{50} 27.13 $\pm$ 0.15 and 41.93 $\pm$ 0.21, respectively, compared to standard ascorbic acid 29.47 $\pm$ 0.17 (Figure 1). Moreover, methylene chloride and ethyl acetate fractions showed promising antimicrobial inhibition activity against the tested microorganisms with comparison with antibacterial ampicillin and antifungal clotrimazole standards (Table S1). The comparison with the standard revealed that the highest activity in terms of percent activity index was observed for the methylene chloride fraction against *S. aureus* (81.6), followed by *C. albicans* (65.4), and *E. coli* (36.26) (Table S1).

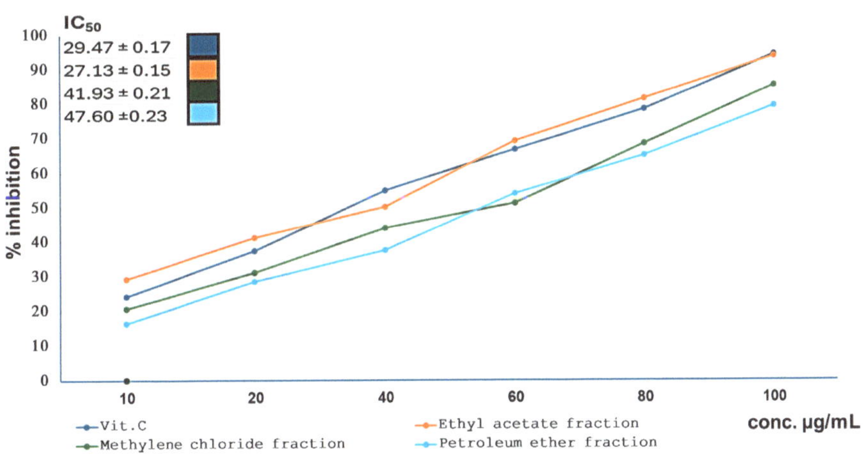

Figure 1. IC_{50} of different fractions of lettuce seedcake extract in ABTS assay using ascorbic acid as standard.

2.3. Bio-Guided Isolation of Active Constituents from L. sativa Seedcake Active Fractions

Using reversed-phase HPLC fitted with a diode array detector, the relevant extract was fractionated to determine the phytochemical components of lettuce seed cake extract that are in charge of the antioxidant and antibacterial actions. It was observed that luteolin glucoside represents the major compound (Figure 2). This could be easily deduced by comparing the retention time and UV spectrum of standard luteolin glucoside (Figure S2). In addition, the less polar compound that was eluted at Rt 26 mint has the same UV spectrum as luteolin glucoside (Figure S2). This compound was tentatively identified as an aglycon part of luteolin glucoside, e.g., luteolin. However, reliable identification of these compounds requires much more information, and this cannot be achieved before the isolation and spectroscopic identification by NMR techniques. In the same manner, the identification of the other peaks displayed in Figure 2 also required much more information.

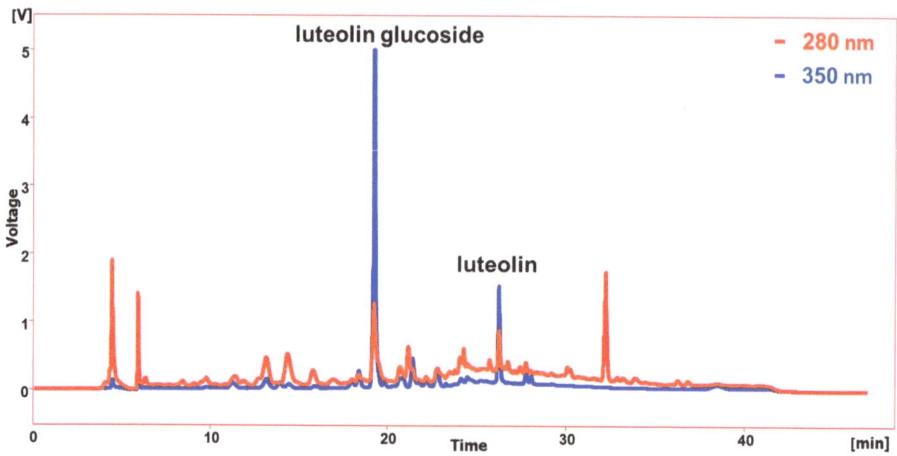

Figure 2. HPLC of the major compounds determined in lettuce seed cake extract. Compounds were monitored using a photo diode array (PDA) detector at 280, 350 nm.

Accordingly, methylene chloride and ethyl acetate fractions were subjected to normal silica column chromatography, which afforded five compounds (Figure 3). The compounds were identified depending on NMR data as 1,3-propanediol-2-amino-1-(3′,4′-methylenedioxyphenyl) (**1**) [19], luteolin (**2**) [20], luteolin 7-O-β-D-glucoside (**3**) [21], apigenin 7-O-β-D-glucoside (**4**) [22], and β-sitosterol 3-O-β-D-glucoside (**5**). Compound (**5**) was characterized upon comparison with an authentic standard sample. For more details of structure elucidation, see the Supporting Information. Luteolin (**2**) and luteolin 7-O-β-D glucoside (**3**) were previously reported to be isolated from lettuce [23], however, up to our knowledge, this is the first time apigenin 7-O-β-D glucoside (**4**) has been reported from *L. sativa* seed cake. It is worth mentioning that compound (**1**) is the first time to be isolated from *Lactuca* species, the second time from the Asteraceae family [19], and the third time from nature [24].

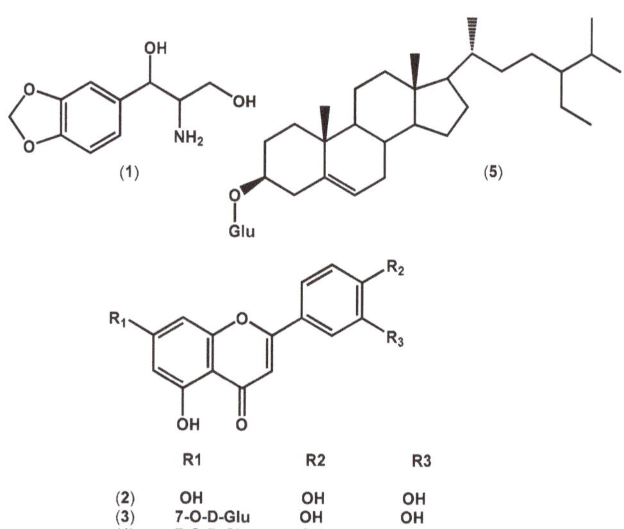

Figure 3. Structures of the isolated compounds from *Lactuca sativa* seedcake active fractions: 1,3-propanediol-2-amino-1-(3′,4′-methylenedioxyphenyl) (**1**), luteolin (**2**), luteolin 7-O-β-D glucoside (**3**), apigenin 7-O-β-D glucoside (**4**), and β-sitosterol 3-O-β-D-glucoside (**5**).

2.4. Biological Evaluation of the Isolated Compounds

The isolated compounds were evaluated for their antioxidant, antimicrobial, and cytotoxicity. Compounds (**3**) and (**4**) showed strong antioxidant activity of IC_{50} 35.18 ± 0.18 and 51.27 ± 0.25, respectively. Meanwhile, compound (**2**) showed moderate activity with an IC_{50} 60.49 ± 0.29. In contrast, compounds (**1**) and (**5**) showed relatively weaker antioxidant activity compared to the ascorbic acid standard (Figure 4). In addition, compounds (**2**) and (**3**) exhibited promising antimicrobial activity throughout the tested compounds, with the highest percent activity index against *S. aureus* (71.7, 61.9), followed by *P. aeruginosa* (56.6, 51.5), *C. albicans* (40.5, 36.7), and *E. coli* (37.8, 33.8), respectively (Table 1). In addition, compounds (**4**) and (**5**) also possess antimicrobial activity but with a much lower percentage, especially against *S. aureus*, with percent activity 50 and 49.6, respectively. On the other hand, compound (**1**) showed weak anti-microbial activity compared to ciprofloxacin and clotrimazole standards (Table 1). The potential antioxidant and antimicrobial activity of methylene chloride and ethyl acetate fractions could be attributed to the content of compounds (**2**) and (**3**) in these active fractions. There are previous studies that reported both the antioxidant and antimicrobial activities of luteolin and its glucoside [25–27]. Accordingly, they are the major compounds responsible for the antioxidant and antimicrobial capacity of *L. sativa* seed cake.

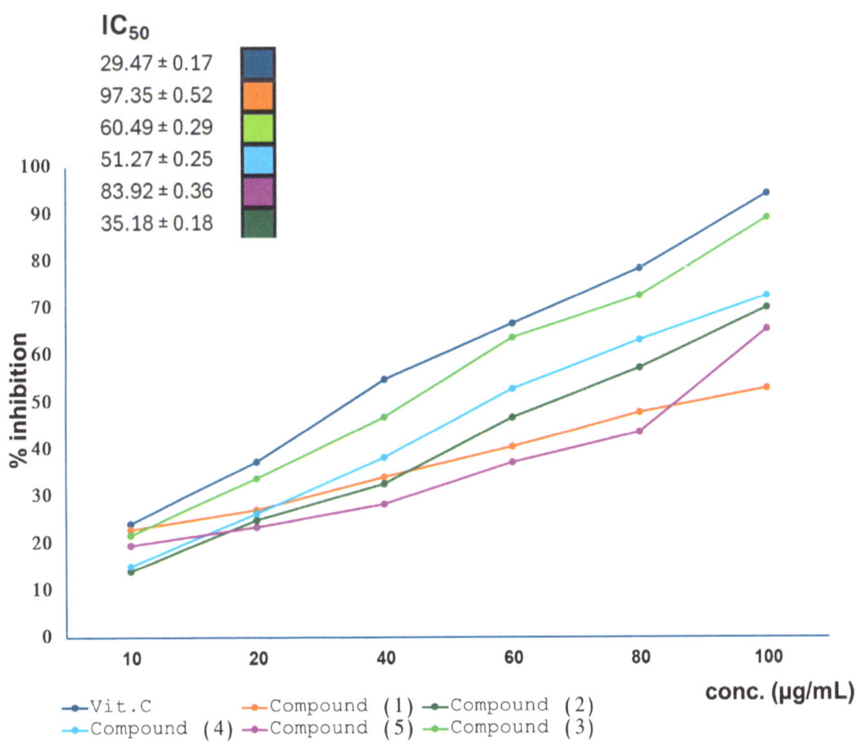

Figure 4. IC_{50} of isolated compounds from lettuce seedcake in ABTS assay using ascorbic acid as standard.

Table 1. Antimicrobial activity index (mean ± SE) of compounds isolated from lettuce seedcake extract (1 mg/mL).

Compound No.	*E. coli*		*P. aeruginosa*		*S. aureus*		*C. Albicans*	
	Diameter of Inhibition Zone (mm)	%Activity Index	Diameter of Inhibition Zone (mm)	%Activity Index	Diameter of Inhibition Zone (mm)	%Activity Index	Diameter of Inhibition Zone (mm)	%Activity Index
1	NA *	----	3.8 ± 0.2	16.5	6.9 ± 0.1	28.9	NA *	----
2	9.8 ± 0.2	37.8	12.9 ± 0.1	56.6	17 ± 0.1	71.7	10.7 ± 0.15	40.5
3	8.8 ± 0.15	33.8	11.77 ± 0.25	51.5	14.83 ± 0.15	61.9	9.87 ± 0.15	36.7
4	6.8 ± 0.15	26.1	9.8 ± 0.15	42.8	11.87 ± 0.15	50	5.95 ± 0.07	22.1
5	5.1 ± 0.1	19.6	7.9 ± 0.1	34.5	10.83 ± 0.2	49.6	4.9 ± 0.1	18.2
Ciprofloxacin	26 ± 0.3	100	22.87 ± 0.15	100	23.9 ± 0.1	100	NA *	----
Clotrimazole	NA *	----	NA *	----	----	----	26.9 ± 0.1	100

* NA = No Activity.

Moreover, compounds (**2**) and (**3**) demonstrated potent cytotoxic action against the cancer cell lines (HepG-2) and (MCF-7) with IC_{50} (19.50 ± 1.3, 11.63 ± 0.9), and (24.83 ± 1.9, 17.79 ± 1.3), respectively. Compared to compounds (**5**) and (**1**), which had minimal anticancer action, compound (**4**) demonstrated moderate efficacy with IC_{50} (43.35 ± 2.5, 39.79 ± 2.2) (Table 2). To investigate the safety and selectivity of the isolated compounds, the normal lung fibroblast cell line (WI-38) was used in the assessment process. In comparison to ordinary doxorubicin, which had selectivity indices of 1.5 for (HepG2) and 1.6 for (MCF-7) cell lines, respectively, compounds (**2**) and (**3**) showed greater selectivity indices (3, 2) for the (HepG2) cell line and (5.1, 2.8) for the (MCF-7) cell line, indicating higher safety profiles as anticancer drugs. On HepG2 and MCF-7 cell lines, compounds (**2**) and (**3**) exhibited cytotoxic action while maintaining a reasonable level of safety. There are also previous studies that reported the anticancer potential of the bioactive molecule luteolin and its analogs [28]. On the other hand, compound (**1**) cytotoxic power was very weak and almost nontoxic, which may lead to the possibility of anti-inflammatory potential. Compound (**1**) was reported before to have moderate inhibitory activities against SARS-CoV-2 main protease Mpro or 3CLpro [19], also showed strong inhibitory against IL-6 production in TNF-a-stimulated MG-63 cells [24], which supports our vision that compound (**1**) might be a potential anti-inflammatory agent.

Table 2. Cytotoxic activity of isolated compounds from lettuce seedcake against (HepG2) and (MCF-7) cancer cell lines and compared with its effect on normal cell lines.

Compound No.	In Vitro Cytotoxicity IC_{50} (μg/mL)			Selectivity Index	
	HepG-2	MCF-7	WI-38	HepG2	MCF-7
1	82.75 ± 4.1	86.85 ± 3.9	61.71 ± 3.4	0.75	0.71
2	19.50 ± 1.3	11.63 ± 0.9	59.33 ± 3.2	3	5.1
3	24.83 ± 1.9	17.79 ± 1.3	49.64 ± 2.7	2	2.8
4	43.35 ± 2.5	39.79 ± 2.2	84.30 ± 4.3	1.9	2.1
5	63.59 ± 3.4	52.76 ± 2.9	34.77 ± 2.1	0.54	0.66
Doxorubicin	4.50 ± 0.2	4.17 ± 0.2	6.72 ± 0.5	1.5	1.6

2.5. *Anti-Inflammatory Activity of Lettuce Seed Cake Extract and Isolated Compound* (**1**) *in Carrageenan-Induced Paw Oedema In Vivo*

The carrageenan-induced paw edema test is widely used to assess the anti-inflammatory effects of new pharmaceutical agents [29–31]. In this study, the subcutaneous injection of carrageenan in rats led to a time-dependent increase in paw edema (Table 3). The inhibitory effects of different treatments are displayed in Table 3 and Figures 5 and S14. Results showed that an oral dose of lettuce seed cake extract (2 g/kg) significantly reduced paw edema size ($p < 0.05$) after 4 h. Compound (**1**) at doses of 40 mg/kg caused significant inhibition of paw edema ($p < 0.001$) starting from the 3rd hour after carrageenan administration. Moreover, the combination of ibuprofen (10 mg/kg) with compound (**1**) (20 mg/kg) further reduced paw edema ($p < 0.001$) after 5 h, showing more promising results than the standard drug.

Table 3. Treatments effect on carrageenan-induced paw oedema.

Treatment	Dose (mg/kg)	Increase in Paw Oedema (mL) and % Inhibition (%I)				
		1 h	2 h	3 h	4 h	5 h
Control	-	1.47± 0.08	3.33 ± 0.05	3.95 ± 0.04	3.99 ± 0.40	4.13 ± 0.46
Ibuprofen	10	0.69 ± 0.16 (15.62)	1.57 ± 0.21 ** (25.62)	1.82 ± 0.38 *** (28.5)	1.49 ± 0.47 *** (62.66)	1.20 ± 0.37 ** (70.9)
Lettuce seed cake extract	2000	1.29 ± 0.11 (5.87)	2.20 ± 0.44 ** (18)	2.59 ± 0.19 *** (19.57)	2.49 ± 0.13 *** (37.5)	2.42 ± 0.10 ** (41.40)
Compound (**1**)	20	1.05 ± 0.20 (10.25)	1.81 ± 0.44 ** (23.51)	2.26 ± 0.28 *** (23.77)	2.18 ± 0.74 *** (45.36)	1.75 ± 0.46 ** (57.63)
Compound (**1**)	40	0.77 ± 0.28 (16.32)	1.59 ± 0.22 ** (27.07)	2.11 ± 0.29 *** (26.1)	1.78 ± 0.45 *** (55.39)	1.49 ± 0.49 ** (63.92)
Ibuprofen + Compound (**1**)	10 + 20	0.58 ± 0.17 (20)	1.30 ± 0.29 ** (31.28)	1.42 ± 0.57 *** (35.35)	1.12 ± 0.40 *** (71.93)	0.83 ± 0.34 ** (79.90)

Values are expressed as mean ± S.D. ** $p < 0.01$, and *** $p < 0.001$, compared to the vehicle control group. Differences between groups were analyzed by analysis of variance (one-way ANOVA) followed by Dunnett's test. The standard anti-inflammatory drug ibuprofen (10 mg/kg) also significantly decreased paw oedema (*** $p < 0.001$) after 3rd and 5th h of carrageenan administration.

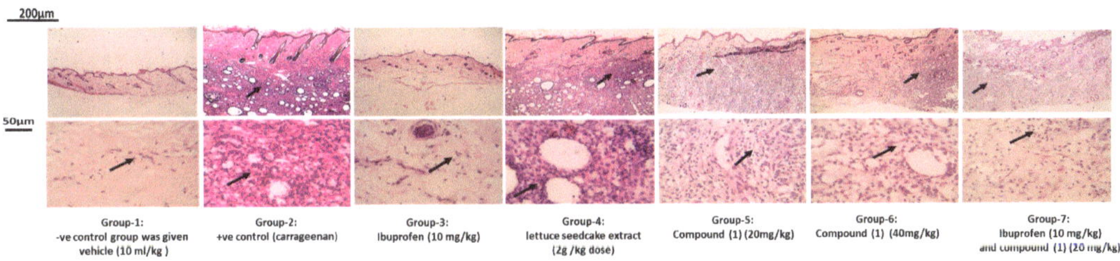

Figure 5. Microscopic pictures of H&E-stained skin sections. Magnifications X: 40 bar 200 are represented in the first column, thin black arrows refer to dermal inflammation and X: 400 bar 50 are represented in the second column where thin black arrows indicate leukocytic cells infiltration in dermis.

Histopathological analysis revealed that skin sections from Group 1 (negative control group) showed normal epidermis and dermis, including hair follicles, sebaceous glands, and collagen bundles, with no signs of inflammation. Microscopic images at higher magnification (Figure 5) displayed very few leukocytes in the dermis. In contrast, Group 2 (carrageenan group) exhibited severe dermal inflammation and significant leukocyte infiltration (thin black arrow). Treatment with lettuce seed cake extract in Group 4 led to a slight reduction in dermal inflammation. An oral dose of compound (**1**) at 20 mg/kg resulted in a moderate decrease in both inflammation and leukocyte infiltration, while Group 6, treated with compound (**1**) at 40 mg/kg, showed a substantial reduction in dermal inflammation. Group 3 (standard treatment) showed mild dermal inflammation, whereas the combination of compound (**1**) with standard ibuprofen (Group 7) demonstrated a synergistic effect, with no signs of inflammation and very few leukocytes (Figure 5).

2.5.1. Impact on Enzymatic Anti-Inflammatory Activity

The effects of lettuce seed cake extract and compound (**1**) on the levels of inflammatory mediators, including IL-1β, TNF-α, and PGE2, in carrageenan-induced paw edema in rats are shown in Figure 6. Treatment with compound (**1**) resulted in a significant reduction in IL-1β and TNF-α levels compared to the vehicle control group at doses of 20 and 40 mg/kg (* $p < 0.05$, ** $p < 0.01$). Notably, the combination of ibuprofen (10 mg/kg) with compound (**1**) (20 mg/kg) yielded particularly promising results, significantly reducing IL-1β, TNF-α, and PGE2 levels (*** $p < 0.001$).

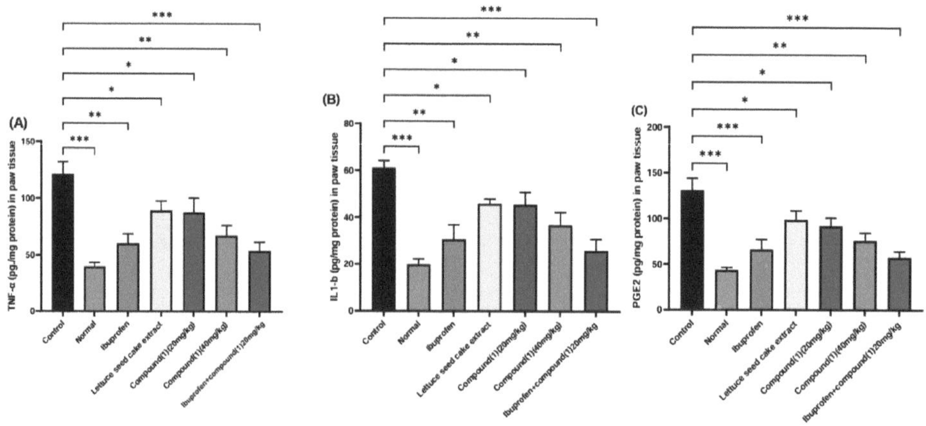

Figure 6. Effect of treatments on TNF-α (**A**), IL-1β (**B**), and PGE2 (**C**) in carrageenan-induced oedema in rat hind paws. Values represent mean ± SD for each group (n = 6). * $p < 0.05$, ** $p < 0.01$, *** $p < 0.001$.

2.5.2. Impact on Enzymatic Antioxidant Status

Carrageenan-induced local inflammation has been linked to the generation of reactive oxygen species (ROS) and is a significant contributor to oxidative stress. The results for SOD, CAT, and GSH levels in the paw edema tissue across different test groups are summarized in Table 4. Treatment with compound (**1**) at 40 mg/kg restored SOD activity by 71.43%, CAT activity by 72.08%, and GPx activity by 73.17% compared to the control group. The combination of ibuprofen (10 mg/kg) and compound (**1**) (20 mg/kg) showed even greater protective effects, with SOD, CAT, and GSH activities increasing by 85.08%, 84.28%, and 84.23%, respectively. In comparison, ibuprofen alone provided protection of 75.17%, 83.39%, and 78.74%, respectively. These findings suggest that both lettuce extract and compound (**1**) enhance antioxidant enzyme activities, potentially by bolstering the cellular antioxidant defense mechanisms. This indicates that they may offer protective benefits by stimulating the expression and activity of antioxidant enzymes during the inflammatory response.

Table 4. Effects of treatments and ibuprofen on CAT, SOD, and GPx activities in carrageenan-induced paw odema.

Treatment	GPx (U/gm Tissue)	CAT (U/gm Tissue)	SOD (U/gm Tissue)
Normal	118.55 ± 10.25 ***	2.83 ± 0.11 ***	189.05 ± 13.79 ***
Control	52.85 ± 7.99	0.87 ± 0.13	76.55 ± 10.68
Ibuprofen	93.35 ± 8.84 **	2.20 ± 0.20 ***	142.1 ± 14.71 **
Lettuce seed cake extract	77.05 ± 3.89 *	1.45 ± 0.08 *	116.65 ± 1.77 *
Compound (**1**) (20 mg/kg)	77.65 ± 3.18 *	1.50 ± 0.19 *	118.35 ± 1.06 *
Compound (**1**) (40 mg/kg)	86.75 ± 6.01 *	2.04 ± 0.26 **	135.05 ± 11.67 *
Combination (Ibuprofen + compound (**1**) (20 mg/kg))	99.85 ± 6.72 **	2.36 ± 0.26 ***	160.85 ± 15.06 **

Values are expressed as mean ± S.D. * $p < 0.05$, ** $p < 0.01$, and *** $p < 0.001$, compared to the vehicle control group. Differences between groups were analyzed by analysis of variance (one-way ANOVA) followed by Dunnett's test.

The phytochemical analysis of lettuce seed cake suggests that the compound 1,3-propanediol-2-amino-1-(3′,4′-methylenedioxyphenyl) (compound **1**) is primarily responsible for its anti-inflammatory activity. Previous studies have shown that compound **1** effectively inhibits IL-6 production in TNF-α-stimulated MG-63 cells [23]. Additionally, flavonoids such as luteolin and luteolin-7-O-*β*-D-glucoside are well recognized for their anti-inflammatory properties [32,33]. As a result, *L. sativa* seed cake extract emerges as a promising candidate for anti-inflammatory applications. This study lays the foundation for further investigation into the mechanism of action of compound **1** and the potential uses of lettuce seed cake extract in the food and cosmetic industries.

2.6. In Vitro COX-1 and COX-2 Inhibition Assay and Molecular Docking for Compound (**1**)

Compound (**1**) demonstrated promising activity against the COX-1 and 2 inhibition assay in comparison to the ibuprofen standard. Additionally, Figure 7 shows that it was more active against COX-2 than COX-1, with IC_{50} values of 4.814 ± 0.24 and 17.31 ± 0.65, respectively. The results were supported by docking as unique interaction patterns of compound (**1**) with COX-2 and COX-1 (Figure 8).

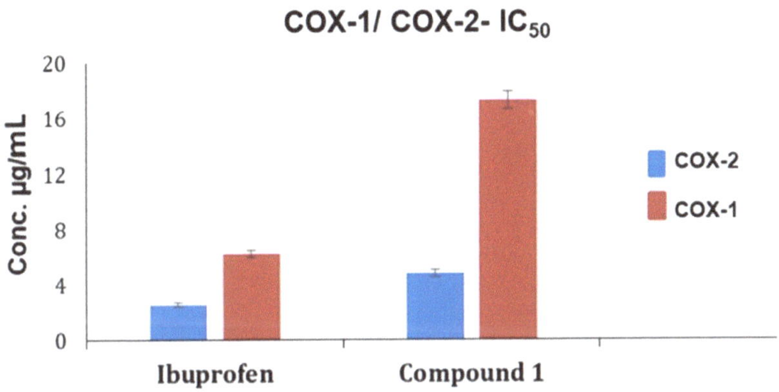

Figure 7. IC_{50} of compound **1** and Ibuprofen standard for COX-1 and COX-2 inhibition assay.

The results of molecular docking and binding affinities between compound (**1**) and COX-1 and COX-2 proteins are shown in Figure 8 and Table 5. Red indicates alpha helices in the protein's tthree-dimensional structure, green shows sheets, and restricted lines show the chemical bonds that have been established between the drug and COX.

Table 5. ΔG (kcal/mol) for each compound with tested proteins (COX1 and COX2).

Compound	COX-1	COX-2
Ibuprofen	−5.4	−4.8
Compound **1**	−6.6	−5.4

In comparison to ibuprofen, compound (**1**) shows stronger binding to both COX-1 and COX-2, according to the binding affinity results, which are displayed in Table 5. Compared to ibuprofen (−5.4 kcal/mol for COX-1 and −4.8 kcal/mol for COX-2), the tested compound's ΔG values (−6.6 kcal/mol for COX-1 and −5.4 kcal/mol for COX-2) indicate a higher binding affinity and possibly a larger inhibitory impact. This enhanced binding profile might point to a more effective anti-inflammatory effect of the investigated substance.

The 2D and 3D visualizations highlight the key residues involved in the interactions and provide a visual confirmation of the tabulated data (Figure 8). For COX-1, ASN59 and ASN77 are significant residues in the binding site that the tested compound makes hydrogen bonds with. The interaction's binding affinity and specificity are greatly influenced by these hydrogen bonds. To further stabilize the binding, there are hydrophobic interactions between PHE81 and the vicinity of GLY55 and GLY56. While the tested chemical exhibited a robust hydrogen bonding network, ibuprofen's interactions with COX-1 are largely hydrophobic.

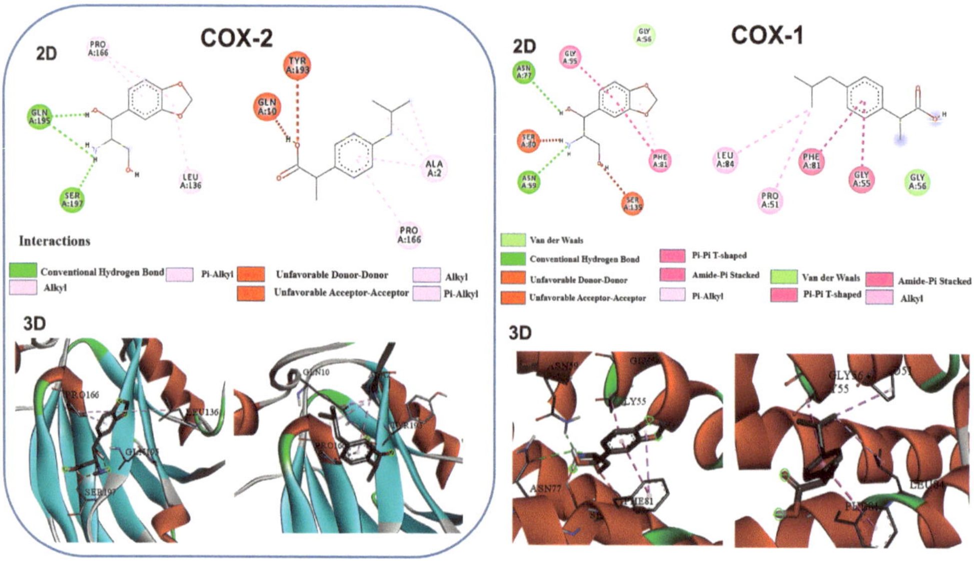

Figure 8. The 3D and 2D representations of compound **1** and ibuprofen interacting with the COX-1 and COX-2 enzymes are shown. The 2D diagrams illustrate color-coded interactions between the ligand and specific amino acid residues, while the 3D models display the ligand within each enzyme's binding pocket, highlighting key structural features. In 3D model of protein (red: alpha helices; cyan: beta sheets; green: loops) complexed with drug (stick model).

Even more noticeable variations can be seen in the interactions with COX-2 (Figure 8). The evaluated substance establishes many hydrogen bonds with GLN195 and SER197, residues that are essential for the selectivity and efficacy of COX-2 inhibitors. Together with the hydrophobic contacts between PRO166 and LEU136, these hydrogen bonds point to a potent and distinct binding mechanism. However, in the COX-2 binding region, ibuprofen only exhibits hydrophobic interactions with ALA2 and PRO166, suggesting a potentially less ideal binding configuration. These differences in binding patterns and affinities have several implications for the biological activity of the investigated substance in comparison to ibuprofen. Compound **1** may have more potency as a COX inhibitor due to its stronger and more frequent interactions with both enzymes. The decreased binding energies (ΔG values) found in the docking studies provide support for this.

More research on the molecule may reveal its greater COX-2 selectivity, which could lead to better gastrointestinal tolerability when compared to non-selective NSAIDs such as ibuprofen. In this context, it is especially interesting that important residues like SER197 are involved in COX-2 binding. It is possible that the tested chemical will spend a longer amount of time at the active site of the enzyme due to its more stable binding mechanism and large hydrogen bonding network. This might result in a longer duration of action than ibuprofen, which would be beneficial for sustaining therapeutic benefits and dose frequency.

Moreover, the presence of crucial residues such as GLN195 and SER197 in COX-2 and ASN59 and ASN77 in COX-1 suggests that the investigated substance may successfully obstruct the entry of arachidonic acid, the natural substrate, or disrupt the catalytic functions of these enzymes. Therefore, while talking about the ADMET profile, ibuprofen has a Log p value of 3.687, which indicates good lipophilicity, which contributes to its ability to cross biological membranes. The high protein binding (PPB) of 94.37% suggests that ibuprofen circulates in the bloodstream primarily bound to plasma proteins. This characteristic can influence its distribution and half-life in the body. The low fraction unbound (Fu) of 3.65%

further supports this high protein binding nature. The tested compound (**1**), in contrast, demonstrates different pharmacokinetic properties. With a lower Log *p* of −0.763, it is more hydrophilic than ibuprofen, which may affect its ability to cross cell membranes. Its protein binding is considerably lower at 31.56%, with a higher fraction unbound (72.23%), suggesting that it may have a different distribution profile in the body compared to ibuprofen (Table S6, Figure S17).

Both compounds show low blood-brain barrier (BBB) penetration probabilities (0.463 for ibuprofen and 0.326 for the tested compound), indicating they may not readily enter the central nervous system. This could be beneficial for reducing potential neurological side effects but might limit their efficacy for CNS-related conditions. Therefore, further investigation of semi-synthetic derivatives could be suggested to produce derivatives with better profiles.

The information also includes the interactions of the compounds with cytochrome P450 enzymes, which are critical for drug metabolism. The information shows that ibuprofen is more likely than the tested chemical (0.171) to be a substrate for CYP2C9 (0.982) (Table S6, Figure S17). This is in line with the enzyme's documented function in ibuprofen metabolism. Being aware of this helps you prepare for potential drug interactions. According to toxicity predictions, there is a minimal likelihood of either chemical having serious negative effects. The studied compound's hERG inhibition probability is 0.07, compared to 0.018 for ibuprofen, suggesting a low risk of cardiotoxicity. Important safety factors, such as genotoxicity (Ames test) and carcinogenicity, both exhibit low probabilities for these substances.

3. Conclusions

Oil seed cakes are hidden treasures for a variety of biologically active substances. The findings revealed that *L. sativa* seed cake extract contains various bioactive substances with potential antibacterial, antioxidant, and anti-inflammatory properties. Compound (**1**), both on its own and in combination with ibuprofen, significantly reduced the production of TNF-α, IL-1β, and IL-6, showed selective inhibition of COX-2, and decreased carrageenan-induced paw oedema in rats. Consequently, *L. sativa* seed cake may serve as a valuable natural resource for developing novel nutraceuticals, functional foods, and applications in the pharmaceutical and food industries.

Ultimately, oilseed cakes represent a promising agro-industrial waste with considerable potential due to their rich phytochemical content, including phenolics and flavonoids [17]. Despite their current use in biodiesel production and animal feed [34], further research is needed to fully explore their phytochemical and biological properties. This study showed the importance of oilseed cakes as a valuable resource for secondary metabolites, which could have significant applications in the medical and commercial sectors. As interest in sustainable practices grows, more attention should be given to utilizing these by-products for their broader economic and environmental benefits.

4. Materials and Methods

4.1. General Experimental

The HPLC was performed in the Institute for Plant Biology, TU Braunschweig, using a Young Lin quaternary pump, vacuum degasser, column oven (40 °C), diode array detector (YL9160 PDA), and a Midas Spark Holland autosampler. UV spectra (λ max) were measured in the central laboratory of the Faculty of Pharmaceutical Science at Mansoura University using spectroscopic methanol with an ultraviolet-visible spectrophotometer (Shimadzu 1601 PC, version TCC-240A, Kyoto, Japan). The NMR unit of Mansoura University's Faculty of Pharmacy obtained nuclear magnetic resonance spectra (^{1}H-NMR, ^{13}C-NMR, DEPT-Q, APT, HSQC, and HMBC) using a Bruker DRX 600 NMR spectrometer (600 and 150 MHz for ^{1}H and ^{13}C-NMR, respectively) and with the Bruker Corporation Avance III 400 spectrometer, which measures ^{1}H and ^{13}C-NMR at 400 and 100 MHz, respectively. The types of solvents that are used are $CDCl_3$, CD_3OD, and DMSO-d6. For normal phase chromatography, silica gel G 60–230 mesh (Merck, Darmstadt, Germany) was packed using a wet or dry method in the specified solvent. The purchase of Normal Human lung fibroblast (WI-38), hepatic cancer cell lines (HePG-2), and mammary gland (MCF-7) cell lines from ATCC was made possible by the Holding firm for biological goods and vaccines (VACSERA), which is based in Cairo, Egypt.

The levels of inflammatory mediators (IL-1β, TNF-α, and PGE2) in rat paw tissue homogenate were measured using the following Kits: Rat TNF-α (Tumor Necrosis Factor Alpha) ELISA Kit from Elabscience® Company, Houston, TX, USA, Interleukin 1 Beta (IL-1b) Organism Species: Rattus norvegicus (Rat) from Cloud Clone Crop Company and Rat PGE2 (Prostaglandin E2) ELISA Kit from Fine test® company Hubei, China. In addition, oxidative stress parameters in rat paw tissue homogenate were measured using superoxide dismutase, glutathione peroxidase, and catalase Kits from the Biodiagnostic company, Birmingham, UK.

4.2. Chemicals

All solvents used were of HPLC grade and supplied by Fisher Scientific. Luteolin-7-glucoside standard was purchased from Carl Roth (Karlsruhe, Germany). The solvents used for extraction, chromatographic separation, and crystallization (i.e., petroleum ether, methylene chloride, ethyl acetate, and methanol) were of reagent grade, obtained from EL-Nasr Company for pharmaceutical chemicals, Egypt. ABTS (Azino-bis-(3-ethyl benzthiazoline-6-sulfonic acid), COX-1, COX-2, RPMI-1640 medium, MTT, DMSO, CMC, doxorubicin, ciprofloxacin, and clotrimazole were purchased from Sigma Co., St. Louis, MO, USA. Ascorbic acid (Cevarol®) and ibuprofen (Brofen) tablets were obtained from Memphis Pharmaceutical Co., Cairo, Egypt, and abbott pharmaceuticals, Abbott Park, IL, USA, respectively.

4.3. Plant Material

Lactuca sativa (lettuce), *Nigella sativa* (black cumin), *Eruca sativa* (rocket), and *Linum usitatissimum* (flaxseed) seed cakes were obtained from Albadawia Oil Factory, Ferdos City, Mansoura, Dakahliya, Egypt, 5 August 2022.

4.4. Extraction of Seed Oil Cakes Derived from Different Plant Species

For preliminary screening to obtain dried methanolic extracts from various oilseed cakes, 200 g each of lettuce, black seed, rockets, and flaxseed seed cakes were extracted three times by overnight maceration using methanol (3× 400 mL each). The methanolic extracts were separated from the crude powders through centrifugation and filtration using Whatman filter paper. The extracts were then concentrated under reduced pressure at 45 °C using a rotary evaporator. The resulting extracts were dried in a desiccator over anhydrous calcium chloride until they reached a constant weight, then stored at room temperature for biological assays and further chromatographic analysis.

4.5. HPLC Analysis of Lettuce Seed Cake Defatted Extract

HPLC separation was performed using a Nucleosil RP-C18 column (5 μm particle size, L × I.D. 25 cm × 3.2 mm). A binary gradient was applied, starting with 85% A (aqueous trifluoroacetic acid 0.05%) and 15% B (acetonitrile). After 5 min, the ratio was adjusted to 80% A and 20% B, followed by the following gradient: 15 min: 70% A, 30% B; 25 min: 20% A, 80% B; 30 min: 20% A, 80% B; 35 min: 10% A, 90% B; 37 min: 85% A, 15% B, and held until 47 min. The flow rate was set at 1 mL/min, with an injection volume of 20 μL. Detection was performed using a photodiode array (PDA) detector at 254 nm and 350 nm.

4.6. Bio-Guided Isolation of Active Constituents from Lactuca sativa Seed Cake Active Fractions

For chromatographic investigation of lettuce seed cake, 5 kg of the dried cake was extracted three times with methanol (10 L each) through overnight maceration in siphon extraction jars. The methanolic extracts were separated from the crude powders and concentrated under reduced pressure at 45 °C using a rotary evaporator. The resulting dried extract of 760 g of residue was suspended in water for fractionation using petroleum ether, methylene chloride, and ethyl acetate. The weights of the resulting fractions were as follows: 590 g for the petroleum ether extract (extremely thick and oily), 11 g for the methylene chloride extract, and 14 g for the ethyl acetate extract.

Column chromatography of regular silica gel (3.5 cm, 300 g) packed with methylene chloride solvent was applied to the methylene chloride fraction extract, eluting it with methylene chloride-methanol (95:5), and progressively raising the polarity of the eluent by 5%. Utilizing thin-layer chromatography, the resulting fractions' homogeneity was assessed. Subfractions were obtained by combining similar fractions. Subfraction was further purified by crystallization from methanol and eluted with various systems to yield compound **1** (560 mg) with an R_f of 0.2 when developed on GF_{254} precoated silica gel plate using solvent system methylene chloride absolute (100%), and compound **2** (8 mg) with an R_f value of 0.60 when 10% methanol in methylene chloride was used. Subfraction 3 was chromatographed on silica to provide compound **5** (18 mg) R_f 0.26 after being eluted with methylene chloride-ethyl acetate (70:30).

Normal silica gel column chromatography packed with methylene chloride (3.5 cm, 350 g) was applied to the extract of ethyl acetate. Methanol was added once elution was completed using methylene chloride-ethyl acetate (95:5) and a 5% increase in eluent polarity up to 100% ethyl acetate. Thin-layer chromatography was employed as a monitoring tool to monitor the collected fractions' homogeneity. By combining similar fractions, compounds **3** and **4** were produced. Chromatographic analysis of compound **3** (15 mg) on TLC-precoated silica gel plates GF_{254} using the solvent system 20% methanol in methylene chloride revealed an R_f value of 0.32 and compound **4** (6 mg) as a yellow amorphous powder, which was further purified by reprecipitation with an R_f value of 0.46 using 15% methanol in the ethyl acetate system on TLC.

4.7. Biological Screening

4.7.1. Antioxidant Screening (ABTS Assay)

The experiment was performed in triplicate according to Zaky et al. [35]. The absorbance of control of the resulting green-blue solution (ABTS* radical solution) was recorded at λ max 734 nm.

4.7.2. Antimicrobial Screening

The procedure was performed according to Stylianakis et al. [36]. A comparison of the extract's zone of inhibition to that of a reference antibiotic's activity index was determined using the following equation:

$$\text{Activity index} = (\text{the inhibition zone of extract}/\text{the inhibition zone of the standard}) \times 100$$

4.7.3. Cytotoxic Activity (MTT Assay)

Using the MTT test, the inhibitory effects of substances on cell growth were ascertained using the cell lines [37] compared to normal lung fibroblast cells (WI-38). The selectivity index was calculated via the Selectivity index (SI) equation: SI equals IC_{50} normal/IC_{50} cancer. IC_{50} normal refers to the concentration of the investigated substance that killed 50% of normal cells, whereas IC_{50} malignancy denotes the concentration that killed 50% of malignant cells [38].

4.7.4. Anti-Inflammatory Test: Carrageenan-Induced Paw Oedema

Adult albino rats 180–220 g of both sexes used in the investigation were acquired from the Mansoura University Faculty of Pharmacy's Animal House. Standard laboratory diet and water were provided, and they were housed in cages maintained at a constant temperature of 22 ± 2 °C. The Animal Care and Use Committee (MU-ACUC) of Mansoura University, Egypt, has approved all procedures involving animals with code number: MU-ACUC (PHARM.R.24.09.40). Throughout the experiment, medications and extracts were given orally in doses of 1 mL/100 g of the rat's body weight after being solubilized in carboxymethyl cellulose (CMC) in order to create a suspension formula.

In this test, rats were divided into six groups, with each group containing 5 animals.

Group-1: The negative control group was given a vehicle, i.e., a 1% aqueous solution of CMC (10 mL/kg).

Group-2: Positive control (carrageenan-induced oedema, Sigma (St. Louis, MO, USA)) without any treatment.

Group-3: Were given the standard drug Ibuprofen at the dose of 10 mg/kg.

Group-4: Seedcake extract of *Lactuca sativa* was administered at a 2 g/kg dose.

Group-5: Compound (**1**) is administered at a dose of 20 mg/kg.

Group-6: Compound (**1**) at a dose of 40 mg/kg.

Group-7: Combination of standard and compound (**1**) at doses of (10 + 20) mg/kg, respectively.

After one hour, rats' right hind paws were subcutaneously injected with 100 μL of carrageenan (1% in normal saline) to cause acute inflammation. Using a digital vernier calliper, the swelling of the paw that had been injected with carrageenan was measured prior to injection as well as 1, 2, 3, 4, and 5 h following the treatment's induction. The percentage inhibition of oedema in the test animals treated with the extract compared to the carrageenan control group was used to calculate the anti-inflammatory activity. The formula $\%I = (dt/dc) \times 100$ was used to calculate the percentage inhibition of inflammation, where "dt" represents the difference in paw volume in the drug-treated group and "dc" represents the difference in paw volume relative to the control group. Moreover, "I" represents the suppression of inflammation [29].

Histopathological Assessment of Skin Tissue

Five hours following the carrageenan injection, tissue specimen samples from the paw skin tissue of every group under study were removed for histological examination. Prior to being submerged in paraffin blocks, they were fixed in a 10% formalin solution and stained with hematoxylin-eosin. Using a microtome, paraffin sections with a thickness of 5 μm were cut, and hematoxylin and eosin were frequently applied to the sections. Microscopically, these tissue slices were inspected. Consequently, the sections were examined under a light microscope; on the other hand, photos taken with a digital camera (Canon Power-shot D10) recorded the progression of paw oedema [30].

Determination of the Levels of Inflammatory Mediators and Oxidative Stress Parameters in Rat Paw Tissue Homogenate

Rats were slaughtered, and their paw tissues were gathered and weighed five hours after the carrageenan was administered. The tissues underwent processing after being fixed in 10% formalin. Rat paw tissue was homogenized, and its oxidative stress indices (SOD, CAT, and GPx) as well as the levels of inflammatory mediators (IL-1β, TNF-α, and PGE2) were measured [31].

4.7.5. Anti-Inflammatory Activity (COX-1 and COX-2) Inhibition Assay

With some modifications, anti-inflammatory activity was carried out following Smith et al. 1998 [39], and ibuprofen served as the benchmark medication.

4.8. *Molecular Docking of Compound* (**1**) *against COX-1 and COX-2*

4.8.1. Protein Preparation

The protein structures of COX-1 and COX-2 were acquired from their respective sources, Uniport, and are identified by the UniProt IDs A0A1B0RKU9 and P00406, respectively. The CB-Dock2 server [40] was utilized to estimate the active sites of these proteins. Using AutoDock Tools 1.5.7 [41], additional processing was performed on the produced protein structures, including the addition of polar hydrogen atoms, the assignment of Gasteiger charges, and the merger of non-polar hydrogen atoms. Molecular docking simulations were performed using the generated PDBQT files as input.

4.8.2. Ligand Preparation

The active material that is extracted from the ligand molecules was obtained in their corresponding SDF formats from the PubChem database. Following this, they were minimized using the Conjugate Gradients technique and the Force Field (MMFF94) in the Avogadro 1.2.0 software [42]. The reduced ligand configurations were transformed into a PDBQT file that was compatible with AutoDock Vina [43], which was used to run molecular docking simulations. BIOVIA Discovery Studio 2020 [BIOVIA 2020] was used to visualize and analyze the docking data.

4.9. In Silico ADMET Prediction for Compound (**1**)

Compound **1**'s SMILES codes were entered into the ADMETlab 2.0 web server [44] to forecast the pharmacokinetic characteristics, such as absorption, distribution, metabolism, and excretion.

4.10. Statistical Analysis

The statistical analyses have been conducted using GraphPad Prism version 10.3.1 software. The data are presented as means ± standard deviations. Every determination was made three times, after which averages were estimated. The significance of the difference between means was determined by one-way ANOVA followed by Dunnett's post hoc test, and the $p < 0.05$ values were considered significant.

Supplementary Materials: The following Supporting Information can be downloaded at: https://www.mdpi.com/article/10.3390/ijms252011077/s1.

Author Contributions: Conceptualization, A.A.G., M.M.A., D.S. and S.A.; Formal analysis, M.M., M.M.A. and S.A.; Methodology, M.M. and S.A.; Resources, M.M. and D.S.; Software, M.M. and S.A.; Validation, A.A.G. and S.A. Writing—original draft, M.M. and S.A.; Writing—review and editing, A.A.G., M.M.A. and D.S. All authors have read and agreed to the published version of the manuscript.

Funding: We acknowledge the support of the Open Access Publication Funds of Technische Universität Braunschweig.

Institutional Review Board Statement: The animal study protocol was approved by the Animal Care and Use Committee (MU-ACUC) of Mansoura University, Egypt, with code number: MU-ACUC (PHARM.R.24.09.40; 28.09.2024).

Informed Consent Statement: Not applicable.

Data Availability Statement: Data are contained within the article and Supplementary Materials.

Conflicts of Interest: The authors declare no conflicts of interest.

References

1. Rakita, S.; Kokić, B.; Manoni, M.; Mazzoleni, S.; Lin, P.; Luciano, A. Cold-pressed oilseed cakes as alternative and sustainable feed ingredients: A review. *Foods* **2023**, *12*, 432. [CrossRef] [PubMed]
2. Vermaak, I.; Kamatou, G.P.P.; Komane-Mofokeng, B.; Viljoen, A.; Beckett, K. African seed oils of commercial importance—Cosmetic applications. *S. Afr. J. Bot.* **2011**, *77*, 920–933. [CrossRef]
3. Zhang, M.; Wang, O.; Cai, S.; Zhao, L.; Zhao, L. Composition, functional properties, health benefits and applications of oilseed proteins: A systematic review. *Int. Food Res.* **2023**, *171*, 113061. [CrossRef] [PubMed]
4. Sielicka, M.; Małecka, M.J. Preservation. Enhancement of oxidative stability of flaxseed oil through flaxseed, evening primrose and black cumin cake extracts. *J. Food Process. Preserv.* **2017**, *41*, e13070. [CrossRef]
5. Tavarini, S.; De Leo, M.; Matteo, R.; Lazzeri, L.; Braca, A.; Angelini, L.G.J.P. Flaxseed and camelina meals as potential sources of health-beneficial compounds. *Plants* **2021**, *10*, 156. [CrossRef]
6. Krimer Malešević, V.; Vaštag, Ž.; Popović, L.; Popović, S.; Peričin-Starčevič, I. Characterisation of black cumin, pomegranate and flaxseed meals as sources of phenolic acids. *Int. J. Food Sci.* **2014**, *49*, 210–216. [CrossRef]

7. Alam, M.S.; Kaur, G.; Jabbar, Z.; Javed, K.; Athar, M. *Eruca sativa* seeds possess antioxidant activity and exert a protective effect on mercuric chloride induced renal toxicity. *Food Chem. Toxicol.* **2007**, *45*, 910–920. [CrossRef]
8. Mariod, A.A.; Ibrahim, R.M.; Ismail, M.; Ismail, N. Antioxidant activity and phenolic content of phenolic rich fractions obtained from black cumin (*Nigella sativa*) seedcake. *Food Chem.* **2009**, *116*, 306–312. [CrossRef]
9. Sayyah, M.; Hadidi, N.; Kamalinejad, M.J. Analgesic and anti-inflammatory activity of *Lactuca sativa* seed extract in rats. *J. Ethnopharmacol.* **2004**, *92*, 325–329. [CrossRef]
10. Gugliandolo, A.; Giacoppo, S.; Ficicchia, M.; Aliquò, A.; Bramanti, P.; Mazzon, E. *Eruca sativa* seed extract: A novel natural product able to counteract neuroinflammation. *Mol. Med. Rep.* **2018**, *17*, 6235–6244. [CrossRef]
11. Kulsum, K.; Syahrul, S.; Hasbalah, K.; Balqis, U. Phytocompounds of Nigella sativa seeds extract and their neuroprotective potential via EGR1 receptor inhibition: A molecular docking study. *Narra J.* **2023**, *3*, e173. [CrossRef] [PubMed]
12. Alawlaqi, M.M.; Al-Rajhi, A.M.; Abdelghany, T.M.; Ganash, M.; Moawad, H. Evaluation of Biomedical Applications for Linseed Extract: Antimicrobial, Antioxidant, Anti-Diabetic, and Anti-Inflammatory Activities In Vitro. *J. Funct. Biomater.* **2023**, *14*, 300. [CrossRef] [PubMed]
13. Prasad, K.J. Suppression of phosphoenolpyruvate carboxykinase gene expression by secoisolariciresinol diglucoside (SDG), a new antidiabetic agent. *Int. Angiol.* **2002**, *11*, 107–109. [CrossRef]
14. Khadem, S.; Marles, R.J. Monocyclic phenolic acids; hydroxy-and polyhydroxybenzoic acids: Occurrence and recent bioactivity studies. *Molecules* **2010**, *15*, 7985–8005. [CrossRef] [PubMed]
15. Bárta, J.; Bártová, V.; Jarošová, M.; Švajner, J.; Smetana, P.; Kadlec, J. Oilseed cake flour composition, functional properties and antioxidant potential as effects of sieving and species differences. *Foods* **2021**, *10*, 2766. [CrossRef] [PubMed]
16. Kim, M.J.; Moon, Y.; Tou, J.C.; Mou, B.; Waterland, N.L. Nutritional value, bioactive compounds and health benefits of lettuce (*Lactuca sativa* L.). *J. Food Compos. Anal.* **2016**, *49*, 19–34. [CrossRef]
17. Ancuța, P.; Sonia, A. Oil press-cakes and meals valorization through circular economy approaches: A review. *Appl. Sci.* **2020**, *10*, 7432. [CrossRef]
18. Haq, A.; Siddiqi, M.; Batool, S.Z.; Islam, A.; Khan, A.; Khan, D.; Khan, S.; Khan, H.; Shah, A.A.; Hasan, F.; et al. Comprehensive investigation on the synergistic antibacterial activities of *Jatropha curcas* pressed cake and seed oil in combination with antibiotics. *Amb Express* **2019**, *9*, 67. [CrossRef]
19. Wahab, G.A.; Aboelmaaty, W.S.; Lahloub, M.F.; Sallam, A. In vitro and in silico studies of SARS-CoV-2 main protease M^{pro} inhibitors isolated from Helichrysum bracteatum. *RSC Adv.* **2022**, *29*, 18412–18424. [CrossRef]
20. Lin, L.C.; Pai, Y.F.; Tsai, T.H. Isolation of luteolin and luteolin-7-O-glucoside from Dendranthema morifolium Ramat Tzvel and their pharmacokinetics in rats. *J. Agric. Food Chem.* **2015**, *63*, 7700–7706. [CrossRef]
21. Salama, M.M.; Ezzat, S.M.; Sleem, A.A. A new hepatoprotective flavone glycoside from the flowers of *Onopordum alexandrinum* growing in Egypt. *Z. Naturforsch. C* **2011**, *66*, 251–259. [CrossRef] [PubMed]
22. Peng, H.Y.; Zhang, X.H.; Xu, J.Z. Apigenin-7-O-β-D-glycoside isolation from the highly copper-tolerant plant *Elsholtzia splendens*. *J. Zhejiang Univ. Sci. B* **2016**, *17*, 447. [CrossRef] [PubMed]
23. Abu-Reidah, I.; Contreras, M.; Arráez-Román, D.; Segura-Carretero, A.; Fernández-Gutiérrez, A. Reversed-phase ultra-high-performance liquid chromatography coupled to electrospray ionization-quadrupole-time-of-flight mass spectrometry as a powerful tool for metabolic profiling of vegetables: *Lactuca sativa* as an example of its application. *J. Chromatogr. A* **2013**, *1313*, 212–227. [CrossRef]
24. Kim, A.R.; Ko, H.J.; Chowdhury, M.A.; Chang, Y.-S.; Woo, E.-R. Chemical constituents on the aerial parts of Artemisia selengensis and their IL-6 inhibitory activity. *Arch. Pharmacal. Res.* **2015**, *38*, 1059–1065. [CrossRef] [PubMed]
25. Abouzeid, S.; Beutling, U.; Elekhnawy, E.; Selmar, D. Antibacterial and antibiofilm effects of allelopathic compounds identified in *Medicago sativa* L. seedling exudate against *Escherichia coli*. *Molecules* **2023**, *28*, 2645. [CrossRef]
26. Ahmadi, S.M.; Farhoosh, R.; Sharif, A.; Rezaie, M. Structure-antioxidant activity relationships of luteolin and catechin. *J. Food Sci.* **2020**, *85*, 298–305. [CrossRef]
27. Wang, W.; Yue, R.-F.; Jin, Z.; He, L.-M.; Shen, R.; Du, D. Efficiency comparison of apigenin-7-O-glucoside and trolox in antioxidative stress and anti-inflammatory properties. *J. Pharm. Pharmacol.* **2020**, *72*, 1645–1656. [CrossRef]
28. Basha, N.J.; Basavarajaiah, S. Anticancer potential of bioactive molecule luteolin and its analogs: An update. *Polycycl. Aromat. Compd.* **2023**, *43*, 3958–3976. [CrossRef]
29. Ismail, H.; Mirza, B. Evaluation of analgesic, anti-inflammatory, anti-depressant and anti-coagulant properties of *Lactuca sativa* (CV. Grand Rapids) plant tissues and cell suspension in rats. *BMC Complement. Altern. Med.* **2015**, *15*, 199. [CrossRef]
30. Zouari Bouassida, K.; Makni, S.; Tounsi, A.; Jlaiel, L.; Trigui, M.; Tounsi, S. Effects of Juniperus phoenicea Hydroalcoholic Extract on Inflammatory Mediators and Oxidative Stress Markers in Carrageenan-Induced Paw Oedema in Mice. *Biomed. Res. Int.* **2018**, *2018*, 3785487. [CrossRef]
31. Karim, N.; Khan, I.; Khan, W.; Khan, I.; Khan, A.; Halim, S.A.; Khan, H.; Hussain, J.; Al-Harrasi, A. Anti-nociceptive and anti-inflammatory activities of asparacosin a involve selective cyclooxygenase 2 and inflammatory cytokines inhibition: An in-vitro, in-vivo, and in-silico approach. *Front. Immunol.* **2019**, *10*, 581. [CrossRef] [PubMed]
32. Harris, G.K.; Qian, Y.; Leonard, S.S.; Sbarra, D.C.; Shi, X. Luteolin and chrysin differentially inhibit cyclooxygenase-2 expression and scavenge reactive oxygen species but similarly inhibit prostaglandin-E2 formation in RAW 264.7 cells. *J. Nutr.* **2006**, *136*, 1517–1521. [CrossRef] [PubMed]

33. Hämäläinen, M.; Nieminen, R.; Asmawi, M.Z.; Vuorela, P.; Vapaatalo, H.; Moilanen, E. Effects of flavonoids on prostaglandin E2 production and on COX-2 and mPGES-1 expressions in activated macrophages. *Planta Med.* **2011**, *77*, 1504–1511. [CrossRef] [PubMed]
34. Sudalai, S.; Prabakaran, S.; Varalakksmi, V.; Kireeti, I.S.; Upasana, B.; Yuvasri, A.; Arumugam, A. A review on oilcake biomass waste into biofuels: Current conversion techniques, sustainable applications, and challenges: Waste to energy approach (WtE). *Energy Convers. Manag.* **2024**, *314*, 118724. [CrossRef]
35. Zaky, R.; Yousef, T.; Ibrahim, K. Co (II), Cd (II), Hg (II) and U (VI) O2 complexes of o-hydroxyacetophenone [N-(3-hydroxy-2-naphthoyl)] hydrazone: Physicochemical study, thermal studies, and antimicrobial activity. *Spectrochim. Acta Part A Mol. Biomol. Spectrosc.* **2012**, *97*, 683–694. [CrossRef]
36. Stylianakis, I.; Kolocouris, A.; Kolocouris, N.; Fytas, G.; Foscolos, G.B.; Padalko, E. Spiro [pyrrolidine-2, 2′-adamantanes]: Synthesis, anti-influenza virus activity and conformational properties. *Bioorganic Med. Chem. Lett.* **2003**, *13*, 1699–1703. [CrossRef]
37. Dutta, A.; Bandyopadhyay, S.; Mandal, C.; Chatterjee, M. Development of a modified MTT assay for screening antimonial resistant field isolates of Indian visceral leishmaniasis. *Parasitol. Int.* **2005**, *54*, 119–122. [CrossRef]
38. Elattar, E.M.; Shaban, M.; Saad, H.E.; Badria, F.A.; Galala, A.A. Evaluation of antimicrobial, antiquorum sensing, and cytotoxic activities of new vanillin 1, 2, 3-triazole derivatives. *Nat. Prod. Res.* **2023**, *37*, 2662–2671. [CrossRef]
39. Smith, C.J.; Zhang, Y.; Koboldt, C.M.; Muhammad, J.; Zweifel, B.S.; Shaffer, A. Pharmacological analysis of cyclooxygenase-1 in inflammation. *Proc. Natl. Acad. Sci. USA* **1998**, *95*, 13313–13318. [CrossRef]
40. Liu, Y.; Yang, X.; Gan, J.; Chen, S.; Xiao, Z.-X.; Cao, Y. CB-Dock2: Improved protein–ligand blind docking by integrating cavity detection, docking and homologous template fitting. *Nucleic Acids Res.* **2022**, *50*, 159–164. [CrossRef]
41. Morris, G.M.; Huey, R.; Lindstrom, W.; Sanner, M.F.; Belew, R.K.; Goodsell, D.S. AutoDock4 and AutoDockTools4: Automated docking with selective receptor flexibility. *J. Comput. Chem.* **2009**, *30*, 2785–2791. [CrossRef] [PubMed]
42. Hanwell, M.D.; Curtis, D.E.; Lonie, D.C.; Vandermeersch, T.; Zurek, E.; Hutchison, G.R. Avogadro: An advanced semantic chemical editor, visualization, and analysis platform. *J. Cheminform.* **2012**, *4*, 17. [CrossRef] [PubMed]
43. Trott, O.; Olson, A.J. AutoDock Vina: Improving the speed and accuracy of docking with a new scoring function, efficient optimization, and multithreading. *J. Comput. Chem.* **2010**, *31*, 455–461. [CrossRef] [PubMed]
44. Xiong, G.; Wu, Z.; Yi, J.; Fu, L.; Yang, Z.; Hsieh, C. ADMETlab 2.0: An integrated online platform for accurate and comprehensive predictions of ADMET properties. *J. Comput. Chem.* **2021**, *49*, 5–14.

International Journal of
Molecular Sciences

MDPI

Article

Citrus Flavanone Effects on the Nrf2-Keap1/GSK3/NF-κB/NLRP3 Regulation and Corticotroph-Stress Hormone Loop in the Old Pituitary

Marko Miler [1,*], Jasmina Živanović [1], Sanja Kovačević [2], Nevena Vidović [3], Ana Djordjevic [2], Branko Filipović [1] and Vladimir Ajdžanović [1]

1 Department of Cytology, Institute for Biological Research "Siniša Stanković"—National Institute of the Republic of Serbia, University of Belgrade, 11060 Belgrade, Serbia; jasminap@ibiss.bg.ac.rs (J.Ž.); brankof@ibiss.bg.ac.rs (B.F.); avlada@ibiss.bg.ac.rs (V.A.)

2 Department of Biochemistry, Institute for Biological Research "Siniša Stanković"—National Institute of the Republic of Serbia, University of Belgrade, 11060 Belgrade, Serbia; sanja.kovacevic@ibiss.bg.ac.rs (S.K.); djordjevica@ibiss.bg.ac.rs (A.D.)

3 Centre of Research Excellence in Nutrition and Metabolism, Institute for Medical Research, National Institute of the Republic of Serbia, University of Belgrade, 11000 Belgrade, Serbia; nevena.vidovic@imi.bg.ac.rs

* Correspondence: marko.miler@ibiss.bg.ac.rs; Tel.: +381-11-2078-321; Fax: +381-11-2761-433

Abstract: Oxidative stress and inflammation are significant causes of aging. At the same time, citrus flavanones, naringenin (NAR), and hesperetin (HES) are bioactives with proven antioxidant and anti-inflammatory properties. Nevertheless, there are still no data about flavanone's influence and its potential effects on the healthy aging process and improving pituitary functioning. Thus, using qPCR, immunoblot, histological techniques, and biochemical assays, our study aimed to elucidate how citrus flavanones (15 mg/kg b.m. *per os*) affect antioxidant defense, inflammation, and stress hormone output in the old rat model. Our results showed that HES restores the redox environment in the pituitary by down-regulating the nuclear factor erythroid 2-related factor 2 (Nrf2) protein while increasing kelch-like ECH-associated protein 1 (Keap1), thioredoxin reductase (TrxR1), and superoxide dismutase 2 (SOD2) protein expression. Immunofluorescent analysis confirmed Nrf2 and Keap1 down- and up-regulation, respectively. Supplementation with NAR increased *Keap1*, *Trxr1*, glutathione peroxidase (*Gpx*), and glutathione reductase (*Gr*) mRNA expression. Decreased oxidative stress aligned with NLRP3 decrement after both flavanones and glycogen synthase kinase-3 (GSK3) only after HES. The signal intensity of adrenocorticotropic hormone (ACTH) cells did not change, while corticosterone levels in serum decreased after both flavanones. HES showed higher potential than NAR in affecting a redox environment without increasing the inflammatory response, while a decrease in corticosterone level has a solid link to longevity. Our findings suggest that HES could improve and facilitate redox and inflammatory dysregulation in the rat's old pituitary.

Keywords: hesperetin; naringenin; ACTH; SOD; CAT; GPx; GR; Interleukin 1 and 6; TNF-α; NLRP3

Citation: Miler, M.; Živanović, J.; Kovačević, S.; Vidović, N.; Djordjevic, A.; Filipović, B.; Ajdžanović, V. Citrus Flavanone Effects on the Nrf2-Keap1/GSK3/NF-κB/NLRP3 Regulation and Corticotroph-Stress Hormone Loop in the Old Pituitary. *Int. J. Mol. Sci.* **2024**, *25*, 8918. https://doi.org/10.3390/ijms25168918

Academic Editors: Vassilis Athanasiadis and Theodoros G. Chatzimitakos

Received: 22 July 2024
Revised: 10 August 2024
Accepted: 11 August 2024
Published: 16 August 2024

1. Introduction

Male aging is a progressive and gradual process characterized by low free testosterone, the main characteristic of the aging male hormonal status [1,2]. Specifically, aging of the pituitary gland is a multi-vector process that includes changes in hormonal and neuronal inputs to the gland itself, in addition to various structural and functional alterations occurring within the pituitary tissue. In the broader context, endocrine deficiency is associated with a progressive decline in pituitary function that, in turn, may contribute to senescence [3]. In parallel, degenerative processes affect pituitary-related regulatory limbic, hippocampal, and hypothalamic neurons and synapses, followed by compensatory gliosis [2,4]. During aging, lesions of the pituitary gland tissue vary and may include loss of pituitary endocrine cells, cyst development, fibrosis associated with chronic inflammation, distended, blood-filled vessels in the pars distalis, hemorrhage, or thrombosis [5]. Aged pituitary glands show an accumulation of oxidative products that further contribute to aging manifestation [6]. These suggest that a local redox imbalance may cause oxidative damage to cells of the hypothalamic–pituitary axis [3,7]. Generated reactive oxygen species (ROS) can lead to a rise in apoptosis of adenohypophysis cells, causing direct cellular damage but also impairment in protein function and overall hormone production with aging [3,8,9]. Studies with old rodents indicate their elevated circulating adrenocorticotropic hormone (ACTH) and glucocorticoid levels and increased release of corticotropin-releasing hormone from the hypothalamus [10]. The hyperplastic corticotrophs (ACTH cells) were observed to be mainly peripherally localized in the pituitary of old male rats [11]. In the present study, we remain focused on pituitary ACTH cells, considering their crucial role as a functional module of the hypothalamic–pituitary–adrenal (HPA) axis in affecting stress response/metabolism in advanced age.

In stress conditions, activation of nuclear factor erythroid 2-related factor 2 (Nrf2) is mediated by various exogenous and endogenous stressors such as electrophilic agents and ROS [12,13]. Nrf2 transits to the nucleus when activated, binding to antioxidant response elements (AREs) in the promoter region of genes encoding antioxidant enzymes (superoxide dismutase (SOD), catalase (CAT), glutathione reductase (GR), and peroxidase (GPx), thioredoxin (Trx) and its reductase (TrxR) system) and cytoprotective proteins. In addition, TrxR and Nrf2 signaling are crucial in regulating inflammation by modulating the transcription of antioxidants and suppressing inflammatory cytokines [14]. Besides its role in maintaining physiological cellular redox homeostasis, Nrf2 is considered a double-edged sword since overexpression of the Nrf2 leads to the development of various cancers, such as breast and prostate cancers [15,16]. Compounds that can control Nrf2 expression by maintaining it in the physiological range in healthy organisms, and preventing its constitutive activation in cancer, are of great importance [17]. Inflammation is intrinsically linked to oxidative stress, as ROS can directly or indirectly activate transcription factors such as nuclear factor kappa-light-chain-enhancer of activated B cells (NF-κB), which can promote inflammation and inflammaging [3,18]. Glycogen synthase kinase-3 (GSK3) has been proven to be a pivotal player within oxidative stress and inflammation networks, inhibiting the antioxidant function of Nrf2 and encouraging the inflammatory role of NF-κB [19]. Indeed, the active form of GSK3 is responsible for suppressing the expression of transcription factors required to overcome oxidative stress and inflammation [20]. GSK3 is a potent regulator of inflammation, while GSK3 inhibition protects from inflammatory conditions in animal models [21]. At the same time, the large varieties of stimuli fueling inflammaging converge on the activation of NF-κB and NLR family pyrin domain containing 3 (NLRP3) inflammasome, responsible for the production of inflammatory molecules. Together, Nrf2-Keap1, GSK3, NF-κB, and NLRP3 constitute a crucial regulatory mechanism in neuroinflammation and inflammaging [18,19].

In our previous papers, we showed that citrus flavanones naringenin (NAR), hesperetin (HES), and lemon extract have great potential in the restoration of the redox environment and alleviation of oxidative stress in old-aged rats [22–24]. Namely, by modulating the redox environment, these polyphenols up-regulated vitagene Sirtuin 1 (Sirt1) expression in thyrotropic cells [25–27], and improved thyroid gland sensitivity to thyroid-stimulating hormone (TSH) and overall liver health. Interestingly, studies regarding oxidative stress in the pituitary gland are scarce and mainly focused on the effects of alcohol, heavy metals, or moderate exercise [28–30]. However, our results suggest that soy isoflavone genistein significantly affected the HPA axis and decreased circulating ACTH and corticosterone in andropausal rats [31]. When it comes to the effects of citrus flavanones in the old pituitary, especially their relation to ACTH cell function, oxidative stress, and inflammation, there are no data so far.

Preliminarily speaking, the significance of the results presented herein regarding HES application in old-aged rats lies in the physiological improvement of the pituitary redox environment reflected in kelch-like ECH-associated protein 1 (Keap1) up-regulation, which consequently quenches excessive Nrf2 and directs its proteasomal degradation. Even the oxidative stress and inflammation are intertwined and contextualized as the Nrf2-Keap1/GSK3/NF-κB/NLRP3 regulatory loop (mechanism); these are the first data about this phenomenon in the old pituitary gland, through which we shed light on a notable potential of the plant-based, economical, and easily accessible nutraceuticals usage in the context of the pituitary health with advanced age.

2. Results

2.1. Treatment with NAR and HES Affects Nrf2 Immunofluorescent Signal, Protein, and Gene Expression in the Old Pituitary

Nrf2 is a master regulator of antioxidant and cytoprotective pathways. Its activation or inhibition leads to a series of alterations in the gene expression of molecules necessary to cope with oxidative stress. What we observed was that HES treatment induced an increase in Nrf2 gene expression by 132% and, at the same time, decreased its protein expression by 68% (Figure 1C,D). To confirm this, we applied the IF analysis of Nrf2 (counterstained with DAPI) to see the exact localization of Nrf2. Our analysis showed that the Nrf2 IF signal in ICON-, CON-, and NAR-treated groups was located in the cytoplasm of pituitary cells (Figure 1A). At the same time, we noticed a reduced signal in HES-treated pituitary sections (Figure 1A). Our IF analysis aligns with Nrf2 protein expression, which supports the thesis of a more reduced environment in the pituitary gland.

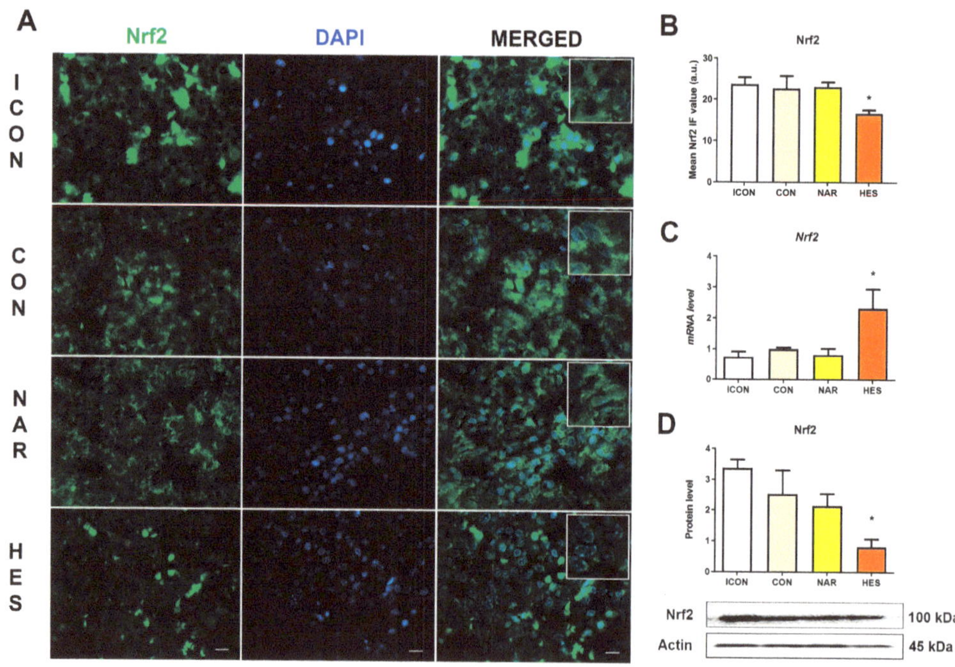

Figure 1. The effects of citrus flavanones on Nrf2 in the pituitary of old rats. Immunofluorescent staining of Nrf2, green and DAPI, blue signal (**A**), the mean Nrf2 IF value (**B**), the relative level of Nrf2 mRNA (**C**), and Nrf2 protein expression (**D**). Insets in the white boxes represent magnified parts of the same image, showing Nrf2 localization. The intense fluorescence spots on the micrographs represent erythrocytes and do not represent Nrf2. 63× magnification, bar = 20 μm. Each value represents mean ± SD, $n = 4$; statistics: one-way ANOVA, Dunett's multiple comparison *post hoc* test, * $p < 0.05$ citrus flavanone vs. CON rats. ICON, intact control; CON, control sunflower oil; NAR, naringenin; HES, hesperetin.

2.2. Treatment with NAR and HES Affects Keap1 Immunofluorescent Signal, Protein, and Gene Expression in the Old Pituitary

To assess the functional status of the Nrf2 protein, we tested the gene and protein expression of its negative regulator, Keap1. In physiological conditions, Keap1 is attached to Nrf2, which targets Nrf2 for proteasomal degradation. At the same time, in the stage of oxidative stress, Nrf2 detaches from Keap1, translocates to the nucleus, and activates the antioxidant enzyme expressions. Our results showed that NAR increased Keap1 gene expression by 46%, while HES treatment increased Keap1 protein expression by 200% (Figure 2C,D). Keap1 IF analysis showed that an intensive Keap1 IF signal was located in the cytoplasm of most cells in the pituitaries after treatment with HES (Figure 2A). In contrast, Keap1 IF signals in ICON, CON, and NAR were moderate and found in several cells' cytoplasm (Figure 2A). A significant increase in Keap1 led to a physiological decrease in Nrf2 protein expression, suggesting a less oxidative environment.

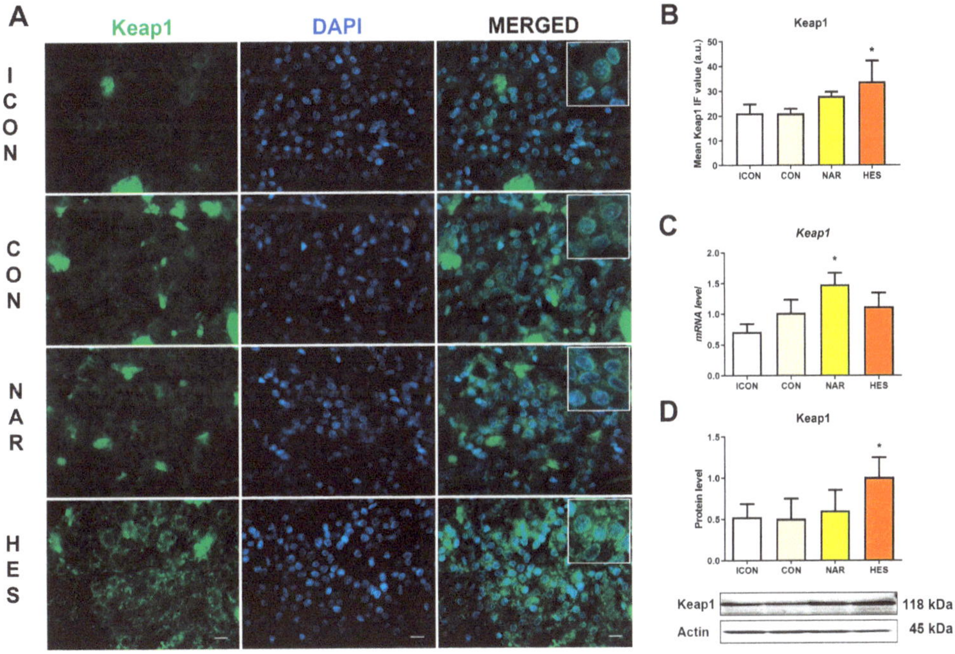

Figure 2. The effects of citrus flavanones on Keap1 in the pituitary of old rats. Immunofluorescent staining of Keap1, green and DAPI, blue signal (**A**), the mean Keap1 IF value (**B**), the relative level of Keap1 mRNA (**C**), and Keap1 protein expression (**D**). Insets in the white boxes represent magnified parts of the same image, showing Keap1 localization. The intense fluorescence spots on the micrographs represent erythrocytes and do not represent Keap1. 63× magnification, bar = 20 μm. Each value represents mean ± SD, $n = 4$; statistics: one-way ANOVA, Dunett's multiple comparison *post hoc* test, * $p < 0.05$ citrus flavanone vs. CON rats. ICON, intact control; CON, control sunflower oil; NAR, naringenin; HES, hesperetin.

2.3. Treatment with NAR and HES Affects Antioxidant Enzyme Gene and Protein Expressions in the Old Pituitary

Gene expression analysis of the antioxidant enzymes can give us information on the mRNA level of the enzymes involved in antioxidant defense. After treatment with NAR, *Trxr1*, *Gpx*, and *Gr* gene expression increased by 55, 109, and 59%, respectively (Figure 3A), while all other targets did not alter (Figure 3A). Gene expression of examined antioxidant enzymes remained unchanged after HES treatment (Figure 3A).

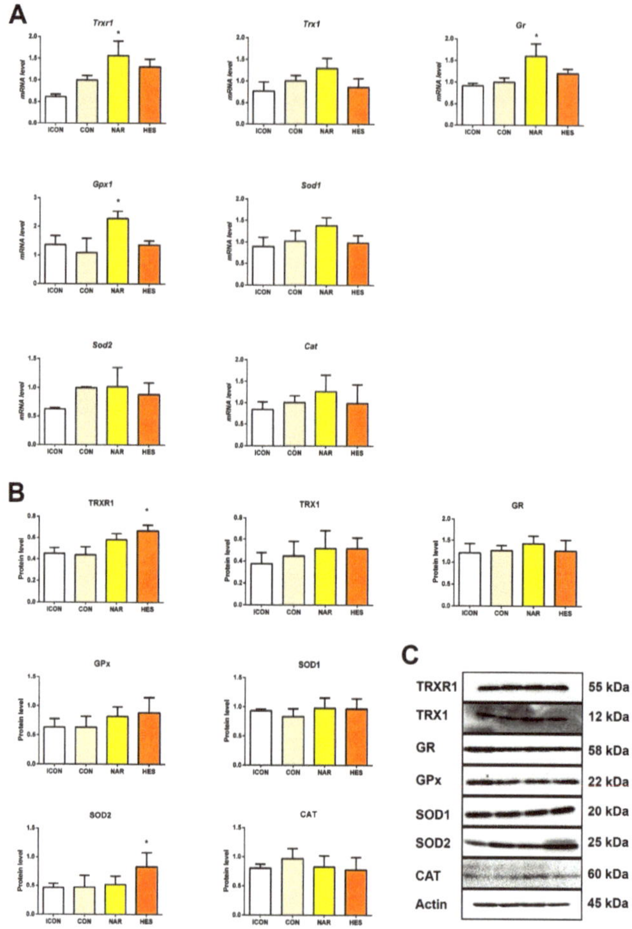

Figure 3. The effects of citrus flavanones on antioxidant enzymes expressions in the pituitary of old rats. Thioredoxin reductase 1 (*Trxr1*), thioredoxin 1 (*Trx1*), glutathione reductase (*Gr*), glutathione peroxidase 1 (*Gpx1*), superoxide dismutase 1 (*Sod1*), superoxide dismutase 2 (*Sod2*), and catalase (*Cat*) gene expression (**A**), TRXR1, TRX1, GR, GPx, SOD1, SOD2, and CAT protein expression (**B**), and protein profile (**C**). Each value represents mean ± SD, $n = 4$; statistics: one way ANOVA, Dunett's multiple comparison post hoc test, * $p < 0.05$ citrus flavanone vs. CON rats. ICON, intact control; CON, control sunflower oil; NAR, naringenin; HES, hesperetin.

Immunoblot analysis is informative regarding the protein abundance of the main actors in the antioxidant defense system. Namely, only after the HES treatment, protein levels of TrxR1 and SOD2 were up-regulated by 51% and 76%, respectively (Figure 3B). All other examined parameters did not change, likewise after NAR (Figure 3B).

2.4. Treatment with NAR and HES Affects Collagen Abundance, While Inflammatory Markers Remain Unchanged in the Old Pituitary

Picro-Sirius Red staining is used to visualize collagen accumulation and fibrosis onset in tissue sections (Figure 4A). In both the ICON and CON control groups of aged rats, some small amount of collagen is distributed in the periphery of the pituitary pars distalis, mainly between the cell clusters and around the blood vessels (Figure 4A). Upon the treatment with citrus flavanones (NAR and HES), collagen accumulation is observed around the dilated blood vessels and between the cell clusters (Figure 4A). Of all examined parameters related to inflammation, neither NAR nor HES affected their gene expression (Figure 4B). Protein expression of GSK3 decreased by 16% (Figure 4C), while protein expression of other examined parameters did not change after the treatment with citrus flavanones (Figure 4C).

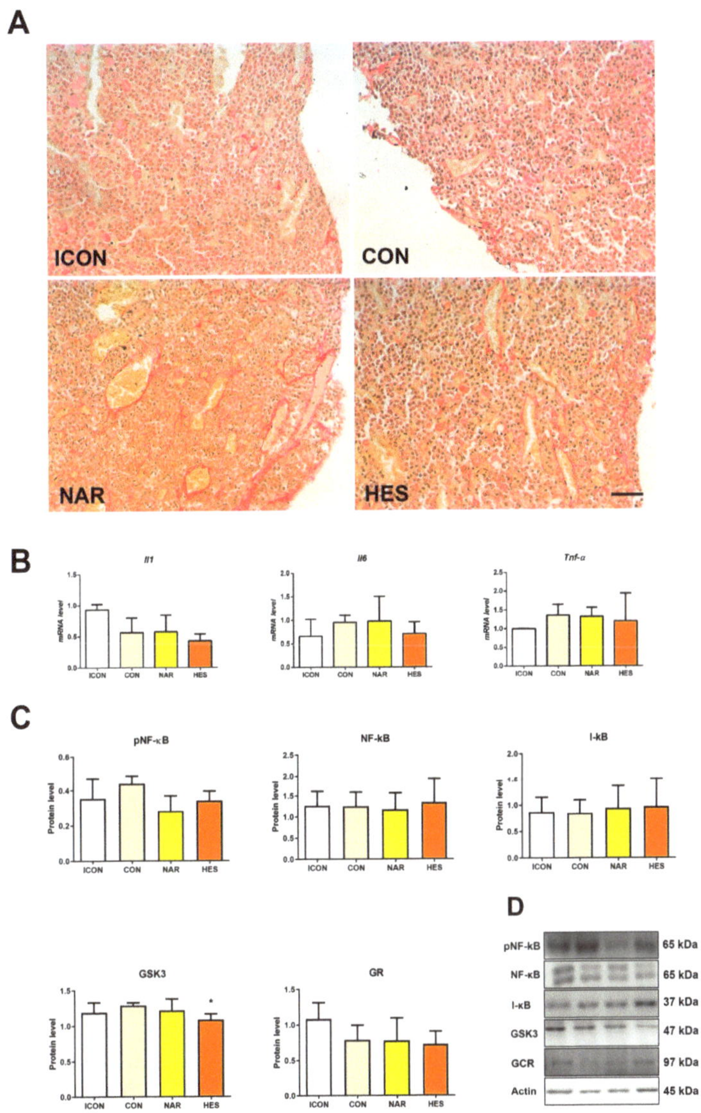

Figure 4. The effects of citrus flavanones on collagen accumulation and inflammatory pathway in the pituitary of old rats. Picro-Sirius Red staining of collagen (**A**), the relative level of *Il1*, *Il6*, and *Tnf-α* mRNA (**B**), protein expression of pNF-κB, NF-κB, I-κB, GSK3, and GCR (**C**), and protein profile (**D**). 20× magnification, bar = 50 µm. Each value represents mean ± SD, $n = 4$; statistics: one-way ANOVA, Dunett's multiple comparison post hoc test, * $p < 0.05$ citrus flavanone vs. CON rats. ICON, intact control; CON, control sunflower oil; NAR, naringenin; HES, hesperetin.

2.5. Treatment with NAR and HES Affects NLRP3 Expression in the Old Pituitary

The NLRP3 inflammasome is a multiprotein complex that plays a pivotal role in regulating the innate immune system and inflammatory signaling, mediating the secretion of proinflammatory cytokines in response to cellular damage. Our results showed that NAR and HES decreased the immunohistochemical optical density of NLRP3 by 28 and 20%, respectively (Figure 5A,B). Even the immunoblot analysis did not show statistically significant values; the NLRP3 protein profile followed the same lowering trend after both flavanones (Figure 5C).

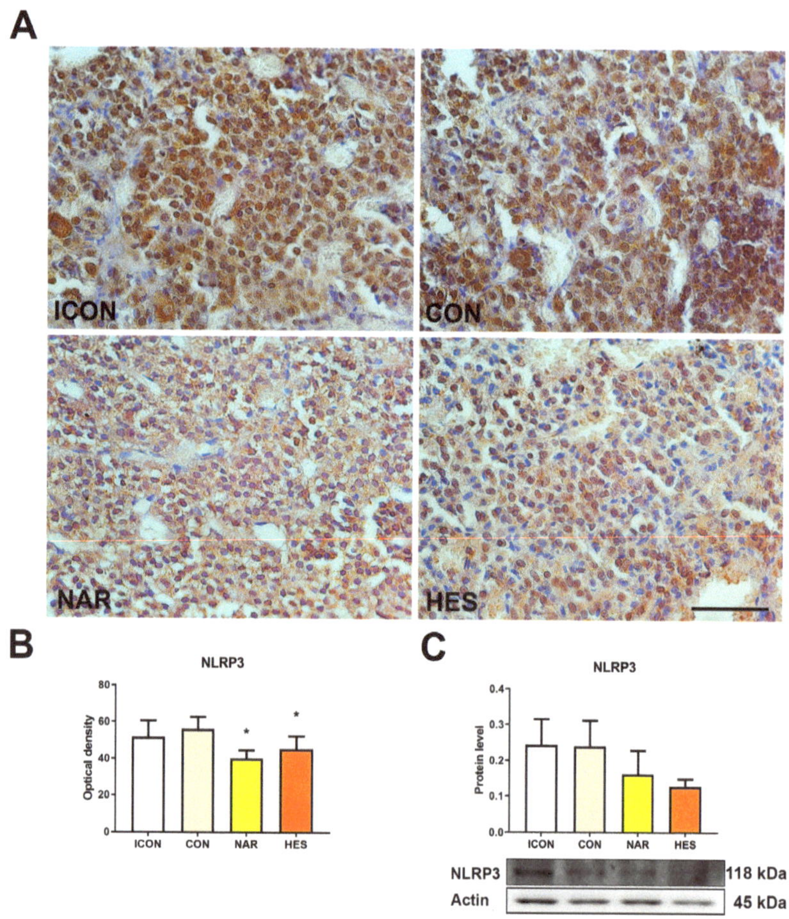

Figure 5. The effects of citrus flavanones on NLRP3 expression in the pituitary of old rats. Immunohistochemical staining of NLRP3 (**A**), optical density of NLRP3 (**B**), and protein expression of NLRP3 (**C**). 40× magnification, bar = 25 μm. Each value represents mean ± SD, $n = 4$; statistics: one-way ANOVA, Dunett's multiple comparison post hoc test, * $p < 0.05$ citrus flavanone vs. CON rats. ICON, intact control; CON, control sunflower oil; NAR, naringenin; HES, hesperetin.

2.6. Treatment with NAR and HES Do Not Affect ACTH Signal Intensity in the Old Pituitary

In control-aged rats (ICON and CON), ACTH-immunopositive cells were mainly distributed in the periphery of the pituitary *pars distalis*, constituting small groups between or close to dilated blood vessels (asterisk, Figure 6A). The ACTH cells were morphologically polygonal or prolate in shape, occasionally with elongated cytoplasmic protrusions penetrating between neighboring cells (Figure 6A). In a naringenin-treated group of old-aged rats, ACTH cells were uniformly distributed throughout the *pars distalis* and resembled the shape of the ACTH cells in control groups (Figure 6A). Upon HES treatment of aged rats, ACTH cells' uniform distribution and characteristic shape were maintained; however, their IHC signal, reflecting the hormonal content in the cells, looked lower (Figure 6A). This may suggest that some cells produce more ACTH after the HES-treatment group (Figure 6A,B). Nevertheless, our quantitative analysis showed no difference in the signal intensity of ACTH cells in treated groups compared to control values (Figure 6B).

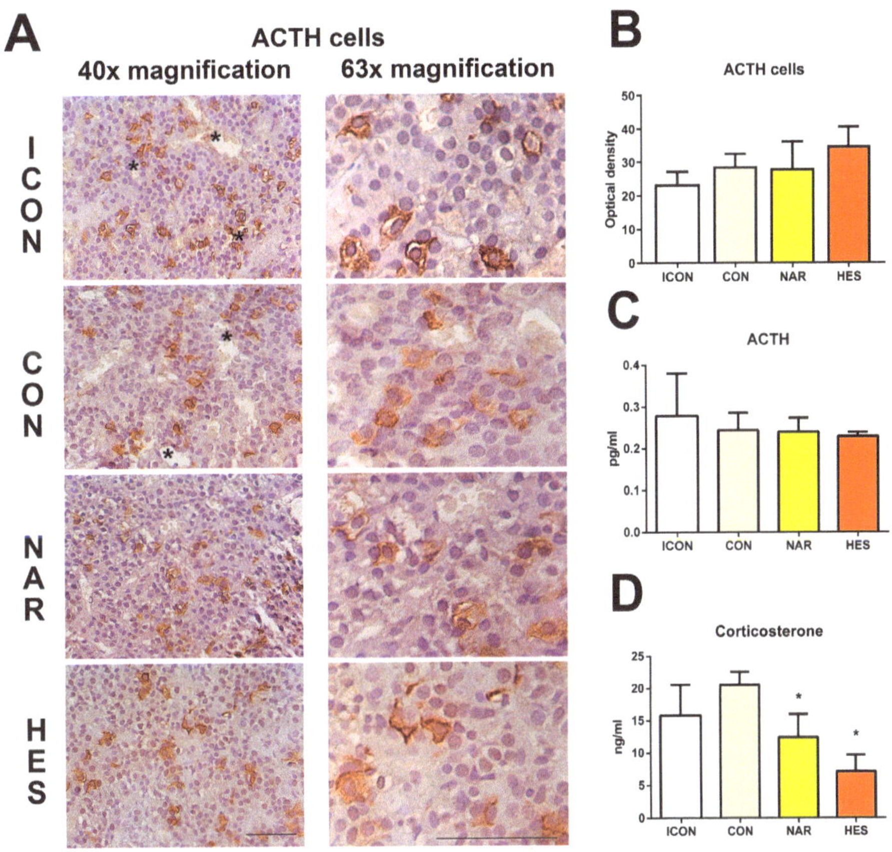

Figure 6. The effects of citrus flavanones on corticotroph cells in the pituitary, ACTH, and corticosterone concentration of old rats. Immunohistochemical staining of ACTH (**A**), optical density of ACTH cells (**B**), plasma level of ACTH (**C**), and corticosterone level in serum (**D**). Asterisk (*) highlights dilated blood vessels with ACTH cells around them. 40× and 63× magnification, bar = 25 and 20 μm. Each value represents mean ± SD, $n = 4$; statistics: one-way ANOVA, Dunett's multiple comparison post hoc test, * $p < 0.05$ citrus flavanone vs. CON rats. ICON, intact control; CON, control sunflower oil; NAR, naringenin; HES, hesperetin.

2.7. Treatment with NAR and HES Affects the Serum Level of Corticosterone but No ACTH Level in Old Rats

Levels of plasma ACTH and serum corticosterone were measured biochemically, and they are essential parameters for testing the pituitary–adrenal axis function (Figure 6C). The level of ACTH in all examined groups was almost the same. The level of corticosterone in the ICON group was 15.81 ng/mL, while in the CON group, the value was 20.58 ng/mL

(Figure 6D). After treatment with NAR or HES, the serum level of corticosterone was 12.41 ng/mL and 7.00 ng/mL, respectively (Figure 6D). So, the corticosterone level after NAR decreased by 40%, while after HES, it was 66% lower than the CON values (Figure 6D).

3. Discussion

Our study has revealed that HES, but not NAR, can improve a reduced redox milieu in the old pituitary. HES showed the most substantial effect on TrxR1 and Keap1 up-regulation, followed by the down-regulation of Nrf2, GSK3, and NLRP3 protein expression. At the same time, NAR affects the gene expression of *Keap1*, *Trxr1*, *Gpx*, and *Gr* enzymes involved in the antioxidant defense and decreases expression of NLRP3 protein in the pituitary glands of old rats. Both citrus flavanones decreased serum corticosterone concentration without affecting the pituitary's ACTH level or inflammation signaling. Accumulation of the collagen around blood vessels was observed after both flavanones. Our results suggest that HES (and NAR), by attuning oxidative stress, could help the pituitary gland, without interruption, maintain its master endocrine regulatory function in old age.

Considering the worldwide consumption of citruses and the lack of available data concerning the effects of NAR and HES on pituitary function, we aimed to evaluate their outcomes in the old rats. We combined molecular biology results with structural data obtained with immuno-histochemical/-fluorescent staining regarding redox parameters, inflammatory signaling, and hormonal levels. The rationale for this study is evidence of the existence of the oxidative and proinflammatory Nrf2-Keap1/GSK3/NF-κB/NLRP3 regulatory loop [19], which is unexplored so far in the model of natural old age.

HES exerted one of the most significant results in the study by increasing Keap1 and decreasing Nrf2 protein expression. This effect is noteworthy, considering Keap1 constantly targets Nrf2 for proteasomal degradation under normal, reduced conditions [32]. Keap1 also acts as a sensor for Nrf2-activating compounds, which target the key Cys151 residues in Keap1, causing the protein to undergo conformational changes [33]. Despite HES's high binding affinity to Keap1 [23], it does not possess prooxidant properties that could activate it, leading to its detachment from Nrf2. This result aligns with unchanged antioxidant enzyme gene expression regulated by Nrf2, observed after HES treatment. Thus, we support the thesis of a more reduced environment in the old pituitary as a positive effect.

The pituitary gland is a master endocrine gland, which develops during week four of fetal development in humans and from the 12th gestational day in rats [34,35]. Such a substantial organ has to have robust redox-sensitive regulation and antioxidant protection like the TrxR1/Trx1 system. An increase in TrxR1 protein expression after HES treatment leads to a higher capacity for Trx1-mediated disulfide reduction, de-nitrosylation, and aldehyde–protein adduct formation [3,33]. This increment helps to suppress alterations in peptide hormone synthetic machinery by decreasing ROS-mediated cellular damage. Trx enzymes recognize the oxidized form of their target proteins with higher selectivity than their corresponding reduced forms [36]. After HES treatment, higher TrxR1 expression keeps Cys151 in Keap1 in reduced form (attached to Nrf2) [33]. Thus, an improved TrxR1/Trx1 system can improve overall pituitary cell health and hormone production with aging [3].

In addition, TrxR1 is an essential negative regulator of Nrf2, meaning that loss of TrxR1 activity leads to Nrf2 activation and vice versa [33]. Interestingly, several polyphenolic compounds have dual but opposite effects on these two proteins. Compounds such as curcumin, quercetin, ellagic acid, isothiocyanates, and sulforaphane possess both inhibitory activities against TrxR1 and, at the same time, the ability to activate Nrf2 by covalently modifying reactive Cys residues in Keap1 in vitro [33]. Nonetheless, our in vivo study suggests this is a less possible scenario due to the up-regulation of Keap1 and TrxR1. In vivo studies differ from in vitro studies due to complex interaction and metabolic transformations of the examined compounds.

Up-regulation in the mitochondrial enzyme SOD2 is another positive outcome after HES treatment. Even though we showed a decrease in Nrf2 and unaltered NF-κB protein expression after HES, a possible explanation could involve increased Sirt1 expression in the same organ and model we showed in our previous study [23,37]. Namely, Sirt1, by deacetylation, enhances SOD2 expression in mitochondria by activating forkhead box protein O (FOXOs), resulting in increased removal of O_2^- or attenuating mitochondrial ROS production [38,39]. Additionally, resveratrol and quercetin have been found to enhance SOD2 expression mediated by Sirt1, thus mitigating oxidative stress in different animal models [40]. SOD2, by eliminating O_2^- generates O_2 and H_2O_2, which GPx and CAT scavenge. Both enzymes are crucial in the host's defense against oxidative stress and cell protection against H_2O_2 toxicity [41]. In our study, GPx and CAT expression did not change after treatment with NAR and HES, proposing no significant H_2O_2 insult in the old pituitary.

Furthermore, our study revealed a decrease in GSK3 protein expression after HES. This result is beneficial since the active form of GSK3 promotes the production of proinflammatory cytokines such as IL-1β, TNF-α, and IL-6 [20,21], whose gene expressions were unchanged after citrus flavanones. In addition, GSK3 down-regulation reduces its proinflammatory function since GSK3 directly stimulates NF-κB signaling [21,42]. Besides inflammation, GSK3 maintains redox equilibrium in the body. It does so by stimulating the production of ROS in the mitochondria and increasing the inner mitochondrial membrane permeability to detrimental molecules [43]. Activated GSK3 down-regulates the expression of Nrf2 and its ability to bind to ARE in cerebral ischemia and reperfusion injury [44]. Overall, down-regulating GSK3 and up-regulating TrxR1 and SOD2 after HES in our study suggest that HES can protect cells from ROS and inflammation, increasing pituitary cell survival.

After treatment with citrus flavanones, we did not observe fibrosis in the aged human adenohypophyses [45]. However, we observed the collagen positivity surrounding the blood vessels, prominently seen after NAR or HES, without affecting the pituitary parenchyma. Fibrosis is a repair mechanism that becomes activated after the body is stimulated by inflammation or physical damage. Activating the NF-κB signaling pathway and increasing IL-6 is probably the critical mechanism in developing pituitary fibrosis and inflammaging process in older adults [46,47]. NAR and HES did not alter inflammatory signaling mediated by NF-κB, or the pNF-κB/NF-κB ratio; thus, this change is probably a result of better blood supply and organ perfusion after our treatment. In general, the anti-inflammatory effect of GCR is attributed to the suppression of inflammatory genes and NFkB [48,49]. However, its protein content remained unchanged under HES and NAR treatment, confirming our assumption that inflammation was not affected and that the observed morphological changes are rather the result of increased blood supply to the pituitary gland. It is important to emphasize that polyphenols can pass the blood–brain barrier. Namely, metabolites of polyphenols are hydrosoluble molecules, unlike their aglycone form [50,51], meaning that after their complete metabolism in the liver, they can be transported all over the organism by blood. Although pituitary capillaries are part of the central nervous system vascularization, their blood vessel components have specific properties that differ from vascular networks in other brain areas due to fenestrated capillaries, allowing the passing of hormones, nutrients, etc. [52]. Polyphenol metabolites can reach pituitary cells, where they exert protective effects by modulating oxidative stress and inflammation [53].

A decrement in NLRP3 expression in the pituitary is in line with the previous results. Inflammasomes have been identified as essential drivers of sterile inflammation, and there is accumulating evidence that NLRP3 activation might also play an important role in the aging process [54]. It has been revealed that the cytosolic ROS induced by NADPH is responsible for the activation of the NLRP3 inflammasome [55], while in our study, lower NLRP3 expression is in line with a better redox environment in the pituitary. Our result is a positive outcome since microglial NLRP3 inflammasome activation up-regulates the pituitary glands' inflammatory cytokines IL1/IL18, and induces prolactinomas [56] or other impairments that could compromise normal pituitary functioning. HES and NAR showed positive neuroprotective effects by down-regulating the NLRP3 inflammasome activation in the brain [57,58]. However, the applied doses were 3–6 times higher than ours. Therefore, this nominates citrus flavanones as nutritionally achievable supplements that can reduce pituitary inflammaging and age-related pathologies at the population level.

The main effect after HES treatment is the activation of two of the three vitagenes essential for the neuronal function and longevity phenomenon [25,59–61], TRXR1, and Sirt1 [23]. This result promotes HES as a desirable supplement in alleviating ROS-induced diseases and improving aging-related malfunction of pituitary-related diseases. NAR also affected examined parameters since it increased *Keap1*, *Trxr1*, *Gpx*, and *Gr* mRNA expression, while there is an increasing trend regarding TrxR1, Trx1, GR, GPx, and Keap1 protein levels. Citrus flavanones in our study exerted different effects since HES, compared to NAR, possesses one hydroxyl less and one methyl group more, making it less prooxidant than NAR. This suggests that even though it has a similar structure, NAR, in comparison with HES, exerts different biological effects, but mainly in the same direction of change. However, low polyphenol doses in our study follow the hormesis concept, meaning preconditioning with HES or NAR can help the old pituitary handle upcoming stress better.

As we described, citrus flavanones (HES more than NAR) protect pituitary gland synthetic/secretory and overall regulatory pathways from ROS-mediated damage of peptide/protein components. Namely, ACTH cell signal intensity in NAR- and HES-treated old rats did not change, but a certain accumulation of hormonal content in these cells upon HES application suggests their synthetic activation. This result may be related to a significantly reduced corticosterone (even more than compared to the NAR group) and the initiation of a feedback mechanism. Generally, ACTH release is mediated by the adenylate cyclase protein kinase (PKA) system [62], and oxidants, if present for a longer time or in higher concentrations, can inhibit PKA phosphorylation by directly oxidizing a reactive cysteine in its catalytic subunit [63]. However, in our study, the serum ACTH and corticotroph IHC signals remained unchanged, suggesting some potential issues in the central regulation at the pituitary level. Namely, it was shown that with aging in rats and humans, there is an impairment in the ACTH cell's response to corticosterone, leading to decreased HPA axis sensitivity to the corticosterone/cortisol feedback loop, which is all accompanied by degenerative processes in the hippocampal and hypothalamic neurons [2,4,64,65]. Therefore, even with lower corticosterone, ACTH remained unaltered in serum and corticotrophs. Additionally, a decrease in corticosterone is in line with the effects of soy isoflavones in the animal model of andropause [66,67]. What deserves attention in future studies that will follow up on this one is examining the effects of citrus flavanones on the activity of adrenocortical steroidogenesis enzymes. Indeed, we did not analyze the potential impact of citrus flavanones on adrenal glands since that was not the main focus of our study. However, by disrupting corticosteroidogenesis, potentially due to inhibiting of hormone-generating machinery, citrus flavanones could modulate and alter stress hormone levels in old rats. Further, we advocate decreasing stress hormones after NAR and HES as a positive outcome since increased cortisol level is associated with metabolic, somatic, and psychiatric conditions related to aging [68,69], while proinflammatory cytokines secreted in many of these conditions may act on the HPA axis, increasing glucocorticoid secretion [69]. Studies showed that lower cortisol in humans and corticosterone in rats positively correlate with mammalian longevity [70–72] due to the ability of these hormones to reduce the cellular production of free radicals, which is observed in long-lived species [72]. Revived thyrotropic activity, which has already been described [23] and showed positive effects on corticosterone, indicates citrus flavanones are potential longevity-modulating molecules.

In general, the results obtained in our study are valuable since we analyzed two processes that can have detrimental effects on the pituitary gland: oxidative stress and inflammation. After HES and NAR, inflammation stayed at a steady state level, and the redox environment even improved. A decline in endocrine function characterizes the old pituitary; still, it remains an essential player in maintaining the organism's homeostasis. However, we showed that our treatments do not influence ACTH cell morphology, IHC signal, and ACTH blood level, even with the provided positive environment. NLRP3 expression was down-regulated, which directly reduced the generation of the local proinflammatory cytokines, which could contribute to HPA axis regulation in a paracrine manner. This is positive since prolonged inflammatory response in aged animals might be linked to dysregulated pituitary cytokine interactions, leading to a rise in serum corticosterone levels [73]. Another benefit of our study is that it opens another new research focus, directed towards the adrenal gland, considering the decline in corticosterone serum levels. In the end, with the other positive outcomes of the present and our previous works [22–24], we are persistently getting closer to elucidating the effects of citrus flavanones on extending the life span.

4. Materials and Methods

4.1. Experimental Animals

Two-year-old male Wistar rats used in the experiment were bred and housed in the Unit for Experimental Animals at the Institute for Biological Research "Siniša Stanković"—National Institute of the Republic of Serbia, Belgrade, Serbia, in cycles of 12 h light and 12 h dark and constant temperature (21 ± 2 °C) conditions, and had access to a chow diet and water *ad libitum*. All animal procedures complied with Directive 2010/63/EU on protecting animals used for experimental and other scientific purposes and were approved by the Ethical Committee for the Use of Laboratory Animals of IBISS, University of Belgrade (No 2-12/12).

4.2. Dosage Regimen

At the beginning of the experiment, we randomly divided the old-aged rats into four experimental groups ($n = 6$ per group). Treated groups of animals received *per os* 15 mg/kg b.m. of citrus flavanones naringenin (NAR) or hesperetin (HES) mixed with sunflower oil. To conduct a nutritionally relevant study, we avoided gavage, and the mixture was applied to the oral cavity, considering the role of oral microbiota in consumed flavanone biotransformation. The applied volume of the mixture was 300 µL *per* animal by syringe directly to the oral cavity. The control group (CON) received the same volume of the vehicle, while physiologically intact controls (ICON) represent physiologically intact animals. The treatments were administered daily for four weeks.

4.3. RNA Isolation, cDNA Transcription and Real-Time PCR

Total pituitary RNA was isolated using TRIzol (Invitrogen, Carlsbad, CA, USA) and purified using the RNeasy mini-kit (74106, QIAGEN, Hilden, Germany) following the manufacturer's instructions. The cDNA was synthesized with a High-Capacity cDNA Reverse Transcription Kit (4368813, Applied Biosystems, Vilnius, Lithuania) with 500 ng of RNA. PCR amplification of cDNAs was performed in a real-time PCR machine ABI Prism 7000 (Applied Biosystems, Waltham, MA, USA) with SYBRGreen PCR master mix (4309155, Applied Biosystems, Waltham, MA, USA). The program included the following conditions: 3 min at 95 °C, followed by 40 cycles of 15 s at 95 °C, 30 s at 60 °C, and 30 s at 72 °C. The list of primers used is shown in Table S1. Melting curve analysis was used to confirm gene-specific amplification. Nuclease-free water was used as a negative control instead of a cDNA template from the individual samples to test if there was no residual genomic DNA. The expression level of each gene was calculated using the formula $2 - (\Delta\Delta^{Ctexp} - \Delta\Delta^{Ctcontrol})$, where ΔCt is different between the cycle threshold value of the gene of interest and the cycle threshold value of *Gapdh* or *Hprt* as a reference gene. All of the data were calculated from duplicate reactions. RNA data are presented as average relative levels vs. *Gapdh* ± SD for antioxidant enzymes or *Hprt* ± SD for inflammatory markers.

4.4. Protein Isolation for Western Blot

Total protein isolation from the rat pituitary was performed using TRIzol following the manufacturer's protocol with slight modifications published by [23]. Briefly, the manufacturer's protocol for protein isolation was followed until the precipitation of proteins from the phenol-ethanol supernatant. At this step, the protein pellet was completely dissolved in 7 M GndCl solution, and then proteins were precipitated again by adding 100% ethanol. This step was repeated, followed by a final wash of the protein pellet with 100% ethanol. After 10 min of air drying, the protein pellet was solubilized in the buffer containing 8 M Urea, 40 mM Tris pH 8, 2% SDS, and 1× Protease G inhibitor cocktail (39101.03, Serva, Heidelberg, Germany). Protein concentration in samples was measured by DC Protein Assay (#5000112, BioRad, Goettingen, Germany), using bovine serum albumin as standard.

4.5. SDS Polyacrylamide Gel Electrophoresis and Western Blot

Proteins were solubilized in 4× Laemmli sample buffer supplemented with 10% β-mercaptoethanol. A quantity of 15 μg of protein *per* lane was subjected to 10% or 12% SDS-polyacrylamide gel electrophoresis and transferred electrophoretically to polyvinylidene difluoride (PVDF) or nitrocellulose membranes with a semidry or wet blotting system (Fastblot B43; Bio-Rad, Goettingen, Germany). The next step was blocking of unbound sites on the membranes with 5% BSA (prior incubation with SOD1, SOD2, CAT, GPx, GR, Keap1, Nrf2, Trx1, TrxR1, pNF-kB, NF-kB, I-kB, GCR, GSK3, NLRP3, and β-Actin; Table S2) for one hour. The membranes were incubated with primary antibodies overnight at 4 °C. The list of used antibodies is shown in Table S2. After washing, blots were incubated with secondary antibodies for one hour at room temperature. Antibody binding was detected using a chemiluminescence detection system (ECL; BioRad, Goettingen, Germany). Signals were quantified by densitometry using ImageJ Image Analysis Software (v1.48).

4.6. Histological Sample Preparation

Pituitaries were excised and fixed in 10% formalin for 48 h, washed in tap water, then dehydrated in a series of increasing concentrations of ethanol (30–100%), enlightened in xylol, and embedded in Histowax® (Histolab Product AB, Göteborg, Sweden). Sections with a thickness of 5 μm were prepared using a rotational microtome (RM 2125RT Leica Microsystems, Wetzlar, Germany).

4.7. Sirius Red Histochemical Staining

Sirius Red staining, used in histological analysis, is often a method of choice when there is a need to distinguish different cells from collagen. In the initial steps, the procedure involved deparaffinization (xylol) and rehydration (100–70% ethanol, distilled water) of the pituitary sections, followed by incubation in Weigert's hematoxylin (8 min). After washing in running tap water, the sections were incubated in a Picro-Sirius Red solution (0.5 g of Sirius Red (Direct Red 80, 365548; Sigma Aldrich, Co., St. Louis, MO, USA) + 500 mL of saturated aqueous solution of picric acid) for 1 h. The next step included double washing pituitary sections in acidified water (5 mL of glacial acetic acid in 1 L of distilled water), followed by vigorous shaking to physically remove most of the water from the slides. Finally, the sections were dehydrated in three changes of 100% ethanol, cleared in xylol, and mounted in DPX (Sigma-Aldrich, Co., St. Louis, MO, USA).

4.8. Immunohistochemical Staining of ACTH and NLRP3

The representative sections were stained with immunohistochemical (IHC) methods according to previously described procedures [31]. After tissue deparaffinization, endogenous peroxidase activity was blocked by sections incubated with 0.3% hydrogen peroxide in methanol for 15 min. Only slides used in NLPR3 IHC analysis were exposed to heat-induced antigen retrieval by placing them in a container with citrate buffer (pH 6.0) and then heated at 750 W in a microwave oven for 10 min. Non-specific background staining was reduced by incubation with normal swine serum (X0901, Dakopatts, Glostrup, Denmark) diluted 1:10 for 45 min. For functional IHC analysis, the rabbit antisera directed against ACTH and NLRP3 (Table S2) were applied overnight at room temperature. The primary antibody was substituted with phosphate buffer saline (PBS) for the negative control of the pituitary sections. Secondary antibodies labeled with HRP were applied for one hour. After secondary antibody incubation, NLRP3 slides were incubated with the VECTASTAIN ABC Kit (PK-4001, Vector Laboratories, Inc., Burlingame, CA, USA) according to the manufacturer's instruction. Visualization was performed using the Dako liquid diaminobenzidine tetrahydrochloride substrate chromogen system (K3468, Dako North America, Inc., Carpinteria, CA, USA) at concentrations suggested by the manufacturer. All washes and dilutions were performed using 0.1 mol/l PBS pH 7.4. Hematoxylin was used as a counterstain, and slides were mounted in DPX medium (Sigma-Aldrich, Barcelona, Spain).

4.9. Immunofluorescent Staining of Nrf2 and Keap1

The representative pituitary sections were stained using the immunofluorescence (IF) method, as previously described [24], with certain modifications. Namely, pituitary sections were exposed to heat-induced antigen retrieval after tissue deparaffinization and rehydration by placing them in a container with Tris-EDTA buffer (pH 9.0) for Nrf2 or citrate buffer (pH 6.0) for Keap 1, and then heated at 750 W in a microwave oven for 10 min. Then, only the samples intended for Nrf2 analysis were incubated for 15 min with PBS containing 0.2% Triton X-100 to permeabilize cells and nuclear membranes. Non-specific background staining was reduced by incubation with normal donkey serum (ab7475, Abcam, Cambridge, UK) diluted 1:10 for 45 min. For functional IF analysis, the rabbit antisera directed against the Nrf2 or Keap1 antigen was applied overnight at room temperature. A secondary antibody (Table S2) was applied for 45 min and washed 5×5 min with PBS 0.05% Tween20 pH 7.4. As counterstain, we used 4′,6-diamidino-2-phenylindole (DAPI; 300 nM) for 5 min and washed with PBS for 3×5 min. Slides were mounted in Mowiol medium (Sigma-Aldrich, St. Louis, MO, USA).

4.10. Image Acquisition

Digital images of the ACTH-, NLRP3-immunopositive, and Sirius Red-stained pituitary sections were taken using a LEITZ DM RB light microscope (Leica Mikroskopie & Systems GmbH, Wetzlar, Germany), a LEICA DFC320 CCD camera (Leica Microsystems Ltd., Heerbrugg, Switzerland), and the Leica DFC Twain Software (4.11.0, Leica, Wetzlar, Germany). A Zeiss Axiovert fluorescent microscope (Zeiss, Graz, Austria) was used to acquire fluorescent images of Nrf2 and Keap1 antibodies.

4.11. Concentration of the ACTH and Corticosterone

Blood was collected from the trunk, and separated plasma and sera samples of all the animals were stored at the same time at −70 °C until assayed. Plasma levels of ACTH were determined without dilution by the IMMULITE method (2500 ACTH, L5KAC2, DPC, Los Angeles, CA, USA), in duplicate samples within a single assay, with an intra-assay CV of 9.6%. The detection limit of the assay, defined as the concentration two standard deviations above the response at zero dose, is approximately 9.0 pg/mL. Serum corticosterone concentrations were measured without dilution by immunoassay (KGE009, R&D Systems Inc., Minneapolis, MN, USA), in duplicate samples within a single assay, with an intra-assay CV of 8.0%. The sensitivity of this corticosterone immunoassay is typically less than 27.0 pg/mL.

4.12. Statistical Analysis

All obtained results were analyzed using GraphPad Prism v.6 for Windows (San Diego, CA, USA). The data for the experimental groups were first tested for distribution normality with the Kolmogorov–Smirnov test. After confirmation of a Gaussian distribution and homogeneity of variance with Bartlett's test, one-way ANOVA was used for further comparative evaluation, followed by Dunett's *post hoc* test. The potential effect of vehicle (sunflower oil; CON) was evaluated versus untreated ICON animals. The effect of NAR or HES treatment on all examined parameters was evaluated in comparison to the values obtained for the CON group. A confidence level of $p < 0.05$ was considered statistically significant. The data are presented as means ± SD.

5. Conclusions

Finally, the lack of research regarding the current topic suggests that more data are needed to make a conclusive decision about how beneficial these molecules could be for pituitary function. However, with data from the recent study (summarized in Figure 7), we opened a new important question: whether citrus flavanones (and other polyphenols) can be used as nutritionally relevant molecules capable of affecting oxidative stress and inflammation in the pituitary. Considering that the pituitary gland is a master endocrine regulator, every molecule that can affect it is worth testing since it could be a new, low-cost, and already accepted food supplement for improving and facilitating pituitary redox and/or endocrine disorders in all ages.

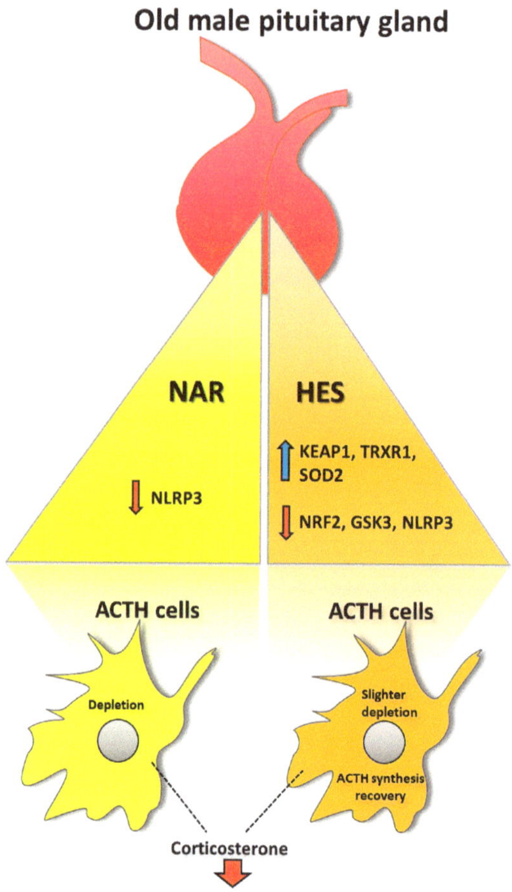

Figure 7. Summarized naringenin (NAR) and hesperetin (HES) effects in the pituitary gland of the old male rats. *Trxr1*, thioredoxin reductase 1; SOD2, superoxide dismutase 2; NRF2, nuclear factor erythroid 2-related factor 2; GSK3, glycogen synthase kinase-3; keap1, kelch-like ECH-associated protein 1; NLRP3, NLR family pyrin domain containing 3. Blue arrow, up-regulation; red arrow, down-regulation.

Supplementary Materials: The following supporting information can be downloaded at: https://www.mdpi.com/article/10.3390/ijms25168918/s1.

Author Contributions: Conceptualization, M.M., J.Ž. and V.A.; investigation, M.M., J.Ž., S.K., N.V. and V.A.; validation, M.M., J.Ž., S.K., V.A., N.V. and A.D.; visualization, M.M., J.Ž., S.K. and N.V.;

formal analysis, A.D. and B.F.; writing—original draft preparation, M.M.; writing—review and editing, J.Ž., V.A., S.K., N.V., A.D. and B.F. All authors have read and agreed to the published version of the manuscript.

Funding: This research was funded by the Ministry of Science, Innovation and Technological Development of the Republic of Serbia, contract number 451-03-66/2024-03/200007.

Institutional Review Board Statement: All animal procedures complied with Directive 2010/63/EU on protecting animals used for experimental and other scientific purposes and was approved by the Ethical Committee for the Use of Laboratory Animals of IBISS, University of Belgrade (No 2-12/12).

Informed Consent Statement: Not applicable.

Data Availability Statement: Data are contained within the article and Supplementary Materials.

Acknowledgments: In this section, you can acknowledge any support given which is not covered by the author contribution or funding sections. This may include administrative and technical support, or donations in kind (e.g., materials used for experiments).

Conflicts of Interest: The authors declare no conflicts of interest.

References

1. Chahal, H.; Drake, W. The endocrine system and ageing. *J. Pathol.* **2007**, *211*, 173–180. [CrossRef]
2. Ajdžanović, V.; Trifunović, S.; Miljić, D.; Šošić-Jurjević, B.; Filipović, B.; Miler, M.; Ristić, N.; Manojlović-Stojanoski, M.; Milošević, V. Somatopause, weaknesses of the therapeutic approaches and the cautious optimism based on experimental ageing studies with soy isoflavones. *EXCLI J.* **2018**, *17*, 279–301. [CrossRef] [PubMed]
3. Vitale, G.; Salvioli, S.; Franceschi, C. Oxidative stress and the ageing endocrine system. *Nat. Rev. Endocrinol.* **2013**, *9*, 228–240. [CrossRef] [PubMed]
4. Ferrari, E.; Cravello, L.; Muzzoni, B.; Casarotti, D.; Paltro, M.; Solerte, S.; Fioravanti, M.; Cuzzoni, G.; Pontiggia, B.; Magri, F. Age-related changes of the hypothalamic-pituitary-adrenal axis: Pathophysiological correlates. *Eur. J. Endocrinol.* **2001**, *144*, 319–329. [CrossRef] [PubMed]
5. Brändli-Baiocco, A.; Balme, E.; Bruder, M.; Chandra, S.; Hellmann, J.; Hoenerhoff, M.J.; Kambara, T.; Landes, C.; Lenz, B.; Mense, M.; et al. Nonproliferative and Proliferative Lesions of the Rat and Mouse Endocrine System. *J. Toxicol. Pathol.* **2018**, *31*, 1S–95S. [CrossRef] [PubMed]
6. Arguelles, S.; Cano, M.; Machado, A.; Ayala, A. Effect of aging and oxidative stress on elongation factor-2 in hypothalamus and hypophysis. *Mech. Ageing Dev.* **2011**, *132*, 55–64. [CrossRef] [PubMed]
7. Rodrigues Siqueira, I.; Fochesatto, C.; Da Silva Torres, I.L.; Dalmaz, C.; Netto, C.A. Aging affects oxidative state in hippocampus, hypothalamus and adrenal glands of Wistar rats. *Life Sci.* **2005**, *78*, 271–278. [CrossRef]
8. Nessi, A.C.; De Hoz, G.; Tanoira, C.; Guaraglia, E.; Consens, G. Pituitary physiological and ultrastructural changes during aging. *Endocrine* **1995**, *3*, 711–716. [CrossRef] [PubMed]
9. Stucker, S.; De Angelis, J.; Kusumbe, A.P. Heterogeneity and Dynamics of Vasculature in the Endocrine System During Aging and Disease. *Front. Physiol.* **2021**, *12*, 624928. [CrossRef]
10. Wang, P.; Lo, M.; Medical, M.K.-J. Glucocorticoids and aging. *J. Formos. Med. Assoc.* **1997**, *96*, 792–801.
11. Attia, M.A. Neoplastic and non-neoplastic lesions in the mammary gland, endocrine and genital organs in aging male and female Sprague-Dawley rats. *Arch. Toxicol.* **1996**, *70*, 461–473. [CrossRef]
12. Tonelli, C.; Chio, I.I.C.; Tuveson, D.A. Transcriptional Regulation by Nrf2. *Antioxid. Redox Signal.* **2018**, *29*, 1727–1745. [CrossRef] [PubMed]
13. Ma, Q. Role of Nrf2 in Oxidative Stress and Toxicity. *Annu. Rev. Pharmacol. Toxicol.* **2013**, *53*, 401. [CrossRef]
14. Ahmed, S.M.U.; Luo, L.; Namani, A.; Wang, X.J.; Tang, X. Nrf2 signaling pathway: Pivotal roles in inflammation. *Biochim. Biophys. Acta Mol. Basis Dis.* **2017**, *1863*, 585–597. [CrossRef] [PubMed]
15. Tossetta, G.; Fantone, S.; Mazzucchelli, R. Role of Natural and Synthetic Compounds in Modulating NRF2/KEAP1 Signaling Pathway in Prostate Cancer. *Cancers* **2023**, *15*, 3037. [CrossRef] [PubMed]
16. Xia, L.; Ma, W.; Afrashteh, A.; Sajadi, M.A.; Fakheri, H.; Valilo, M. The nuclear factor erythroid 2-related factor 2/p53 axis in breast cancer. *Biochem. Med.* **2023**, *33*, 030504. [CrossRef]
17. Rojo de la Vega, M.; Chapman, E.; Zhang, D.D. NRF2 and the Hallmarks of Cancer. *Cancer Cell.* **2018**, *34*, 21–43. [CrossRef] [PubMed]
18. Franceschi, C.; Garagnani, P.; Parini, P.; Giuliani, C.; Santoro, A. Inflammaging: A new immune–metabolic viewpoint for age-related diseases. *Nat. Rev. Endocrinol.* **2018**, *14*, 576–590. [CrossRef]
19. Yousef, M.H.; Salama, M.; El-Fawal, H.A.N.; Abdelnaser, A. Selective GSK3β Inhibition Mediates an Nrf2-Independent Anti-inflammatory Microglial Response. *Mol. Neurobiol.* **2022**, *59*, 5591–5611. [CrossRef]
20. Rana, A.K.; Singh, D. Targeting glycogen synthase kinase-3 for oxidative stress and neuroinflammation: Opportunities, challenges and future directions for cerebral stroke management. *Neuropharmacology* **2018**, *139*, 124–136. [CrossRef]

21. Jope, R.S.; Yuskaitis, C.J.; Beurel, E. Glycogen Synthase Kinase-3 (GSK3): Inflammation, Diseases, and Therapeutics. *Neurochem. Res.* **2007**, *32*, 577. [CrossRef] [PubMed]
22. Miler, M.; Živanović, J.; Ajdžanović, V.; Oreščanin-Dušić, Z.; Milenković, D.; Konić-Ristić, A.; Blagojević, D.; Milošević, V.; Šošić-Jurjević, B. Citrus flavanones naringenin and hesperetin improve antioxidant status and membrane lipid compositions in the liver of old-aged Wistar rats. *Exp. Gerontol.* **2016**, *84*, 49–60. [CrossRef] [PubMed]
23. Miler, M.; Živanović, J.; Ajdžanović, V.; Milenkovic, D.; Jarić, I.; Šošić-Jurjević, B.; Milošević, V. Citrus Flavanones Upregulate Thyrotroph Sirt1 and Differently Affect Thyroid Nrf2 Expressions in Old-Aged Wistar Rats. *J. Agric. Food Chem.* **2020**, *68*, 8242–8254. [CrossRef] [PubMed]
24. Miler, M.; Živanović, J.; Ajdžanović, V.; Milenkovic, D.; Cesar, T.; Filipović, M.R.; Milošević, V. Lemon extract reduces the hepatic oxidative stress and persulfidation levels by upregulating the Nrf2 and Trx1 expression in old rats. *Biofactors* **2024**, *50*, 756–771. [CrossRef] [PubMed]
25. Calabrese, V.; Cornelius, C.; Dinkova-Kostova, A.T.; Calabrese, E.J.; Mattson, M.P. Cellular stress responses, the hormesis paradigm, and vitagenes: Novel targets for therapeutic intervention in neurodegenerative disorders. *Antioxid. Redox Signal.* **2010**, *13*, 1763–1811. [CrossRef] [PubMed]
26. Calabrese, V.; Mancuso, C.; Calvani, M.; Rizzarelli, E.; Butterfield, D.A.; Giuffrida Stella, A.M. Nitric oxide in the central nervous system: Neuroprotection versus neurotoxicity. *Nat. Rev. Neurosci.* **2007**, *8*, 766–775. [CrossRef]
27. Calabrese, V.; Giordano, J.; Signorile, A.; Laura Ontario, M.; Castorina, S.; De Pasquale, C.; Eckert, G.; Calabrese, E.J. Major pathogenic mechanisms in vascular dementia: Roles of cellular stress response and hormesis in neuroprotection. *J. Neurosci. Res.* **2016**, *94*, 1588–1603. [CrossRef] [PubMed]
28. Mercau, M.E.; Repetto, E.M.; Perez, M.N.; Calejman, C.M.; Puch, S.S.; Finkielstein, C.V.; Cymeryng, C.B. Moderate Exercise Prevents Functional Remodeling of the Anterior Pituitary Gland in Diet-Induced Insulin Resistance in Rats: Role of Oxidative Stress and Autophagy. *Endocrinology* **2016**, *157*, 1135–1145. [CrossRef] [PubMed]
29. Nudler, S.I.; Quinteros, F.A.; Miler, E.A.; Cabilla, J.P.; Ronchetti, S.A.; Duvilanski, B.H. Chromium VI administration induces oxidative stress in hypothalamus and anterior pituitary gland from male rats. *Toxicol. Lett.* **2009**, *185*, 187–192. [CrossRef]
30. Ren, J.C.; Banan, A.; Keshavarzian, A.; Zhu, Q.; LaPaglia, N.; McNulty, J.; Emanuele, N.V.; Emanuele, M.A. Exposure to ethanol induces oxidative damage in the pituitary gland. *Alcohol* **2005**, *35*, 91–101. [CrossRef]
31. Ajdžanović, V.Z.; TŠošič-Jurjević, B.; Filipović, B.; Trifunović, S.L.; Brkic, D.D.; Sekulić, M.I.; Milosević, V.L.J. Genistein affects the morphology of pituitary ACTH cells and decreases circulating levels of ACTH and corticosterone in middle-aged male rats. *Biol. Res.* **2009**, *42*, 13–23. [CrossRef]
32. Yamamoto, M.; Kensler, T.W.; Motohashi, H. The KEAP1-NRF2 system: A thiol-based sensor-effector apparatus for maintaining redox homeostasis. *Physiol. Rev.* **2018**, *98*, 1169–1203. [CrossRef] [PubMed]
33. Cebula, M.; Schmidt, E.E.; Arnér, E.S.J. TrxR1 as a Potent Regulator of the Nrf2-Keap1 Response System. *Antioxid. Redox Signal.* **2015**, *23*, 823. [CrossRef] [PubMed]
34. Larkin, S.; Ansorge, O. Development And Microscopic Anatomy of the Pituitary Gland. *Endotext* **2017**, *1625*, 206–217.
35. Szabó, K.; Csányi, K. The vascular architecture of the developing pituitary-median eminence complex in the rat. *Cell Tissue Res.* **1982**, *224*, 563–577. [CrossRef] [PubMed]
36. Palde, P.B.; Carroll, K.S. A universal entropy-driven mechanism for thioredoxin-target recognition. *Proc. Natl. Acad. Sci. USA* **2015**, *112*, 7960–7965. [CrossRef] [PubMed]
37. Kim, Y.S.; Vallur, P.G.; Phaëton, R.; Mythreye, K.; Hempel, N. Insights into the Dichotomous Regulation of SOD2 in Cancer. *Antioxidants* **2017**, *6*, 86. [CrossRef] [PubMed]
38. Ungvari, Z.; Labinskyy, N.; Mukhopadhyay, P.; Pinto, J.T.; Bagi, Z.; Ballabh, P.; Zhang, C.; Pacher, P.; Csiszar, A. Resveratrol attenuates mitochondrial oxidative stress in coronary arterial endothelial cells. *Am. J. Physiol. Heart Circ. Physiol.* **2009**, *297*, H1876. [CrossRef]
39. Li, H. Sirtuin 1 and oxidative stress. In *Systems Biology of Free Radicals and Antioxidants*; Laher, I., Ed.; Springer: Berlin/Heidelberg, Germany, 2014; pp. 417–435. [CrossRef]
40. Hsu, M.-Y.; Hsiao, Y.-P.; Lin, Y.-T.; Chen, C.; Lee, C.-M.; Liao, W.-C.; Tsou, S.-C.; Lin, H.-W.; Chang, Y.-Y.; Hsu, C.; et al. Quercetin Alleviates the Accumulation of Superoxide in Sodium Iodate-Induced Retinal Autophagy by Regulating Mitochondrial Reactive Oxygen Species Homeostasis through Enhanced Deacetyl-SOD2 via the Nrf2-PGC-1α-Sirt1 Pathway. *Antioxidants* **2021**, *10*, 1125. [CrossRef]
41. Rijal, S.; Jang, S.H.; Cho, D.H.; Han, S.K. Hydrogen peroxide suppresses excitability of gonadotropin-releasing hormone neurons in adult mouse. *Front. Endocrinol.* **2022**, *13*, 939699. [CrossRef]
42. Maixner, D.W.; Weng, H.-R. The Role of Glycogen Synthase Kinase 3 Beta in Neuroinflammation and Pain. *J. Pharm. Pharmacol.* **2013**, *1*, 001. [CrossRef]
43. Wang, L.; Li, J.; Di, L. jun Glycogen synthesis and beyond, a comprehensive review of GSK3 as a key regulator of metabolic pathways and a therapeutic target for treating metabolic diseases. *Med. Res. Rev.* **2022**, *42*, 946. [CrossRef]
44. Chen, X.; Liu, Y.; Zhu, J.; Lei, S.; Dong, Y.; Li, L.; Jiang, B.; Tan, L.; Wu, J.; Yu, S.; et al. GSK-3β downregulates Nrf2 in cultured cortical neurons and in a rat model of cerebral ischemia-reperfusion. *Sci. Rep.* **2016**, *6*, 20196. [CrossRef]
45. Sano, T.; Kovacs, K.T.; Scheithauer, B.W.; Young, W.F. Aging and the human pituitary gland. *Mayo Clin. Proc.* **1993**, *68*, 971–977. [CrossRef] [PubMed]

46. Vennekens, A.; Laporte, E.; Hermans, F.; Cox, B.; Modave, E.; Janiszewski, A.; Nys, C.; Kobayashi, H.; Malengier-Devlies, B.; Chappell, J.; et al. Interleukin-6 is an activator of pituitary stem cells upon local damage, a competence quenched in the aging gland. *Proc. Natl. Acad. Sci. USA* **2021**, *118*, e2100052118. [CrossRef]
47. Mao, J.; Huang, H.; Liu, F.; Mai, Y.; Liao, X.; Qiu, B.; Mei, F.; Bao, Y. Activation of Age-Related Nuclear Factor-κB Signaling Pathway Leads to Chronic Inflammation and Pituitary Fibrosis. *World Neurosurg.* **2022**, *157*, e417–e423. [CrossRef]
48. Escoter-Torres, L.; Greulich, F.; Quagliarini, F.; Wierer, M.; Uhlenhaut, N.H. Anti-inflammatory functions of the glucocorticoid receptor require DNA binding. *Nucleic Acids Res.* **2020**, *48*, 8393–8407. [CrossRef] [PubMed]
49. Rao, N.A.S.; McCalman, M.T.; Moulos, P.; Francoijs, K.J.; Chatziioannou, A.; Kolisis, F.N.; Alexis, M.N.; Mitsiou, D.J.; Stunnenberg, H.G. Coactivation of GR and NFKB alters the repertoire of their binding sites and target genes. *Genome Res.* **2011**, *21*, 1404–1416. [CrossRef]
50. Liu, Z.; Hu, M. Natural polyphenol disposition via coupled metabolic pathways. *Expert Opin. Drug Metab. Toxicol.* **2007**, *3*, 389–406. [CrossRef]
51. Bešlo, D.; Golubić, N.; Rastija, V.; Agić, D.; Karnaš, M.; Šubarić, D.; Lučić, B. Antioxidant Activity, Metabolism, and Bioavailability of Polyphenols in the Diet of Animals. *Antioxidants* **2023**, *12*, 1141. [CrossRef]
52. Anbalagan, S.; Gordon, L.; Blechman, J.; Matsuoka, R.L.; Rajamannar, P.; Wircer, E.; Biran, J.; Reuveny, A.; Leshkowitz, D.; Stainier, D.Y.R.; et al. Pituicyte Cues Regulate the Development of Permeable Neuro-Vascular Interfaces. *Dev. Cell* **2018**, *47*, 711–726.e5. [CrossRef]
53. Figueira, I.; Garcia, G.; Pimpão, R.C.; Terrasso, A.P.; Costa, I.; Almeida, A.F.; Tavares, L.; Pais, T.F.; Pinto, P.; Ventura, M.R.; et al. Polyphenols journey through blood-brain barrier towards neuronal protection. *Sci. Rep.* **2017**, *7*, 11456. [CrossRef]
54. Latz, E.; Duewell, P. NLRP3 inflammasome activation in inflammaging. *Semin. Immunol.* **2018**, *40*, 61–73. [CrossRef]
55. Wang, Z.; Zhang, S.; Xiao, Y.; Zhang, W.; Wu, S.; Qin, T.; Yue, Y.; Qian, W.; Li, L. NLRP3 Inflammasome and Inflammatory Diseases. *Oxid. Med. Cell. Longev.* **2020**, *2020*, 4063562. [CrossRef]
56. Wang, X.; Ma, L.; Ding, Q.Y.; Zhang, W.Y.; Chen, Y.G.; Wu, J.H.; Zhang, H.F.; Guo, X.L. Microglial NLRP3 inflammasome activation-mediated inflammation promotes prolactinoma development. *Endocr. Relat. Cancer* **2021**, *28*, 433–448. [CrossRef]
57. Jing, S.; Wang, X.; Zhang, Z.; Cao, D.; Huang, K.; Wang, Y.; Liu, Z.; Su, S.; Wang, Q. Hesperetin attenuates cognitive dysfunction via SIRT6/NLRP3 pathway in scopolamine-induced mice. *Metab. Brain Dis.* **2023**, *38*, 2443–2456. [CrossRef]
58. Chen, C.; Wei, Y.Z.; He, X.M.; Li, D.D.; Wang, G.Q.; Li, J.J.; Zhang, F. Naringenin produces neuroprotection against LPS-induced dopamine neurotoxicity via the inhibition of microglial NLRP3 inflammasome activation. *Front. Immunol.* **2019**, *10*, 936. [CrossRef]
59. Calabrese, V.; Cornelius, C.; Dinkova-Kostova, A.T.; Calabrese, E.J. Vitagenes, cellular stress response, and acetylcarnitine: Relevance to hormesis. *Biofactors* **2009**, *35*, 146–160. [CrossRef]
60. Calabrese, V.; Giordano, J.; Crupi, R.; Di Paola, R.; Ruggieri, M.; Bianchini, R.; Ontario, M.L.; Cuzzocrea, S.; Calabrese, E.J. Hormesis, cellular stress response and neuroinflammation in schizophrenia: Early onset versus late onset state. *J. Neurosci. Res.* **2017**, *95*, 1182–1193. [CrossRef]
61. Trovato Salinaro, A.; Pennisi, M.; Di Paola, R.; Scuto, M.; Crupi, R.; Cambria, M.T.; Ontario, M.L.; Tomasello, M.; Uva, M.; Maiolino, L.; et al. Neuroinflammation and neurohormesis in the pathogenesis of Alzheimer's disease and Alzheimer-linked pathologies: Modulation by nutritional mushrooms. *Immun. Ageing* **2018**, *15*, 8. [CrossRef]
62. Aguilera, G.; Harwood, J.P.; Wilson, J.X.; Morell, J.; Brown, J.H.; Catt, K.J. Mechanisms of action of corticotropin-releasing factor and other regulators of corticotropin release in rat pituitary cells. *ASBMB* **1983**, *258*, 8039–8045. [CrossRef]
63. Humphries, K.M.; Pennypacker, J.K.; Taylor, S.S. Redox Regulation of cAMP-dependent Protein Kinase Signaling. *J. Biol. Chem.* **2007**, *282*, 22072–22079. [CrossRef]
64. Wilkinson, C.W. Human Glucocorticoid Feedback Inhibition Is Reduced in Older Individuals: Evening Study. *J. Clin. Endocrinol. Metab.* **2001**, *86*, 545–550. [CrossRef]
65. Boscaro, M.; Paoletta, A.; Scarpa, E.; Barzon, L.; Fusaro, P.; Fallo, F.; Sonino, N. Age-Related Changes in Glucocorticoid Fast Feedback Inhibition of Adrenocorticotropin in Man. *J. Clin. Endocrinol. Metab.* **1998**, *83*, 1380–1383. [CrossRef] [PubMed]
66. Ajdžanović, V.; Miler, M.; Živanović, J.; Filipović, B.; Šošić-Jurjević, B.; Popovska-Perčinić, F.; Milošević, V. The adrenal cortex after estradiol or daidzein application in a rat model of the andropause: Structural and hormonal study. *Ann. Anat.* **2020**, *230*, 151487. [CrossRef]
67. Ajdžanović, V.; Šošić-Jurjević, B.; Filipović, B.; Trifunović, S.; Manojlović-Stojanoski, M.; Sekulić, M.; Milošević, V. Genistein-Induced Histomorphometric and Hormone Secreting Changes in the Adrenal Cortex in Middle-Aged Rats. *Exp. Biol. Med.* **2009**, *234*, 148–156. [CrossRef]
68. Yiallouris, A.; Tsioutis, C.; Agapidaki, E.; Zafeiri, M.; Agouridis, A.P.; Ntourakis, D.; Johnson, E.O. Adrenal aging and its implications on stress responsiveness in humans. *Front. Endocrinol.* **2019**, *10*, 435083. [CrossRef]
69. Moffat, S.D.; An, Y.; Resnick, S.M.; Diamond, M.P.; Ferrucci, L. Longitudinal Change in Cortisol Levels Across the Adult Life Span. *J. Gerontol. Ser. A* **2020**, *75*, 394–400. [CrossRef]
70. Noordam, R.; Jansen, S.W.M.; Akintola, A.A.; Oei, N.Y.L.; Maier, A.B.; Pijl, H.; Slagboom, P.E.; Westendorp, R.G.J.; van der Grond, J.; de Craen, A.J.M.; et al. Familial Longevity Is Marked by Lower Diurnal Salivary Cortisol Levels: The Leiden Longevity Study. *PLoS ONE* **2012**, *7*, e31166. [CrossRef]

71. Noordam, R.; Gunn, D.A.; Tomlin, C.C.; Rozing, M.P.; Maier, A.B.; Slagboom, P.E.; Westendorp, R.G.J.; van Heemst, D.; De Craen, A.J.M. Cortisol serum levels in familial longevity and perceived age: The Leiden Longevity Study. *Psychoneuroendocrinology* **2012**, *37*, 1669–1675. [CrossRef] [PubMed]
72. Krøll, J. Correlations of plasma cortisol levels, chaperone expression and mammalian longevity: A review of published data. *Biogerontology* **2010**, *11*, 495–499. [CrossRef] [PubMed]
73. Koenig, S.; Bredehöft, J.; Perniss, A.; Fuchs, F.; Roth, J.; Rummel, C. Age dependent hypothalamic and pituitary responses to novel environment stress or lipopolysaccharide in rats. *Front. Behav. Neurosci.* **2018**, *12*, 343528. [CrossRef] [PubMed]

International Journal of *Molecular Sciences*

MDPI

Review

The Impact of Thermal Treatment Intensity on Proteins, Fatty Acids, Macro/Micro-Nutrients, Flavor, and Heating Markers of Milk—A Comprehensive Review

Yi Wang [1,2,†], Ran Xiao [2,†], Shiqi Liu [2], Pengjie Wang [2], Yinhua Zhu [2], Tianjiao Niu [2,*] and Han Chen [1,2,*]

1 Food Laboratory of Zhongyuan, China Agricultural University, Beijing 100083, China; wangyi922217@126.com

2 Department of Nutrition and Health, China Agricultural University, Beijing 100083, China; xiaoran881230@163.com (R.X.); liushiqi97@bjfu.edu.cn (S.L.); wpj1019@cau.edu.cn (P.W.); zhuyinhua@cau.edu.cn (Y.Z.)

* Correspondence: tn2020036@zyfoodlab.com (T.N.); h.chen@cau.edu.cn (H.C.)

† These authors contributed equally to this work.

Abstract: Milk thermal treatment, such as pasteurization, high-temperature short-time processing, and the emerging ultra-short-time processing (<0.5 s), are crucial for ensuring milk safety and extending its shelf life. Milk is a nutritive food matrix with various macro/micro-nutrients and other constituents that are possibly affected by thermal treatment for reasons associated with processing strength. Therefore, understanding the relationship between heating strength and milk quality is vital for the dairy industry. This review summarizes the impact of thermal treatment strength on milk's nutritional and sensory properties, the synthesizing of the structural integrity and bioavailability of milk proteins, the profile and stability of fatty acids, the retention of macro/micro-nutrients, as well as the overall flavor profile. Additionally, it examines the formation of heat-induced markers, such as Maillard reaction products, lactulose, furosine, and alkaline phosphatase activity, which serve as indicators of heating intensity. Flavor and heating markers are commonly used to assess the quality of pasteurized milk. By examining former studies, we conclude that ultra-short-time-processing-treated milk is comparable to pasteurized milk in terms of specific parameters (such as whey protein behavior, furosine, and ALP contents). This review aims to better summarize how thermal treatments influence the milk matrix, guiding the dairy industry's development and balancing milk products' safety and nutritional value.

Keywords: direct steam; indirect steam; pasteurization; milk; heating markers; heat-induced gelation; vitamins; minerals

Citation: Wang, Y.; Xiao, R.; Liu, S.; Wang, P.; Zhu, Y.; Niu, T.; Chen, H. The Impact of Thermal Treatment Intensity on Proteins, Fatty Acids, Macro/Micro-Nutrients, Flavor, and Heating Markers of Milk—A Comprehensive Review. *Int. J. Mol. Sci.* **2024**, *25*, 8670. https://doi.org/10.3390/ijms25168670

Academic Editors: Vassilis Athanasiadis and Theodoros G. Chatzimitakos

Received: 21 June 2024
Revised: 2 August 2024
Accepted: 7 August 2024
Published: 8 August 2024

1. Introduction

Milk is comprised of a variety of compounds, including proteins, carbohydrates, lipids, minerals, vitamins, and other newly found bioactive compounds [1]. Due to these compounds, milk is regarded as a natural whole food with a unique flavor and various health benefits, such as antioxidant [2], anti-inflammatory [3], anti-osteoporosis [4] benefits. With consumers' increasing health awareness, the consumption of dairy products has grown over the past few decades [5]. However, these nutritional compounds are also good substrates for harmful microorganisms, which are not only detrimental to human health but also make the milk susceptible to spoilage [6]. Thus, microorganism control is an indispensable step before commercialization to guarantee milk safety and prolong its shelf life [5].

Of all the methods of microorganism control, thermal processing is the most commonly used, with the advantages of low cost and high efficiency. However, adverse changes occur in milk when exposed to excessive heating conditions, such as the denaturation of proteins and oxidation of lipids, and Maillard reactions are observed in milk after thermal treatment, which further causes some unfavorable flavors, such as cooked, sulfurous, cabbage, and caramelized flavors. Moreover, the loss of heat-sensitive bioactive compounds during heating results in a reduction in the nutritional value of milk [7,8]. In addition, plasmin, a key protease of the fibrinolysis system that regulates sedimentation and age gelation in milk, shows different enzymatic kinetics after various thermal treatments [7]. Research has reported that pasteurization processes, usually referring to heating at 63 °C for 30 min (low temperature long time, LTLT), 72 °C for 15 s, or 85 °C for 2 to 4 s (high temperature short time, HTST), provide minor damage on the flavor, color, bioactive compounds, and rheology properties of milk due to the low processing intensity applied [9,10]. The pasteurization procedure is able to effectively eliminate vegetative microbes but is less feasible against spores and may germinate these aerobic spores in some cases [11], which leads to a short shelf life, requires strict cold-chain transportation, and therefore, limits the supply of milk products with high quality, especially in developing countries [10].

Direct steam heating (DSH), a thermal technology introduced to the dairy industry in recent years, is defined as mixing superheated steam directly with the milk matrix to raise the temperature to the targeted value rapidly [12]. This method causes less heat damage compared to indirect heating systems because of the more rapid heating rate and the absence of a heat transfer surface [13]. The processing intensity of DSH is controllable and DSH treatment with a certain temperature and duration [9] might balance the adverse effects of heating on quality and extend the shelf life of milk to produce high-quality dairy products.

This article briefly summarizes the recent findings of thermal treatments in the dairy industry and compares the impacts of different thermal processes on the structure and content of proteins (whey protein and casein) and lipids, flavor, macro/micro-nutrients (minerals and vitamins), and commonly used heating markers of traditionally pasteurized milk. Afterward, we propose our perspectives on the future optimizations of thermal treatment in milk processing.

2. Impact of Thermal Treatment on Structure and Behavior of Main Milk Constituents

Based on heating principles, steam heating treatments can be either composed of an "indirect" system or a "direct" system (Figure 1). In direct heating systems, products are directly mixed with steam under pressure, significantly improving heat transfer efficiency [14]. After this, products are cooled to the targeted temperature, and excess water is removed by vacuum cooling [15]. The rapid heating and cooling processing in the DSH system reduces heat-induced physical and chemical changes [16] compared to indirect steam heating (ISH)-treated milk [17]. Understanding changes in the structure of the main milk constituents, including whey proteins, casein, and milk fat, caused by direct/indirect steam heating, is essential for enhancing its processing stability and controlling its functionalities. Consequently, this section inspects the impact of DSH on the structure and function behavior of the main milk constituents. Meanwhile, the differences in the effects of different heating methods on these substances are also discussed.

2.1. Whey Proteins

Whey proteins, highly ordered proteins in a globular shape, represent the second most abundant protein group (~20% of milk proteins) after caseins. β-Lactoglobulin (β-Lg) and α-lactalbumin (α-La) are the most abundant in whey proteins, which account for approximately half and one-fifth of whey protein, respectively. At the same time, immunoglobulins (IgGs), bovine serum albumin (BSA), and lactoferrin are minor whey proteins. Whey proteins contain several biological functions, such as antioxidant [17], anti-obesity [18], antitumor [19], and immunomodulation [20] functions. Milk needs to undergo heat treatment to guarantee microbiological safety during its shelf life before commercialization. Whey proteins are more sensitive to heating than caseins. HTST pasteurization (73 °C/15 s) showed no significant changes in β-Lg and α-La compared to raw milk, while denaturation of β-Lg and α-La was observed after a 135 °C/2 s steam infusion treatment or a 151 °C/4 s steam injection treatment, which was identified by SDS-PAGE [21,22]. On the other hand, IgGs are the most heat-sensitive among whey proteins, followed by BSA, LF, β-Lg, and α-La [23]. In most cases, ISH treatment with a higher thermal load can significantly denature whey proteins [23,24]. A previous study investigated directly/indirectly heated milk products from 10 plants in six countries and found that denaturation of β-Lg was lower in milk treated with DSH (35–80%) than in milk treated with ISH (79–100%) due to a more rapid heating rate in direct systems [25,26]. In addition to ISH and DSH, the denaturation of whey protein is nearly negligible during pasteurization (72 °C for 15 s) [14]. Wang et al. [27] compared the differences in the thermal damage of proteins (LF, α-La, and β-Lg) in milk treated with different heat treatments, including ISH treatment, direct steam heating treatment (direct steam infusion, DSI, and direct steam injection, DSIJ), and pasteurization. Steam is injected into the food matrix through apertures on the vessel body or through a pipe positioned inside the vessel in DSIJ technology. It involves discharging a series of steam bubbles into the product. Steam injectors are engineered to create a turbulent zone within the steam injector to help mix the steam and the product. DSI technology is used to accelerate steam at speeds of up to 1000 m/s into the food matrix. The steam disrupts the fluid flow and breaks it into small droplets, resulting in an increased contacting surface and higher heat exchange rate. [25,26]. They found that the DSI heating treatment (139–156 °C/0.116–5 s) provided lower thermal damage to α-La, β-Lg, and LF than the ISH treatment, equivalent to pasteurization [27–29]. Mi et al. [28] investigated the effects of DSIJ treatment on heat-sensitive serum proteins in yak milk (YM). They found that this heat treatment could reduce the content of LF (about 40%) and β-Lg (about 30%) but had little effect on α-La. However, in contrast, previous studies reported 79–100% β-Lg denaturation, 100% LF denaturation, and 18–54% α-La denaturation after ISH treatment [26,29,30]. Therefore, regarding the denaturation of whey proteins, DSH-treated milk could be more similar to pasteurized milk. Advances in analytical methods and proteomics approaches have revealed the abundance and complexity of whey protein compositions [31–33]. Thus, this approach can comprehensively analyze and compare the effects of different heat treatments on the composition of whey proteins in milk. Meanwhile, bioinformatics analysis also helps elucidate these proteins' potential biological functions.

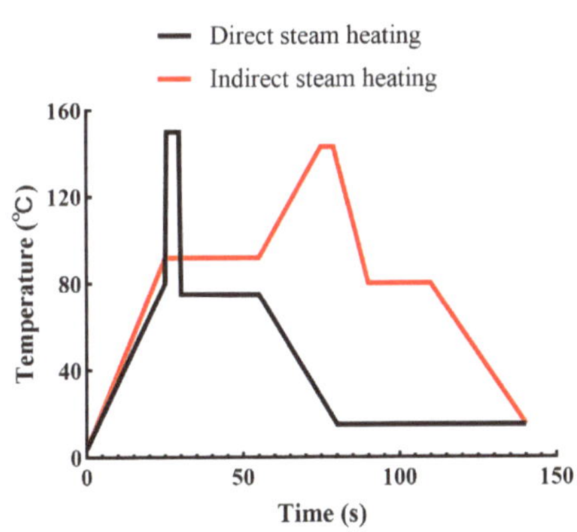

Figure 1. Temperature–time profiles of DSH and ISH.

Compared to whey proteins with a higher molecular weight (Mw), such as immunoglobulins and bovine serum albumin, there are more complexes formed by the participation of β-Lg and α-La induced by heating, especially β-Lg [14,34]. β-Lg and α-La denature at 75–80 °C and 65 °C, respectively [14]. After an 85 °C/30 min treatment, β-Lg and α-La are irreversibly denatured, exposing the reactive hydrophobic and free cysteine residues of the two whey proteins [35]. They consequently interact with themselves or κ-casein on the surface of the casein micelles in milk through disulfide bonds, thiol/disulfide exchange, and hydrophobic interactions to form whey protein aggregates and β-Lg-κ-casein complexes [36,37]. In the formed aggregates, β-lg interacts with α-La and other β-Lgs [34]. β-Lg unfolds when induced by heating and can start association reactions with other β-Lgs via disulfide bonds. α-La has no free sulfhydryl group, but it associates with β-Lg via thiol/disulfide exchange [23]. The release of reactivated free sulfhydryl groups at temperatures higher than 70 °C induced by the disruption of one of the disulfide bonds triggers and spreads non-reversible aggregation in the same pattern as the natural free sulfhydryl in β-Lg [23]. Yun et al. [38] developed a multistep model for the denaturation of β-lg and its following reaction with κ-casein. They found that the β-Lg denaturation mechanism was expanded to three steps: dimerization (at 63 °C), polymerization (at 80 °C), and reaction with κ-casein (at ~90 °C). In addition, the association of denatured whey protein with the casein micelles depends on pH. At a pH of 6.5, 75–80% of the denatured whey protein was associated with micelles, and this association level decreased to approximately 30% at a pH of 6.7 [39].

2.2. Casein

Caseins in cow milk account for approximately 80% of milk proteins, which are present in the form of roughly globular aggregates named micelles with different particle sizes between 50 and 600 nm [40,41]. Casein micelles consist of αS1-casein, αS2-casein, β-casein, and κ-casein (the ratio in cow milk: 4:1:3.5:1.5), colloidal calcium phosphate, and water [42]. Over the last few decades, researchers have proposed different models to illustrate the structure of casein micelles, including the submicelle model [43], the dual binding model [44], and the nanocluster model [41,45]. Although the nanocluster model can best characterize the structural changes of casein micelles in milk caused by different dairy processing methods, the role of β-casein and water is still unclear [45,46]. Therefore, its exact structure in raw milk is still under debate. Dalgleish et al. [41] stated that calcium phosphate nanoclusters linked the αS-caseins and β-caseins; some β-caseins hydrophobically bind to other caseins and stabilize the water channel and pores within the micelles, allowing water, peptides, or other small Mw compounds to pass through the micelles, but these β-caseins can be released from the micelles to the serum phase in milk by cooling [41]. In addition, κ-casein was located on the surface of micelles, providing steric and electrostatic stabilization against aggregation.

Casein micelles resist severe heat treatments because of their lack of a well-defined tertiary structure due to disulfide binding and a large amount of propyl residues, and only temperatures above 120 °C cause the denaturation of casein proteins, which explains the low denaturation rate in pasteurized milk [46]. Upon severe heating, the most critical change in casein micelles is an increase in micellar size due to the association of denatured whey proteins with κ-casein on the surface of micelles [47,48]. The degree of protein particle size change is not the same during different heating conditions. Previous studies found that ISH treatment induces the formation of large-sized aggregates compared to DSH [14,27]. This is because ISH treatment has a higher thermal load than DSH under the same heating conditions. In addition, a previous study compared the effects of two DSH systems (injection and infusion) on milk particle size, and they found that the particle size of steam-infusion-treated milk was lower than that of steam-injection-treated milk [49]. Two possible explanations for this phenomenon are as follows. (a) DSI was gentler. (b) DSIJ induces more self-aggregation of β-Lg, leading to less β-Lg available to associate with κ-casein [9,50,51]. InfusionPlusTM (SPX FLOW, Charlotte, NC, USA) can achieve an extreme temperature (157 °C) in both DSI and DSIJ systems. Thus, our lab used this machine to process skim milk samples over a 139–156 °C temperature range for 0.116–5 s [52]. The results showed that the milk protein particle size was related to the C* value, and it provides a measure of the processing extent of the chemical components of a product [52]. Compared to a low C* value (<0.1; holding times $\leq$ 0.25 s), milk samples with a high C* value (>0.1; holding times were 3–6 s) had smaller particle sizes, suggesting extensive heat-induced partial dissociation of the β-Lg-κ-casein complex from the casein micelle upon DSI treatment [52]. It should be noted that this complex has been considered the key element that contributes to the formation of age gelation in DSH-treated milk and ISH-treated milk [53,54]. Besides proteins, alterations in salt balance were also observed. DSH induces the calcium precipitation and solubilization of colloidal calcium phosphate, but these changes are invertible at heating temperatures lower than 95 °C for a few minutes [55].

The above discussion about casein is based on the milk system, but the impact of DSH on the casein system (without whey proteins, lactose, and salts) needs to be clarified. Therefore, our lab studied the structural changes in casein micelle dispersions upon different DSI treatments (157 °C/0.116 s, 155 °C/3 s, 150 °C/3 s, 145 °C/3 s, and 140 °C/3 s) [44]. As the temperature increased, the particle size of casein micelles reduced, and turbidity decreased significantly [56]. In addition, calcium bridges were deconstructed by DSI and increased electrostatic repulsion inside casein micelles, resulting in the dissociation of casein clusters, and therefore, a looser, smaller, and more porous shape of the casein micelles [56].

2.3. Milk Fat

Fat in milk is present in the form of fat globules. The core of milk fat globules is triacylglycerols (TAGs), which are surrounded by three-layered biofilms called milk fat globule membranes (MFGMs) [57]. These membranes consist of some membrane-specific proteins, triglycerides, cholesterol, phospholipids, and enzymes [34].

MFGM protein occupies 1–2% of total dairy protein, which is much lower than casein and whey protein [34,58]. The major MFGM proteins include fatty acid-binding protein (FABP), xanthine oxidase (XO), cluster of differentiation 36 (CD36), periodic acid Schiff glycoprotein III (PAS III), adipophilin (ADPH), butyrophilin (BTN), periodic acid Schiff glycoprotein 6/7 (PAS 6/7), and Mucin-1 (MUC 1) [58]. These proteins are presented in different layers of MFGMs [59]. ADPH appears in the inner monolayer, while XO is located in between both layers [15]. MUC 1, PAS III, CD 36, and BTN are in the outer layer [15]. PAS 6/7 is only loosely bound to the outer layer of MFGMs [15]. Thermal treatment at >65 °C accelerates the binding of the denatured whey proteins to MFGM proteins, as well as the dissociation of natural MFGM proteins. The denatured β-Lg can undergo complexation with α-La, other β-Lgs, and minor whey proteins with a higher Mw in milk [59]. These complexes can interact with MFGM proteins mainly through β-Lg as a "connector" [59]. In addition, α-La, IgGs, LF, and BSA also directly bind to MFGM proteins [34]. MFGM protein dissociation is not observed in pasteurized milk and the release of BTN (at >85 °C, 3 min), XO (at >85 °C, 3 min), as well as PAS 6/7 (at >80 °C, 3 min) can be observed above certain temperatures [60,61]. According to the SDS-PAGE patterns regarding major MFGM proteins in milk treated with different heat treatments (pasteurization, DSH, and ISH), we found that pasteurization and DSH retain more major MFGM proteins. However, the significant denaturation of major MFGM proteins in milk treated with ISH treatment is observed (up to 50.05% MFGM denatured after a 135 °C/5 s treatment) because of the higher thermal load exerted by ISH treatment [15,34,62]. The condensation and collapse of steam bubbles during DSIJ lead to homogenization and cavitation effects. In addition to these changes, it should be noticed that the homogenization effect can influence the reconstruction of the MFGM [14].

The polar lipids on the MFGM account for 0.1–0.3% of total milk lipids, which can be divided into glycerophospholipids (phosphatidylcholine, PC; phosphatidylinositol, PI; phosphatidylserine, PS; and phosphatidylethanolamine, PE) and sphingolipids (mainly sphingomyelin, SM) [59]. Polar lipids endow the MFGM with fluidity, but the distribution of polar lipids is non-homogeneous. Liquid-ordered domains (SM+cholesterol) with low fluidity lead to a more fragile membrane structure, which can affect the properties of fat globules. Heat treatment causes the membrane to become "brittle". Huang et al. [63] investigated the effects of different pasteurization treatments on the physical properties and interfacial composition of bovine milk fat globules. For polar lipids, they found that the phospholipid layer of the MFGM was incomplete upon heating [63]. Meanwhile, the liquid-ordered domains were more easily shed into the aqueous phase. As a result, the nano-mechanical properties of the MFGM decreased [63]. However, the effect of DSH treatment on MFGM phospholipids is still unclear.

3. Impact of Thermal Treatment on Flavor of Milk

It has been reported that high processing temperatures influence interactions between dairy compounds in the milk matrix or with the environment, such as the Maillard reaction between proteins and carbohydrates [64,65], lipid oxidation [66], and protein degradation [67]. The results of these heat-induced reactions are eventually reflected in the flavor of milk.

Zhao et al. [68] reported that traditional pasteurized milk (63 °C/30 min or 72 °C/15 s) has a higher content of low Mw constituents, such as 2-butanone and dimethyl ketone, while ultra-high temperature treatment with DSI (150 °C, 0.1 s) increases the content of ketones and aldehydes with a higher Mw, such as hexanal, pentanal, heptanal, benzaldehyde, and nonanal. Of these, sulfur-containing compounds such as dimethyl sulfone are related to the cooked taste, hexanal is a compound that has a grassy off flavor, and benzaldehyde is an aromatic substance that emanates a pleasant flavor formed by the Maillard reaction [69]; nonanal is a common flavor substance which emits floral and citrus aromas [70]. UHT milk normally has stronger flavors than HTST milk, especially regarding the cooked flavor and a sulfur or eggy flavor, which are not liked by US consumers who typically prefer light or medium milk flavor intensity [71]. This bias can be dependent on the type of milk most commonly consumed. Liem et al. [72] reported that consumers from China, where 60% of the milk consumed is long-life milk (LLM), preferred the flavor of LLM, whereas consumers from Australia, where only 10% of milk consumed is LLM, preferred HTST milk.

Meng et al. [73] comprehensively compared both DSI and DSIJ treatment (140–155 °C, 0.5–2 s) to HTST treatment in terms of the flavors of milk obtained with two extraction methods, namely solvent-assisted flavor evaporation (SAFE) and solid-phase microextraction (SPME). There were 59 volatile compounds discovered in DSI milk and DSIJ milk. There were 19 and 54 substances detected by SPME and SAFE, respectively, of which 14 substances were detected by both extraction methods. The 14 compounds include hexanal, nonanal, 2-undecanione, δ-decanolactone, dimethyl sulfone, and some short-chain acids. SAFE is effective for the extraction of alcohols, fatty acids, and aldehydes as reported, while SPME is better for the extraction of esters and sulfur-containing compounds [74].

The number of volatile compounds in DSI milk is a bit less than in DSIJ milk (50 vs. 52 substances). Flavor compounds, such as dimethyl sulfone, nonanal, 2-decanone, and 3-hydroxy-2-butanone, were detected in DSIJ milk but not in DSI milk. In terms of content, combining the results of two extraction methods, the content of 2-undecanone and decanoic acid in DSI milk is higher, while the content of dimethyl sulfone is lower.

4. Impact of Thermal Treatment on Macro/Micro-Nutrients in Milk

4.1. Minerals

It is well known that temperature fluctuations can change the distribution of minerals between the colloidal and aqueous phases [75]. The processing conditions, such as temperature, dilution ratio, pH, and the presence of chelating agents or salts, lead to an imbalance of the minerals in milk [76]. The increase in temperature facilitates the disassembly of the inorganic phosphate group (Pi) in the aqueous phase and breaks the balance between H_2PO^{4-} and $HPO4^{2-}$ in milk. Calcium phosphate becomes supersaturated in the aqueous phase with lower concentrations of ionic calcium and Pi. The changes in mineral balance depend on the intensity of the heat treatment between the colloidal and aqueous phases [75,77]. The $HPO_4{}^{2-}$ concentration is more affinitive for calcium, and it reacts at high temperatures with calcium to form calcium phosphate salts. Heat treatment causes the aggregation of calcium phosphate because its solubility decreases at higher temperatures. The content of P and Ca in serum decreases by 35% and 60%, respectively, after being heated from 20 to 90 °C. Moreover, magnesium concentrations in the colloidal phase increase from 1.53 to

1.97 mM but there is little change in potassium and sodium [78,79]. Pouliot et al. [80] also reported that the recovery ability of the heat-induced (85 °C/40 min) changes in P and Ca was 93–99% and 90–95%, respectively.

Boiani et al. [81] reported that Pi and Ca within micellar casein played vital roles as inorganic calcium phosphate salts or in forming casein phosphate nanoclusters at low concentrations of micellar casein during a low heating strength (25 to 60 °C). With more intensive heating (60 to 80 °C), the higher temperature fortified the negative charge of Pi, leading to an interaction between micellar casein and the casein phosphate nanocluster. It has been reported that treatment at 100 °C for 15 min molded a new type of calcium phosphate with the nature of an alkaline salt other than the native colloidal calcium phosphate in fresh milk, while the structure of the freshly formed calcium phosphate due to treatments at below 90 °C for 15 min was identical with native colloidal calcium phosphate nanoclusters [82].

The alterations in milk's mineral balance triggered by thermal treatment are of great importance to the dairy industry [83]. It has been reported by Jeurnink et al. [84] that up to 45% of minerals, including 15.7% calcium, presented in the residues on the heater of a plate heat exchanger with an operating temperature range of 69–85 °C. Moreover, only 0.2% of colloidal minerals in milk from the aqueous phase were adhered to the heater surface after treatment at 85 °C for 1 min. The shifting balance of the minerals results in the poor coagulation of the milk during cheese production [84]. Additionally, severe treatment (115 °C/ 40 s) could cause some extent of mineral losses in milk because of the fouling mechanism related to the severity of thermal treatment.

For the bioavailability of minerals, Seiquer et al. [85] found that overheating (three cycles of 116 °C/16 min heating) decreased the amount of soluble calcium by about 23.6% in milk; which was its bioaccessibility after an in vitro digestion compared to that of 150 °C/6 s treatment. The reduction in solubility of Ca could be due to dephosphorylation triggered by the thermal treatment of casein phosphopeptides, which prevents it from aggregation and the formation of chelates between calcium and Maillard reaction products. No significant effect of ultra-heating on the absorption of Ca was also determined using a rat model, in which food intake was administrated.

4.2. Vitamins

Vitamin B_1 (thiamin), is normally abundant in products from animal sources in the form of phosphorylation and is likely to be denatured by oxidizing, reducing agents, and thermal conditions (as seen in Table 1) [86,87] The circumstances of the thermal treatment module, e.g., plate or tubular energy exchanger in thermal treatment and DSI or DSIJ treatments, showed limited effects on vitamin B_1 levels in milk [88]. Contrarily, for in-bottle sterilized milk, rare in the dairy industry nowadays, the loss of vitamin B_1 might be up to 40% [89]. No significant decrease in vitamin B_1 is observed when opened LLM milk was refrigerated for 10 days [90]. Conversely, as for evaporated milk, the vitamin B_1 loss can be up to 50% [91].

Vitamin B_2 (riboflavin) in milk is insensitive to thermal treatment [92,93], but its stability is weakened when exposed simultaneously to light [94]. It has been reported that skim milk treated from 80 to 120 °C showed a declining trend in photolysis of vitamin B_2, possibly due to the denaturation of whey protein and/or an increase in the size of casein micelle, which blocks light [95]. Aminoreductone, a kind of Maillard-reaction product, is the primary compound promoting the stability of vitamin B_2 against light in LLM [96]. Vitamins B_3 (niacin), B_5 (pantothenic acid), and B_7 (biotin) in milk were insensitive against prolonged pasteurization below 72 °C and following storage at 4 °C [93,97]. A reduction of less than 10% in vitamins B_3, B_5, and B_7 in pasteurized milk was also reported in former studies [91,98,99].

Table 1. Influences of different thermal treatments on vitamins in milk.

Vitamins		Heating Conditions (Whole Milk Unless Stated Otherwise)	Raw Milk	After Treatment	Ref.
water-soluble vitamins	Vitamin B_1 (thiamin) (μg/mL)	77 °C/15 s	0.45 ± 0.02	No loss	[87]
		94 °C/420 s	0.43 ± 0.03	6.34% loss	
		129 °C/2 s, indirect	0.44 ± 0.02	No loss	
		140 °C/4 s, direct	0.46 ± 0.03	No loss	
		141 °C/4 s, indirect	0.50 ± 0.03	No loss	
	Vitamin B_2 (riboflavin)	77 °C/15 s	1.42 ± 0.01 μg/mL	No loss	[87]
		94 °C/420 s	1.43 ± 0.01 μg/mL	3.68% loss	
		129 °C/2 s, indirect	1.39 ± 0.01 μg/mL	1.80% loss	
		140 °C/4 s, direct	1.42 ± 0.01 μg/mL	1.41% loss	
		141 °C/4 s, indirect	1.35 ± 0.01 μg/mL	No loss	
		75 °C/6 s	0.17 ± 0.01 μmol/L	5.9% loss	[98]
		85 °C/6 s	0.17 ± 0.01 μmol/L	No loss	
		92 °C/6 s	0.17 ± 0.01 μmol/L	No loss	
	Vitamin B_3 (niacin)	62.5 °C/30 min	218 μg/100 g	No loss	[97]
		75 °C/15 min	238 μg/100 g	No loss	
		75 °C/6 s	13.28 ± 0.44 μmol/L	No loss	[98]
		85 °C/6 s	13.28 ± 0.44 μmol/L	No loss	
		92 °C/6 s	13.28 ± 0.44 μmol/L	No loss	
	Vitamin B_5 (pantothenic acid) (μg/100 g)	62.5 °C/30 min	469	No loss	[97]
		75 °C/15 min	610	4.8% loss	
	Vitamin B_6 (pyridoxamine) (μmol/L)	75 °C/6 s	2.06 ± 0.60	No loss	[98]
		85 °C/6 s	2.06 ± 0.60	No loss	
		92 °C/6 s	2.06 ± 0.60	No loss	
	Vitamin B_7 (biotin) (μg//100 g)	62.5 °C/30 min	0.73	No loss	[97]
		75 °C/15 min	0.95	No loss	
	Vitamin B_9 (folic acid) (μmol/L)	75 °C/6 s	0.75 ± 0.06	No loss	[98]
		85 °C/6 s	0.75 ± 0.06	No loss	
		92 °C/6 s	0.75 ± 0.06	No loss	
	Vitamin B_{12} (cyanocobalamin) (ng/mL)	77 °C/15 s	3.60 ± 0.40	No loss	[87]
		94 °C/420 s	3.30 ± 0.20	23.30% loss	
		129 °C/2 s, indirect	3.40 ± 0.60	No loss	
		140 °C/4 s, direct	4.40 ± 0.30	No loss	
		141 °C/4 s, indirect	3.30 ± 0.10	No loss	
		100 °C/1 h	100 μM (standardized)	10.2% loss	[100]
		100 °C/1 h (2% fat)	100 μM (standardized)	12.5% loss	
		100 °C/1 h (fat-free)	100 μM (standardized)	14.4% loss	
		76 °C/16 s	0.31 μg 100 g^{-1}	No loss	[101]
		96 °C/5 min	0.31 μg 100 g^{-1}	3.2% loss	
	Vitamin C (ascorbic acid) (μg/mL)	50 & 63 °C/10–60 min	100	~15% loss at 10 min and up to 40% loss at 60 min	[99]
		75 & 90 °C/0.25–10 min	100	~12% loss at 0.25 min and up to ~35% loss at 10 min	

Table 1. *Cont.*

Vitamins		Heating Conditions (Whole Milk Unless Stated Otherwise)	Raw Milk	After Treatment	Ref.
fat-soluble vitamins	Vitamin A (retinol) (IU/L)	63 °C/30 min	325 ± 19.2	2% loss	[102]
		121 °C/15 min	325 ± 19.2	36.6% loss	
	Vitamin D_2 (ergocalciferol) (IU/L)	63 °C/30 min	594.28 ± 2.22 (fortified)	No loss	[103]
		121 °C/15 min	594.28 ± 2.22 (fortified)	No loss	
	Vitamin E (tocopherol) (μg/mL)	77 °C/15 s	0.97 ± 0.03	No loss	[87]
		94 °C/420 s	0.96 ± 0.03	No loss	
		129 °C/2 s, indirect	0.94 ± 0.04	No loss	
		140 °C/4 s, direct	1.00 ± 0.06	No loss	
		141 °C/4 s, indirect	0.98 ± 0.04	No loss	

The effect of a reduction in vitamin B_6 (pyridoxamine) in pasteurized (<8%), evaporated (35–50%), sterilized (20–50%), and LLM (<10%) milk after storage on shelf-life duration was addressed [104]. Interconversion of different forms of vitamin B_6 may occur because of heat treatment, being more pronounced in evaporated milk. A more recent study reported that thermal treatment of 120 °C/400 s can alter the distribution of vitamin B_6 forms. It proposed this as a discriminator of high-temperature treated milk for a novel integrator for time–temperature combination from other thermally processed counterparts [105].

Vitamin B_9 (folic acid) (5-methyltetrahydrofolate) is presented as part of folate-binding proteins (FBPs) in fresh milk [106]. FBPs are partly denatured in pasteurized milk and extensively denatured in exceeded-heated milk. Thus, folate presents as free folate in ultra-heated milk. It remains unclear how FBPs affect the bioavailability of folate. It was earlier reported that FBPs play a role in the direct transport of folate via intestinal mucosa. However, studies published more recently showed that FBPs were less effective in the promotion of the bioavailability of folate [107].

Vitamin B_{12} (cyanocobalamin) shows good stability against pasteurization and ultra-high-temperature treatments [108], but thermal treatment of milk at a higher strength may cause up to a 20% loss in vitamin B_{12} [91]. Despite vitamin B_{12} being insensitive to heat in plain pasteurized or ultra-heated milk, a remarkable loss (~1/3) was observed in chocolate milk processed at 100 °C for 1 h compared to no-chocolate-added milk (<10%), as reported by Johns et al. [100]. These results may be due to the fact that cocoa powder is physically affinitive to vitamin B_{12} and/or has a high capacity for peroxide generation of cocoa polyphenols (i.e., (+)-catechin) and related products such as caffeic acid, (−)-epigallocatechin, and gallic acid. Vitamin B_{12} loss is inappreciable in milk treated at 96 °C for 5 min [101] to 100% in ultra-heated milk stored for 20 weeks at room temperature [109]. Heat treatment might not be a significant factor in the instability of vitamin B_{12}. However, the presence of oxygen-sensitive ingredients such as copper and ascorbic acid as well as oxygen exposure accelerate the destabilization of vitamin B_{12} during thermal treatment.

Various factors could alter the capacity of vitamin C (ascorbic acid) against thermal treatment controlled by a group of factors including thermal treatment patterns, de-airing processes as well as storage conditions and periods. The loss of vitamin C in pasteurized cow milk refrigerated at 4 °C for 6–9 d may be up to 45% [110,111]. As the temperature increases from 63 °C to 100 °C for 30 min, the vitamin C losses in cow milk and camel milk (CM) increase from 18 to 48% and 27 to 67%, respectively [112]. Contrarily, the loss of vitamin C is inappreciable in LLM milk when milk was kept at low oxygen concentrations (<3.3 ppm) [111]. The food matrix considerably affects the rate constant of vitamin C degradation and a higher rate constant for the denaturation of vitamin C is observed in milk than in polyphenol-rich drinks. The duration of thermal treatment and storage, the oxygen concentration, and the presence of antioxidants are essential for the stability of vitamin C [113].

In light and air, there may be a loss of vitamin A (retinol) with a prolonged holding time during thermal treatment [114] or the storage of processed milk at ambient temperatures [115]. Fortified vitamin A is stable against pasteurization but remarkably decreases up to 23% and 37% in boiled and sterilized milk, respectively [102]. Additionally, the fortification of milk with ferrous gluconate hydrate or ferric pyrophosphate soluble adversely affects the thermal stability of vitamin A. The evaporation of pasteurized milk may also cause a reduction in vitamin A, but not carotene [116,117]. Vitamin D (ergocalciferol) is more sensitive to light degradation than thermal treatment, similar to vitamin A [118,119]. Contrarily, the presence of oxygen shows little effect on the stability of milk fortified with vitamin D_3 (cholecalciferol) [103].

The pasteurization or evaporation process shows less effect on Vitamin E (tocopherol) degradation [116,120]. Moreover, minor damage of either δ- or γ-tocotrienol in LLM after storage was observed [120]. A partial degradation of vitamin E (especially α-tocopherol) is demonstrated when exposed to air, being more profound in unesterified tocopherols due to free phenolic hydroxyl groups [114,121]. Milk proteins protect vitamin E against heat in dairy products, like vitamins A and D. Vitamin K in milk is unlikely to be affected by heat treatments, just like vitamin E [104].

Overall, thermal treatment widely utilized in the dairy industry (i.e., HTST, ESL or LLM milk, direct/indirect heat treatment) shows limited effects on the denaturation of milk vitamins, except for vitamin C, which is more sensitive to heating in water-soluble vitamins. Usually, milk is not preferable to supplement water-soluble vitamins except for vitamin B_2 and vitamin B_{12}, and there is minor damage to most vitamins caused by thermal treatment. Interactions between vitamins and other milk components, such as casein and lipids, triggered by heat might be a protective barrier, promoting the thermal stability of vitamins in milk.

5. Influences of Heat Treatment on Heating Markers in Milk

5.1. Furosine and Lactulose

The Maillard reaction is commonly seen during food processing when heating is involved, and it generates multiple products, including precursor substances of furosine, which can cause kidney damage [122]. Furosine has been observed to have a higher level with higher dissolved air/oxygen contents, possibly due to reactions on the oxidative side [123]. Elliott et al. [124] compared heat treatment directly or indirectly (147 °C/6 s) and found that the DSH process achieved a lower level of furosine by 76.6% on week 1 and 27.9% on week 24.

Furosine is typically less in direct-heated milk than in indirect-heated counterparts for 50–70% [125–127]. Lee et al. [128] demonstrated that ultra-pasteurization treatment accelerated the generation of furosine. However, DSI samples had a much lower content than that of indirect heat treatment samples (43.81 mg vs. 168.72 mg/100 g protein) while the HTST samples had the lowest content of 6.95 mg/100 g protein. Furosine formation was only remarkably influenced by the extreme processing conditions (i.e., holding for 60 s at temperatures > 100 °C) when pasteurized either directly or indirectly [127]. Furosine levels for raw and pasteurized milk are 4–5 mg and 4–7 mg/100g protein, respectively, while furosine levels for milk processed under extreme conditions could be up to 372 mg/100 g protein (150 °C/20 s).

As for extreme short-time DSH (<0.5 s), Rauh et al. [129] compared two DSH methods with different preheating treatments (95 °C/5 s or 95 °C/180 s followed by 150 °C/0.2 s) to ISH (140 °C/4 s). They found that both pretreatments followed by 150 °C/0.2 s treatments generated dramatically less furosine than 4 s indirect heating (11.6 mg/100 g protein or 7.6 mg/100 g protein vs. 119.9 mg /100 g) of protein. Similar results were reported in a study on DSIJ treatment with an extremely short timeframe (154~156 °C/0.116 s), and only a slight influence was found in the level of furosine in YM (~11.1 mg for processed milk vs. ~8.7 mg for raw milk) [28,130].

Lactulose is an isomer of lactose that is formed during the thermal processing of milk, also used as a marker for the thermal treatment of milk. Marconi, E. et al. [131] conducted a comprehensive study testing lactulose content in milk samples after different thermal treatments, labeled as follows: pasteurized (PAST) milk, high-temperature pasteurized (HT PAST) milk, direct UHT-treated milk using an injection system (INJ UHT), direct UHT-treated milk using an infusion system (INF UHT), indirect UHT-treated milk using a plate or a tubular heat-exchange system (IND UHT), and in-container sterilized (STER) milk, respectively. The content of lactulose is closely related to thermal treatment intensity as follows: 744 mg/L for STER milk, 348 mg/L for IND UHT milk, 165 mg/L for INJ UHT milk, 107 mg/L for INF UHT milk, 58 mg/L for HT PAST milk, and 3.5 mg/L for PAST milk (the European Union limit for milk is 600 mg/L).

5.2. Alkaline Phosphatase (ALP)

Alkaline phosphatase (ALP) inactivation is commonly regarded as a maker of sufficient pasteurization, and its inactivation is well known to occur even for short-time pasteurization of at least 70 °C for 16 s [132]. Therefore, ALP activities in milk samples are regarded as a marker of heat treatment severity [133].

Dickow et al. [134] measured changes in ALP activity with heat treatment of 63 °C/30 s, 72 °C/15 s, or 150 °C/0.1 s and found that ALP was completely inactivated when heated at 72 °C or above for all three methods. Only from the perspective of ALP activity, thermal treatment at or above 72 °C can be regarded as standard HTST treatment, which agrees with Fox and Kelly [132]. Lorenzen et al. [88] compared ESL milk samples, heated indirectly or directly, to HTST and ultra-heated milk samples regarding indigenous milk enzymes. The activities of alkaline phosphatase (ALP), lipase (LIP), and lactoperoxidase (LPO) were described. The results of LPO proved that a sufficient thermal treatment was applied to directly and indirectly heated ESL milk along with ultra-heated milk as LPO activities of all samples were below 5 U/L, which is 300–2100 U/L for HTST milk samples. As for ALP and LIP activities, all samples were in a similar range (0.01–0.50 U/L).

5.3. Lactoperoxidase (LPO)

Lactoperoxidase (LPO), one of the most abundant enzymes in bovine milk, exerts its function as a nature-sourced antibacterial agent, and it has been reported that lactoperoxidase can alleviate symptoms of asthma and reduce the damaging effects of hydrogen peroxide [133,134]. LPO activity can be used as an indicator of whether the milk was heated over 78 °C for 15 s [135]. It has been reported that the LPO activity of fresh milk decreased by 58% and 92% in milk after being heated at 72.5 °C for 15 s and 25 s, respectively [136]. This finding was reviewed by Lan et al. [137], with the LPO activity dramatically reducing after heating at 75 °C for 15 s and completely deactivating after heating at 85 °C or above for 15 s.

5.4. Plasmin

Treatment with a low temperature (63 °C/30 min) significantly augments plasmin activity in bovine casein fractions. The results of plasminogen are similar to plasmin; 63 °C/30 min treatment has similar plasminogen-derived activity compared to its raw counterpart, but an extended time heating treatment reduces plasminogen-derived activity up to 100%. Similar results were reported by van Asselt et al. [138] that heat treatment of 80 °C/300 s with a 45 s pre/post-treatment could eliminate up to 99.96% plasmin activity. Leite et al. [139] reported that extended-time heating (85 °C/5 min) treatment could cause a significant decline (~100%) of plasmin activity in both whey and casein fractions of bovine milk. Ultra-high temperature treatment exhibits lower efficiency on plasmin inactivation; 135 °C /4 s treatment could decrease plasmin activity by about 70% compared to raw milk [140]. An other study by Leite et al. [141] described the impact of conventional pasteurization (63 °C/30 min or 75 °C/15 s) on cathepsin D and elastase activities. Both heat treatments decreased the activity of cathepsin D by ~45%. As for elastase, 63 °C/30 min treatment showed no effect, and 75 °C/15 s treatment reduced elastase activity by about 25%.

6. Impact of Heat Treatment on Other Properties of Milk

6.1. Age Gelation

Ultra-instantaneous direct steam heating (UI-DSH) treatment has emerged based on the heating system of steam infusion, which can achieve more extreme heating conditions (>155 °C for <0.1 s) [15]. Compared to ISH-treated whole milk, the levels of bio-active proteins and flavor in UI-DSH whole milk are similar to pasteurized whole milk [15] (Figure 2). However, some detrimental changes, including bitterness, creaming, age gelation, and sedimentation/clarification, can occur [53]. Of these, creaming is the first observed defect, seriously affecting the quality of milk and consumers' purchase intentions.

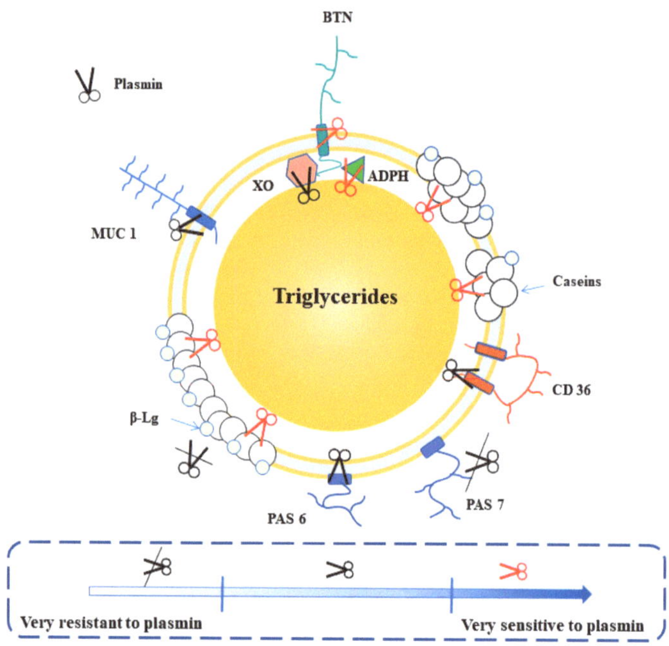

Figure 2. Molecular mechanisms of fat destabilization in UI-DSH whole milk induced by plasmin.

To remain stable for at least 6 months, commercial ISH-treated whole milk undergoes sufficient homogenization; however, an obvious creaming can occur in UI-DSH milk in a short time, because of residual plasmin attacks on interfacial proteins of fat globules. Our previous study determined that the activity of residual plasmin was 1.25 mU/mL in commercial UI-DSH whole milk [54]. Unfortunately, residual plasminogen was gradually converted to plasmin by an activator during storage [55]. Thus, at 25 °C and 37 °C, plasmin activity increased rapidly within 15 days and reached a maximum value (7–8 mU/mL) between days 15 and 45 [142].

The natural MFGM is a three-layer structure; membrane proteins are either integrally or peripherally attached to the membrane [59]. After homogenization disrupts the membrane, caseins are involved in reconstructing the fat globule interface, forming a mixed of three-layer and monolayer structure [34]. This complex structure affects the hydrolysis behaviors of plasmin. In this case, our lab revealed the mechanisms underlying fat destabilization in UI-DSH whole milk at the molecular level induced by added plasmin, and further investigated the impacts of the MFGM structure on the hydrolysis of major MFGM proteins. The results showed that plasmin decreased the zeta-potential value as well as the amount and coverage of interfacial proteins through hydrolysis, causing electrostatic and steric destabilization of fat globules, and thus the fat globules flocculated and coalesced. It is worth noticing that the co-hydrolysis of membrane-anchoring proteins (X0-BTN-ADPH) and caseins was the direct cause of fat globule destabilization [15]. Moreover, at the interface, caseins, BTN, and ADPH were very sensitive to plasmin, while it was difficult for PAS 7 to be hydrolyzed by plasmin [15]. Overall, the complex interfacial structure reduced the susceptibility of some major MFGM proteins to plasmin and provided protective effects [15].

Three methods should be considered to minimize creaming in whole UI-DSH milk during storage, including plasmin activity and the modulation of the components of interfacial proteins [15]. Plasmin activity causes fat destabilization directly in whole UI-DSH milk. Minimizing this issue should involve making a trade-off among some factors, including thermal load, residual plasmin activity, and the retention of bioactive components. Although increasing homogenization pressure within a specific range can reduce the particle size of fat globules, it will provide more caseins at the newly created water–oil interface. Thus, the modulation of the constitution of interfacial proteins will be a new direction to minimize the rise of fat; namely, the proteins involved in the reconstruction of the fat globule membrane are replaced with milk proteins that are resistant to plasmin.

6.2. *Effect of Thermal Treatment on Unconventional Milk*

Unconventional milk lactated by minor mammalian species, such as goats, yaks, and camels, is often accompanied by unique nutritional, textural and medicinal virtues. Ultra-high temperature processing and pasteurization are widely utilized to guarantee milk quality and safety. Goat milk (GM) has been gaining attention recently due to its unique nutritional value, digestive virtues, and less allergenicity than cow milk due to low β-lactoglobulin and αs1-casein content [143]. Moreover, the easy digestibility of GM fat is possibly attributed to its smaller diameter of 3.2–3.6 μm and its higher concentration of short to medium-chain fatty acids, which are used to treat cholesterol disorders, anemia, metabolic disorders, and low mineral bone density [143]. YM is economically essential for herders in the Chinese Qinghai–Tibet Plateau, with 1.2 million tons of annual production. Additionally, YM has different nutritional value than cow milk since the contents of protein, fat, and dry matter in YM are ~5, ~6 g, and ~17 g/100 mL, respectively [144,145], so there is an excellent opportunity for high quality, unique YM products. CM exerts various medicinal and therapeutic values, including anti-cancerous, anti-diabetic, hypo-allergenic, ACE-inhibitory, and cholesterol-lowering properties [146]. These unique features of unconventional milk lead to certain limitations related to intensive thermal treatments, such as nutritional losses, off-flavors, and color modifications.

GM is comparatively sensitive to thermal treatment compared to bovine milk due to its lower concentration of citrates and micellar solvation but its higher concentration in ionic calcium, making it more fragile against heat [147]. Heilig et al. [148] compared GM to bovine milk after heating using different temperature–time combinations (72 °C, 32 s; 120 °C, 4 s (DSIJ/standard tubular)/10 s (DSI); 140 °C, 4 s (DSIJ/standard tubular)/10 s (DSI)). The sensory quality (descriptive sensory analysis and triangle test), casein micelle diameter (dynamic light scattering), and storage stability (precipitation at 4 and 20 °C) of differently treated milk were analyzed. While GM heated at 72 °C showed neither processing nor sensory deficiencies, GMs heated at 120 and 140 °C were observed to be slightly to more clearly grainy and precipitated after 21–90 days of storage. As for storage stability, sensory quality, and process suitability, skim milk showed overall better results than whole milk, and DSH techniques gave better overall results than indirect ones. They also reported in-process coagulation due to casein micelle aggregation in the case of indirectly heated GM.

A study focused on the digestive kinetics of YM after pasteurization (65 °C, 30 min), autoclaving (121 °C, 20 min), microwaving (2.45 GHz at 640 W, 140 s), and non-heated treatment (raw milk) using an in vitro model [149]. Due to the formation of clots with denser structures, a shorter gastric emptying half-time for the pasteurized (36.2 min) and microwaved (34.2 min) milk was determined compared to that of the autoclaved (41.1 min) and fresh (38.4 min) milk. The highest protein digestibility was found in pasteurized milk (92.5%) at the end of digestion, followed by milk treated by microwaving (87.8%), autoclaving (86.1%), and raw milk (80.8%). Mi et al. [28] comprehensively studied the influence of DSI treatment on YM with temperatures of 138 (5 s)–156 °C (0.116 s). They found that DSI treatment effectively sterilized *G. stearothermophilus* spores without extensively degrading serum proteins. Serum protein denaturation in milk caused by DSI was lower than that normally found in traditional ultra-high-temperature treated milk, with losses of 30 and 40% of native β-Lg and lactoferrin, respectively, but not for α-La, which was more similar to pasteurization. Moreover, the heating makers, such as lactulose and furosine in DSI milk, were both below the limits for pasteurized milk (50 mg L^{-1} and 0.12 mg g^{-1} of protein), which indicates the potential that DSI-treated milk might be marketable as a pasteurized product.

Understanding milk's heat stability is important to the dairy industry since thermal processing is commonly used. The coagulation time caused by thermal treatment (CTT) of CM at 130, 120, and 100 °C was investigated in the pH range of 6.3–7.1 [150]. At 130 and 120 °C, the milk was unstable in a wide pH range, with a CTT of 2–3 min. The CTT initially increased to 12 min at 100 °C, then remained constant in a pH range between 6.4 and 6.7 and increased significantly with the increasing pH until approximately 33 min was reached at a pH of 7.1. CM seemed much less stable than bovine milk during heat processing [151]. Similar results were reported by Sagar et al. [152]; the CTTs at 140 °C for CM, buffalo, and bovine milk were determined to be 133.6, 1574.6, and 1807.4 s, respectively. The presence of κ -casein and β-lactoglobulin and the interaction between them during heating is critical for retaining milk stability [150]. Therefore, the lower level of κ-casein (5% of casein in CM vs. 13.6% of casein in bovine milk) and the shortage of β-lactoglobulin may explain CM's instability at high temperatures. Genene et al. [153] compared the denaturation of whey protein and the resultant rennetability of CM to those of bovine milk under heat treatment from 65 °C to 90 °C with different durations, and found that α-La showed less denaturation in CM than bovine milk. β-Lg was not found in CM, which could be one of the main differences compared to bovine milk, while in bovine milk, both β-Lg and α-La denatured along with increasing heat treatment. The gelation properties of CM were significantly affected by the severity of the thermal treatment applied. Zhao et al. [154] analyzed the changes in products regarding the Millard reaction after fresh CM was heated at different temperature–time combinations as well as volatile components.

Markers in CM showed different changes compared to bovine milk. A former study showed that camel ALP is heat-resistant, still showing activity at 90 °C [155], which is typically inactivated at around 72 °C [103]. On the other hand, Tayefi-Nasrabadi et al. [156] determined that camel LPO was more heat-sensitive than bovine LPO. There are insufficient in-depth studies on convenient markers of pasteurization for CM, which limits the establishment of an international standard. Thus, the pasteurization of CM may be incorrect and needs to be revised. Heat-induced changes in milk could affect the properties of reconstituted powders for applications in end-products.

7. Conclusions

Milk is rich in nutrition and is an easy-to-obtain food material; therefore, it is vital to explore its thermal behavior, texturally and nutritively. Although micronutrients and bioactive compounds are normally present at low levels, they present massive influences on the health of humans, especially infants. Developing more sensitive and precise analysis techniques has improved our understanding of the importance and functions of these minor components. Thermal treatment is an essential tool for the dairy industry to ensure microbiological safety and maintain milk quality. On the other hand, heating can cause a loss of macro/micro-nutrients by decomposition or inducing oxidation.

However, pasteurization (<85 °C/ 60 s) has a low-level effect on the stability of nutrients in milk; however, partial degradation of some nutrients, including water-soluble vitamins B_1, B_6, B_{12}, and C, some hormones, and unsaturated fatty acids occurs along with the increased degree of heat treatment. The heating duration and manner could be vital for the degradation of nutrients or the level of age gelation in milk when heated with ultra-high temperatures; a treatment of 156 °C/0.116 s showed much less impact on the quality of milk than a treatment of 138 °C/5s, and direct heating showed less impact than indirect heating. Additionally, evaporation operations can harm oxygen-sensitive components, including unsaturated fatty acids and most water- and fat-soluble vitamins. Moreover, higher processing temperatures alter the viscosity of milk. This is particularly important for products requiring the addition of milk (coffee or baked food) or where viscosity is a key quality parameter of end products (yogurt).

Although heat-induced changes on the macro/micro-nutrients and other constituents of milk have been revealed to a certain extent, knowledge of the differences between direct steam heating and indirect steam heating, the characteristics of DSI technology, and the understanding of the bioavailability and bioaccessibility of these components after heat treatment remains insufficient. More in vitro and in vivo studies should be conducted to explore how heating influences milk's benefits for human health, providing a foundation for optimizing thermal treatment parameters. In addition, the different influences among DSI, DSIJ, and ISH on the food matrix should be further studied to benefit the dairy industry.

Author Contributions: Y.W.: writing—original draft. R.X.: conceptualization, advising, and editing. S.L.: writing—original draft. P.W.: conceptualization, and supervision. Y.Z.: supervision. T.N.: reviewing and editing. H.C.: conceptualization, writing—original draft and writing—review and editing. All authors have read and agreed to the published version of the manuscript.

Funding: This research was funded by key projects of the National Natural Science Foundation of China, Grant No. 32130081; Hohhot Science and Technology Plan Project: 2024-Announcement-Agriculture-3.

Institutional Review Board Statement: Not applicable.

Data Availability Statement: Data sharing is not applicable to this article.

Conflicts of Interest: The authors declare no conflicts of interest.

References

1. Syed, Q.A.; Hassan, A.; Sharif, S.; Ishaq, A.; Saeed, F.; Afzaal, M.; Hussain, M.; Anjum, F.M. Structural and functional properties of milk proteins as affected by heating, high pressure, Gamma and ultraviolet irradiation: A review. *Int. J. Food Prop.* **2021**, *24*, 871–884. [CrossRef]
2. Tonolo, F.; Folda, A.; Cesaro, L.; Scalcon, V.; Marin, O.; Ferro, S.; Bindoli, A.; Rigobello, M.P. Milk-derived bioactive peptides exhibit antioxidant activity through the Keap1-Nrf2 signaling pathway. *J. Funct. Foods* **2020**, *64*, 103696–103703. [CrossRef]
3. Chatterton, D.E.W.; Nguyen, D.N.; Bering, S.B.; Sangild, P.T. Anti-inflammatory mechanisms of bioactive milk proteins in the intestine of newborns. *Int. J. Biochem. Cell Biol.* **2013**, *45*, 1730–1747. [CrossRef] [PubMed]
4. Zhang, J.; Mamet, T.; Guo, Y.; Li, C.; Yang, J. Yak milk promotes renal calcium reabsorption in mice with osteoporosis via the regulation of TRPV5. *J. Dairy Sci.* **2023**, *106*, 7396–7406. [CrossRef] [PubMed]
5. Coutinho, N.M.; Silveira, M.R.; Rocha, R.S.; Moraes, J.; Ferreira, M.V.S.; Pimentel, T.C.; Freitas, M.Q.; Silva, M.C.; Raices, R.S.L.; Ranadheera, C.S.; et al. Cold plasma processing of milk and dairy products. *Trends Food Sci. Technol.* **2018**, *74*, 56–68. [CrossRef]
6. Munir, M.; Nadeem, M.; Qureshi, T.M.; Leong, T.S.H.; Gamlath, C.J.; Martin, G.J.O.; Ashokkumar, M. Effects of high pressure, microwave and ultrasound processing on proteins and enzyme activity in dairy systems—A review. *Innov. Food Sci. Emerg.* **2019**, *57*, 102192–102205. [CrossRef]
7. Jo, Y.; Benoist, D.M.; Barbano, D.M.; Drake, M.A. Flavor and flavor chemistry differences among milks processed by high-temperature, short-time pasteurization or ultra-pasteurization. *J. Dairy Sci.* **2018**, *101*, 3812–3828. [CrossRef]
8. Nikmaram, N.; Keener, K.M. The effects of cold plasma technology on physical, nutritional, and sensory properties of milk and milk products. *LWT-Food Sci. Technol.* **2022**, *154*, 112729–112736. [CrossRef]
9. Eisner, M.D. Direct and indirect heating of milk–A technological perspective beyond time–temperature profiles. *Int. Dairy J.* **2021**, *122*, 105145–105162. [CrossRef]

10. Gao, T.; Sun, D.-W.; Tian, Y.; Zhu, Z. Gold–Silver Core-Shell Nanorods Based Time-Temperature Indicator for Quality Monitoring of Pasteurized Milk in the Cold Chain. *J. Food Eng.* **2021**, *306*, 110624. [CrossRef]
11. Hanson, M.L.; Wendorff, W.L.; Houck, K.B. Effect of Heat Treatment of Milk on Activation of Bacillus Spores. *J. Food Prot.* **2005**, *68*, 1484–1486. [CrossRef] [PubMed]
12. Roberts, W.M.; Dill, C.W. Direct-steam injection system for processing fluid milk products. *J. Dairy Sci.* **1962**, *45*, 937–940. [CrossRef]
13. Wang, X.; Zhao, Z. A mini-review about direct steam heating and its application in dairy and plant protein processing. *Food Chem.* **2023**, *408*, 135233–135239. [CrossRef] [PubMed]
14. Rauh, V. Impact of Plasmin Activity on the Shelf Life and Stability of UHT Milk. Ph.D. Thesis, Aarhus University, Aarhus, Denmark, 2014.
15. Wang, Y.; Guo, M.; Wu, P.; Fan, K.; Zhang, W.; Chen, C.; Ren, F.; Wang, P.; Luo, J.; Yu, J.J.C.; et al. New insights into the destabilization of fat globules in ultra-instantaneous UHT milk induced by added plasmin: Molecular mechanisms and the effect of membrane structure on plasmin action. *Colloids Surf. B Biointerfaces* **2024**, *240*, 113987. [CrossRef] [PubMed]
16. Kelleher, C.M.; O'Mahony, J.A.; Kelly, A.L.; O'Callaghan, D.J.; Kilcawley, K.N.; McCarthy, N.A. The effect of direct and indirect heat treatment on the attributes of whey protein beverages. *Int. Dairy J.* **2018**, *85*, 144–152. [CrossRef]
17. Elias, R.J.; Kellerby, S.S.; Decker, E.A. Antioxidant activity of proteins and peptides. *Crit. Rev. Food Sci. Nutr.* **2008**, *48*, 430–441. [CrossRef] [PubMed]
18. Shi, J.; Ahlroos-Lehmus, A.; Pilvi, T.K.; Korpela, R.; Tossavainen, O.; Mervaala, E.M. Metabolic effects of a novel microfiltered native whey protein in diet-induced obese mice. *J. Funct. Foods* **2012**, *4*, 440–449. [CrossRef]
19. Attaallah, W.; Yilmaz, A.M.; Erdogan, N.; Yalçin, A.S.; Aktan, A. Whey protein versus whey protein hydrolyzate for the protection of azoxymethane and dextran sodium sulfate induced colonic tumors in rats. *Pathol. Oncol. Res.* **2012**, *18*, 817–822. [CrossRef]
20. Badr, G.; Ebaid, H.; Mohany, M.; Abuelsaad, A.S. Modulation of immune cell proliferation and chemotaxis towards CC chemokine ligand (CCL)-21 and CXC chemokine ligand (CXCL)-12 in undenatured whey protein-treated mice. *J. Nutr. Biochem.* **2012**, *23*, 1640–1646. [CrossRef]
21. Arena, S.; Renzone, G.; D'Ambrosio, C.; Salzano, A.M.; Scaloni, A. Dairy products and the Maillard reaction: A promising future for extensive food characterization by integrated proteomics studies. *Food Chem.* **2017**, *219*, 477–489. [CrossRef]
22. Wada, Y.; Lönnerdal, B. Effects of different industrial heating processes of milk on site-specific protein modifications and their relationship to in vitro and in vivo digestibility. *J. Agric. Food Chem.* **2014**, *62*, 4175–4185. [CrossRef] [PubMed]
23. Krishna, T.C.; Najda, A.; Bains, A.; Tosif, M.M.; Paplinski, R.; Kaplan, M.; Chawla, P. Influence of ultra-heat treatment on properties of milk proteins. *Polymers* **2021**, *13*, 3164. [CrossRef] [PubMed]
24. Pizzano, R.; Manzo, C.; Nicolai, M.A.; Addeo, F. Occurrence of major whey proteins in the pH 4.6 insoluble protein fraction from UHT-treated milk. *J. Agric. Food Chem.* **2012**, *60*, 8044–8050. [CrossRef] [PubMed]
25. Datta, N.; Elliott, A.J.; Perkins, M.L.; Deeth, H.C. Ultra-high-temperature (UHT) treatment of milk: Comparison of direct and indirect modes of heating. *Aust. J. Dairy Technol.* **2002**, *57*, 211–227.
26. Burton, H. *Ultra-High-Temperature Processing of Milk and Milk Products*; Springer Science & Business Media: Berlin, Germany, 2012.
27. Wang, X.J.; Zhao, Z.T. Structural and colloidal properties of whey protein aggregates produced by indirect tubular heating and direct steam injection. *Food Struct.-Neth.* **2023**, *35*, 100301. [CrossRef]
28. Mi, L.; Liang, Q.; Zhang, W.B.; Zhen, S.B.; Song, X.M.; Wen, P.C.; Zhang, Y. Sterilization of yak milk by direct steam injection (DSI) and effects on milk quality. *LWT-Food Sci. Technol.* **2023**, *189*, 115450. [CrossRef]
29. Paulsson, M.A.; Svensson, U.; Kishore, A.R.; Naidu, A.S. Thermal behavior of bovine lactoferrin in water and its relation to bacterial interaction and antibacterial activity. *J. Dairy Sci.* **1993**, *76*, 3711–3720. [CrossRef] [PubMed]
30. Dannenberg, F.; Kessler, H.G. Reaction kinetics of the denaturation of whey proteins in milk. *J. Food Sci.* **1988**, *53*, 258–263. [CrossRef]
31. Yang, Y.X.; Bu, D.P.; Zhao, X.W.; Sun, P.; Wang, J.Q.; Zhou, L.Y. Proteomic analysis of cow, yak, buffalo, goat and camel milk whey proteins: Quantitative differential expression patterns. *J. Proteome Res.* **2013**, *12*, 1660–1667. [CrossRef]
32. Yang, M.; Cao, X.Y.; Wu, R.N.; Liu, B.A.; Ye, W.H.; Yue, X.Q.; Wu, J.R. Comparative proteomic exploration of whey proteins in human and bovine colostrum and mature milk using iTRAQ-coupled LC-MS/MS. *Int. J. Food Sci. Nutr.* **2017**, *68*, 671–681. [CrossRef]
33. Li, W.X.; Li, M.H.; Cao, X.Y.; Han, H.J.; Kong, F.H.; Yue, X.Q. Comparative analysis of whey proteins in donkey colostrum and mature milk using quantitative proteomics. *Food Res. Int.* **2020**, *127*, 108741. [CrossRef]
34. Wang, Y.; Liu, M.X.; Qu, X.N.; Wang, S.C.; Ma, Z.N.; Zhang, R.M.; Li, H.J.; Liu, X.H.; Yu, J.H. Changes in the fat globule membrane protein components of pasteurized milk caused by different homogenization conditions determined using a label-free proteomic approach. *LWT-Food Sci. Technol.* **2019**, *115*, 108430. [CrossRef]
35. Lucey, J.A.; Wilbanks, D.J.; Horne, D.S. Impact of heat treatment of milk on acid gelation. *Int. Dairy J.* **2022**, *125*, 105222. [CrossRef]
36. Anema, S.G. On heating milk, the dissociation of κ-casein from the casein micelles can precede interactions with the denatured whey proteins. *J. Dairy Res.* **2008**, *75*, 415–421. [CrossRef] [PubMed]
37. Raikos, V. Effect of heat treatment on milk protein functionality at emulsion interfaces. A review. *Food Hydrocolloids* **2010**, *24*, 259–265. [CrossRef]

38. Yun, G.U.; Gillies, G.; Ripberger, G.; Hashemizadeh, I.; Whitby, C.P.; Bronlund, J. Modelling the reaction kinetics of β-lactoglobulin and K-casein heat-induced interactions in skim milk. *J. Food Eng.* **2023**, *344*, 111391. [CrossRef]
39. Anema, S.G.; Li, Y.M. Effect of pH on the association of denatured whey proteins with casein micelles in heated reconstituted skim milk. *J. Agric. Food Chem.* **2003**, *51*, 1640–1646. [CrossRef]
40. Nicolai, T.; Chassenieux, C. Heat-induced gelation of casein micelles. *Food Hydrocolloids* **2021**, *118*, 106755. [CrossRef]
41. Dalgleish, D.G. On the structural models of bovine casein micelles-review and possible improvements. *Soft Matter* **2011**, *7*, 2265–2272. [CrossRef]
42. Dalgleish, D.G.; Corredig, M. The structure of the casein micelle of milk and its changes during processing. *Annu. Rev. Food Sci. Technol.* **2012**, *3*, 449–467. [CrossRef] [PubMed]
43. Walstra, P. Casein sub-micelles: Do they exist? *Int. Dairy J.* **1999**, *9*, 189–192. [CrossRef]
44. Home, D.S. Casein micelle structure: Models and muddles. *Curr. Opin. Colloid. In.* **2006**, *11*, 148–153. [CrossRef]
45. Holt, C.; de Kruif, C.G.; Tuinier, R.; Timmins, P.A. Substructure of bovine casein micelles by small-angle X-ray and neutron scattering. *Colloid Surf. A* **2003**, *213*, 275–284. [CrossRef]
46. Bhat, M.Y.; Dar, T.A.; Rajendrakumar Singh, L. Casein Proteins: Structural and Functional Aspects. In *Milk Proteins—From Structure to Biological Properties and Health Aspects*; IntechOpen: London, UK, 2016; ISBN 978-953-51-2537-2.
47. Anema, S.G. Role of κ-casein in the association of denatured whey proteins with casein micelles in heated reconstituted skim milk. *J. Agric. Food Chem.* **2007**, *55*, 3635–3642. [CrossRef]
48. Anema, S.G.; Li, Y.M. Association of denatured whey proteins with casein micelles in heated reconstituted skim milk and its effect on casein micelle size. *J. Dairy Res.* **2003**, *70*, 73–83. [CrossRef]
49. Kelleher, C.M.; Tobin, J.T.; O'Mahony, J.A.; Kelly, A.L.; O'Callaghan, D.J.; McCarthy, N.A. A Comparison of Pilot-Scale Supersonic Direct Steam Injection to Conventional Steam Infusion and Tubular Heating Systems for the Heat Treatment of Protein-Enriched Skim Milk-Based Beverages. *Innov. Food Sci. Emerg. Technol.* **2019**, *52*, 282–290. [CrossRef]
50. Oldfield, D.J.; Singh, H.; Taylor, M.W. Association of β-lactoglobulin and α-lactalbumin with the casein micelles in skim milk heated in an ultra-high temperature plant. *Int. Dairy J.* **1998**, *8*, 765–770. [CrossRef]
51. Oldfield, D.J.; Singh, H.; Taylor, M.W.; Pearce, K.N. Kinetics of denaturation and aggregation of whey proteins in skim milk heated in an ultra-high temperature (UHT) pilot plant. *Int. Dairy J.* **1998**, *8*, 311–318. [CrossRef]
52. Wu, P.P.; Guo, M.Y.; Wang, P.J.; Wang, Y.; Fan, K.; Zhou, H.; Qian, W.T.; Li, H.L.; Wang, M.H.; Wei, X.J.; et al. Age gelation in direct steam infusion ultra-high-temperature milk: Different heat treatments produce different gels. *Foods* **2024**, *13*, 1236. [CrossRef] [PubMed]
53. Rauh, V.M.; Johansen, L.B.; Ipsen, R.; Paulsson, M.; Larsen, L.B.; Hammershoj, M. Plasmin activity in UHT milk: Relationship between proteolysis, age gelation, and bitterness. *J. Agric. Food Chem.* **2014**, *62*, 6852–6860. [CrossRef]
54. Zhang, C.Y.; Bijl, E.; Hettinga, K. Destabilization of UHT milk by protease AprX from Pseudomonas fluorescens and plasmin. *Food Chem.* **2018**, *263*, 127–134. [CrossRef] [PubMed]
55. Beliciu, C.M.; Sauer, A.; Moraru, C.I. The effect of commercial sterilization regimens on micellar casein concentrates. *J. Dairy Sci.* **2012**, *95*, 5510–5526. [CrossRef] [PubMed]
56. Guo, M.Y.; Wang, Y.; Wang, P.J.; Luo, J.; Qian, W.T.; Li, H.L.; Wang, M.H.; Yang, J.H.; Ren, F.Z. Multiscale structure analysis reveals changes in the structure of casein micelles treated with direct-steam-infusion UHT. *Food Hydrocolloids* **2024**, *153*, 110033. [CrossRef]
57. Wang, M.Q.; Cao, C.J.; Wang, Y.; Li, H.B.; Li, H.J.; Yu, J.H. Comparison of bovine milk fat globule membrane protein retention by different ultrafiltration membranes using a label-free proteomic approach. *LWT-Food Sci. Technol.* **2021**, *144*, 111219. [CrossRef]
58. Wang, Y.; Guo, M.Y.; Ren, F.Z.; Wang, P.J.; Li, H.J.; Li, H.B.; Li, Y.X.; Luo, J.; Yu, J.H. A novel strategy to construct stable fat globules with all major milk fat globule membrane proteins to mimic breast milk fat emulsions at the protein level. *Food Res. Int.* **2023**, *173*, 113351. [CrossRef] [PubMed]
59. Wiking, L.; Gregersen, S.B.; Hansen, S.F.; Hammershoj, M. Heat-induced changes in milk fat and milk fat globules and its derived effects on acid dairy gelation—A review. *Int. Dairy J.* **2022**, *127*, 105213. [CrossRef]
60. Lee, S.J.E.; Sherbon, J.W. Chemical changes in bovine milk fat globule membrane caused by heat treatment and homogenization of whole milk. *J. Dairy Res.* **2002**, *69*, 555–567. [CrossRef] [PubMed]
61. Yang, Y.X.; Zheng, N.; Zhao, X.W.; Yang, J.H.; Zhang, Y.D.; Han, R.W.; Qi, Y.X.; Zhao, S.G.; Li, S.L.; Wen, F.; et al. Changes in bovine milk fat globule membrane proteins caused by heat procedures using a label-free proteomic approach. *Food Res. Int.* **2018**, *113*, 1–8. [CrossRef]
62. Yan, D.; Zhang, L.; Zhu, Y.; Han, M.; Wang, Y.; Tang, J.; Zhou, P.J.F. Changes in caprine milk fat globule membrane proteins after heat treatment using a label-free proteomics technique. *Foods* **2022**, *11*, 2705. [CrossRef]
63. Huang, Y.C.; Wei, T.; Chen, F.; Tan, C.L.; Gong, Z.Q.; Wang, F.X.; Deng, Z.Y.; Li, J. Effects of various thermal treatments on interfacial composition and physical properties of bovine milk fat globules. *Food Res. Int.* **2023**, *167*, 112580. [CrossRef]
64. Kokkinidou, S.; Peterson, D.G. Control of Maillard-type off-flavor development in ultrahigh-temperature-processed bovine milk by phenolic chemistry. *J. Agric. Food Chem.* **2014**, *62*, 8023–8033. [CrossRef]
65. Zhang, Y.; Yi, S.; Lu, J.; Pang, X.; Xu, X.; Lv, J.; Zhang, S. Effect of different heat treatments on the Maillard reaction products, volatile compounds and glycation level of milk. *Int. Dairy J.* **2021**, *123*, 105182. [CrossRef]

66. Zhang, H.; Xu, Y.; Zhao, C.; Xue, Y.; Tan, D.; Wang, S.; Jia, M.; Wu, H.; Ma, A.; Chen, G. Milk lipids characterization in relation to different heat treatments using lipidomics. *Food Res. Int.* **2022**, *157*, 111345. [CrossRef] [PubMed]
67. Chen, J.; He, J.; Zhao, Z.; Li, X.; Tang, J.; Liu, Q.; Wang, H. Effect of heat treatment on the physical stability, interfacial composition and protein-lipid co-oxidation of whey protein isolate-stabilised O/W emulsions. *Food Res. Int.* **2023**, *172*, 113126. [CrossRef] [PubMed]
68. Zhao, J.; Zhou, B.; Wang, P.; Ren, F.; Mao, X. Physicochemical properties of fluid milk with different heat treatments and HS-GC-IMS identification of volatile organic compounds. *Int. Dairy J.* **2023**, *142*, 105654. [CrossRef]
69. Cerny, C. The aroma side of the Maillard reaction. *Ann. N. Y. Acad. Sci.* **2008**, *1126*, 66–71. [CrossRef] [PubMed]
70. Feng, D.; Wang, J.; Ji, X.; Min, W.; Yan, W. HS-GC-IMS detection of volatile organic compounds in yak milk powder processed by different drying methods. *LWT-Food Sci. Technol.* **2021**, *141*, 110855. [CrossRef]
71. Jo, Y. Flavor and Flavor Chemistry of Fluid Milk. Ph.D. Thesis, North Carolina State University, Raleigh, NC, USA, 2019.
72. Liem, D.G.; Bolhuis, D.P.; Hu, X.; Keast, R.S.J. Short communication: Influence of labeling on Australian and Chinese consumers' liking of milk with short (pasteurized) and long (UHT) shelf life. *J. Dairy Sci.* **2016**, *99*, 1747–1754. [CrossRef] [PubMed]
73. Meng, F.Y.; Han, Z.S.; Zhang, Z.T.; Ding, H.; Lu, X.L.; Lu, C.; Ma, L.G.; Kang, Z.Y.; Wang, B.; Li, Y. Effect of steam infusion and steam injection ultra-high temperature treatment on active proteins and flavor compounds in milk. *J. Food Sci. Technol.* **2023**, *41*, 70–80. [CrossRef]
74. Li, Y.; Wang, Y.; Yuan, D.; Li, Y.; Zhang, L. Comparison of SDE and SPME for the analysis of volatile compounds in butters. *Food Sci. Biotechnol.* **2019**, *29*, 55–62. [CrossRef]
75. Gaucheron, F. Milk salts | distribution and analysis. In *Encyclopedia of Dairy Sciences*, 2nd ed.; Fuquay, J.W., Ed.; Academic Press: San Diego, CA, USA, 2011; pp. 908–916. ISBN 978-0-12-374407-4.
76. Kilic-Akyilmaz, M.; Ozer, B.; Bulat, T.; Topcu, A. Effect of heat treatment on micronutrients, fatty acids and some bioactive components of milk. *Int. Dairy J.* **2022**, *126*, 105231. [CrossRef]
77. Fox, P.F.; Uniacke-Lowe, T.; McSweeney, P.L.H.; O'Mahony, J.A. Salts of milk. In *Dairy Chemistry and Biochemistry*; Fox, P.F., Uniacke-Lowe, T., McSweeney, P.L.H., O'Mahony, J.A., Eds.; Springer International Publishing: Cham, Switzerland, 2015; pp. 241–270. ISBN 978-3-319-14892-2.
78. Pouliot, Y.; Boulet, M.; Paquin, P. Observations on the heat-induced salt balance changes in milk I. Effect of heating time between 4 and 90 °C. *J. Dairy Res.* **1989**, *56*, 185–192. [CrossRef]
79. Pouliot, Y.; Boulet, M.; Paquin, P. Experiments on the heat-induced salt balance changes in cow's milk. *J. Dairy Res.* **1989**, *56*, 513–519. [CrossRef]
80. Pouliot, Y.; Boulet, M.; Paquin, P. Observations on the heat-induced salt balance changes in milk II. Reversibility on cooling. *J. Dairy Res.* **1989**, *56*, 193–199. [CrossRef]
81. Boiani, M.; Fenelon, M.; FitzGerald, R.J.; Kelly, P.M. Use of ^{31}P NMR and FTIR to investigate key milk mineral equilibria and their interactions with micellar casein during heat treatment. *Int. Dairy J.* **2018**, *81*, 12–18. [CrossRef]
82. Wang, Q.; Ma, Y. Effect of temperature and pH on salts equilibria and calcium phosphate in bovine milk. *Int. Dairy J.* **2020**, *110*, 104713. [CrossRef]
83. Britten, M.; Giroux, H.J. Rennet coagulation of heated milk: A review. *Int. Dairy J.* **2022**, *124*, 105179. [CrossRef]
84. Jeurnink, T.; Walstra, P.; Kruif, D. Mechanisms of fouling in dairy processing. *Neth. Milk Dairy J.* **1996**, *50*, 407–426.
85. Seiquer, I.; Delgado-Andrade, C.; Haro, A.; Navarro, M.P. Assessing the effects of severe heat treatment of milk on calcium bioavailability: In vitro and in vivo studies. *J. Dairy Sci.* **2010**, *93*, 5635–5643. [CrossRef]
86. Hrubša, M.; Siatka, T.; Nejmanová, I.; Vopršalová, M.; Kujovská Krčmová, L.; Matoušová, K.; Javorská, L.; Macáková, K.; Mercolini, L.; Remião, F.; et al. Biological properties of vitamins of the B-complex, Part 1: Vitamins B1, B2, B3, and B5. *Nutrients* **2022**, *14*, 484. [CrossRef]
87. Lalwani, S.; Lewerentz, F.; Håkansson, A.; Löfgren, R.; Eriksson, J.; Paulsson, M.; Glantz, M. Impact of Thermal Processing on Micronutrients and Physical Stability of Milk and Cream at Dairy Production Scale. *Int. Dairy J.* **2024**, *153*, 105901. [CrossRef]
88. Lorenzen, P.C.; Clawin-Rädecker, I.; Einhoff, K.; Hammer, P.; Hartmann, R.; Hoffmann, W.; Martin, D.; Molkentin, J.; Walte, H.G.; Devrese, M. A survey of the quality of extended shelf life (ESL) milk in relation to HTST and UHT milk. *Int. J. Dairy Technol.* **2011**, *64*, 166–178. [CrossRef]
89. Ryley, J.; Kajda, P. Vitamins in thermal processing. *Food Chem.* **1994**, *49*, 119–129. [CrossRef]
90. Lalić, J.; Denić, M.; Sunarić, S.; Kocić, G.; Trutić, N.; Mitić, S.; Jovanović, T. Assessment of thiamine content in some dairy products and rice milk. *CyTA-J. Food* **2014**, *12*, 203–209. [CrossRef]
91. Ottaway, P.B. 10—The stability of vitamins during food processing. In *The Nutrition Handbook for Food Processors*; Henry, C.J.K., Chapman, C., Eds.; Woodhead Publishing Series in Food Science, Technology and Nutrition; Woodhead Publishing: Cambridge, UK, 2002; pp. 247–264. ISBN 978-1-85573-464-7.
92. Guneser, O.; Karagul Yuceer, Y. Effect of ultraviolet light on water- and fat-soluble vitamins in cow and goat milk. *J. Dairy Sci.* **2012**, *95*, 6230–6241. [CrossRef] [PubMed]
93. Zhu, D.; Kebede, B.; Chen, G.; McComb, K.; Frew, R. Effects of the vat pasteurization process and refrigerated storage on the bovine milk metabolome. *J. Dairy Sci.* **2020**, *103*, 2077–2088. [CrossRef] [PubMed]
94. Damodaran, S.; Parkin, K.L. (Eds.) *Fennema's Food Chemistry*, 5th ed.; CRC Press: Boca Raton, FL, USA, 2017; ISBN 978-1-315-37291-4.

95. Saidi, B.; Warthesen, J.J. Effect of heat and homogenization on riboflavin photolysis in milk. *Int. Dairy J.* **1995**, *5*, 635–645. [CrossRef]
96. Thu Trang, V.; Kurogi, Y.; Katsuno, S.; Shimamura, T.; Ukeda, H. Protective effect of aminoreductone on photo-degradation of riboflavin. *Int. Dairy J.* **2008**, *18*, 344–348. [CrossRef]
97. Friend, B.A.; Shahani, K.M.; Long, C.A.; Agel, E.N. Evaluation of freeze-drying, pasteurization, high-temperature heating and storage on selected enzymes, B-vitamins and lipids of mature human milk. *J. Food Prot.* **1983**, *46*, 330–334. [CrossRef]
98. Matera, A.; Altieri, G.; Genovese, F.; Polidori, P.; Vincenzetti, S.; Perna, A.; Simonetti, A.; Rashvand Avei, M.; Calbi, A.; Di Renzo, G.C. Effect of Continuous Flow HTST Treatments on Donkey Milk Nutritional Quality. *LWT-Food Sci. Technol.* **2022**, *153*, 112444. [CrossRef]
99. Bendicho, S.; Espachs, A.; Arántegui, J.; Martín, O. Effect of High Intensity Pulsed Electric Fields and Heat Treatments on Vitamins of Milk. *J. Dairy Res.* **2002**, *69*, 113–123. [CrossRef] [PubMed]
100. Johns, P.W.; Das, A.; Kuil, E.M.; Jacobs, W.A.; Schimpf, K.J.; Schmitz, D.J. Cocoa polyphenols accelerate vitamin B12 degradation in heated chocolate milk. *Int. J. Food Sci. Technol.* **2015**, *50*, 421–430. [CrossRef]
101. Arkbåge, K.; Witthöft, C.; Fondén, R.; Jägerstad, M. Retention of vitamin B12 during manufacture of six fermented dairy products using a validated radio protein-binding assay. *Int. Dairy J.* **2003**, *13*, 101–109. [CrossRef]
102. Sachdeva, B.; Kaushik, R.; Arora, S.; Khan, A. Effect of processing conditions on the stability of native vitamin A and fortified retinol acetate in milk. *Int. J. Vitam. Nutr. Res.* **2021**, *91*, 133–142. [CrossRef] [PubMed]
103. Kaushik, R.; Sachdeva, B.; Arora, S. Vitamin D2 stability in milk during processing, packaging and storage. *LWT—Food Sci. Technol.* **2014**, *56*, 421–426. [CrossRef]
104. Fox, P.F.; Uniacke-Lowe, T.; McSweeney, P.L.H.; O'Mahony, J.A. Vitamins in milk and dairy products. In *Dairy Chemistry and Biochemistry*; Fox, P.F., Uniacke-Lowe, T., McSweeney, P.L.H., O'Mahony, J.A., Eds.; Springer International Publishing: Cham, Switzerland, 2015; pp. 271–297. ISBN 978-3-319-14892-2.
105. Schmidt, A.; Mayer, H.K. Milk process authentication by vitamin B6 as a novel time temperature integrator. *Food Control* **2018**, *91*, 123–127. [CrossRef]
106. Wigertz, K.; Hansen, I.; Høier-Madsen, M.; Holm, J.; Jägerstad, M. Effect of milk processing on the concentration of folate-binding protein (FBP), folate-binding capacity and retention of 5-methyltetrahydrofolate. *Int. J. Food Sci. Nutr.* **1996**, *47*, 315–322. [CrossRef] [PubMed]
107. de Jong, R.J.; Verwei, M.; West, C.E.; van Vliet, T.; Siebelink, E.; van den Berg, H.; Castenmiller, J.J.M. Bioavailability of folic acid from fortified pasteurised and UHT-treated milk in humans. *Eur. J. Clin. Nutr.* **2005**, *59*, 906–913. [CrossRef]
108. Riaz, M.N.; Asif, M.; Ali, R. Stability of vitamins during extrusion. *Crit. Rev. Food Sci. Nutr.* **2009**, *49*, 361–368. [CrossRef]
109. Godoy, H.T.; Amaya-Farfan, J.; Rodriguez-Amaya, D.B. Chapter 8—Degradation of vitamins. In *Chemical Changes During Processing and Storage of Foods*; Rodriguez-Amaya, D.B., Amaya-Farfan, J., Eds.; Academic Press: Cambridge, MA, USA, 2021; pp. 329–383. ISBN 978-0-12-817380-0.
110. Khan, I.T.; Nadeem, M.; Imran, M.; Ayaz, M.; Ajmal, M.; Ellahi, M.Y.; Khalique, A. Antioxidant capacity and fatty acids characterization of heat treated cow and buffalo milk. *Lipids Health Dis.* **2017**, *16*, 163. [CrossRef]
111. Andersson, I.; Öste, R. Loss of ascorbic acid, folacin and vitamin B12, and changes in oxygen content of UHT milk. I: Introduction and methods. *Milchwiss.-Milk Sci. Int.* **1992**, *47*, 223–224.
112. Mehaia, M.A. Vitamin C and riboflavin content in camels milk: Effects of heat treatments. *Food Chem.* **1994**, *50*, 153–155. [CrossRef]
113. Burch, R. 26—The stability and shelf life of vitamin-fortified foods. In *Food and Beverage Stability and Shelf Life*; Kilcast, D., Subramaniam, P., Eds.; Woodhead Publishing Series in Food Science, Technology and Nutrition; Woodhead Publishing: Cambridge, UK, 2011; pp. 743–754. ISBN 978-1-84569-701-3.
114. Berry Ottaway, P. 19—Stability of vitamins during food processing and storage. In *Chemical Deterioration and Physical Instability of Food and Beverages*; Skibsted, L.H., Risbo, J., Andersen, M.L., Eds.; Woodhead Publishing Series in Food Science, Technology and Nutrition; Woodhead Publishing: Cambridge, UK, 2010; pp. 539–560. ISBN 978-1-84569-495-1.
115. Lau, B.L.; Kakuda, Y.; Arnott, D.R. Effect of milk fat on the stability of vitamin A in ultra-high temperature milk. *J. Dairy Sci.* **1986**, *69*, 2052–2059. [CrossRef]
116. Bezie, A. The effect of different heat treatment on the nutritional value of milk and milk products and shelf-life of milk products. A review. *J. Dairy Vet. Sci.* **2019**, *11*, 1–8. [CrossRef]
117. Bonrath, W.; Gao, B.; Houston, P.; McClymont, T.; Müller, M.A.; Schäfer, C.; Schweiggert, C.; Schütz, J.; Medlock, J.A. 75 Years of vitamin A production: A historical and scientific overview of the development of new methodologies in chemistry, formulation, and biotechnology. *Org. Process Res. Dev.* **2023**, *27*, 1557–1584. [CrossRef]
118. Gordon, M.H. Fat-soluble vitamin assays in food analysis. *Food Chem.* **1990**, *36*, 87. [CrossRef]
119. Fennema, O. Chemical changes in food during processing–An overview. In *Chemical Changes in Food during Processing*; Richardson, T., Finley, J.W., Eds.; Springer: Boston, MA, USA, 1985; pp. 1–16. ISBN 978-1-4613-2265-8.
120. Pandya, J.K.; DeBonee, M.; Corradini, M.G.; Camire, M.E.; McClements, D.J.; Kinchla, A.J. Development of vitamin E-enriched functional foods: Stability of tocotrienols in food systems. *Int. J. Food Sci. Technol.* **2019**, *54*, 3196–3204. [CrossRef]
121. García-Martínez, C.; Holgado, F.; Velasco, J.; Márquez-Ruiz, G. Effect of classic sterilization on lipid oxidation in model liquid milk-based infant and follow-on formulas. *Eur. J. Lipid Sci. Technol.* **2012**, *114*, 1373–1380. [CrossRef]

122. Li, H.; Yang, H.; Li, P.; Li, M.; Yao, Q.; Min, L.; Zhang, Y.; Wang, J.; Zheng, N. Maillard reaction products with furan ring, like furosine, cause kidney injury through triggering ferroptosis pathway. *Food Chem.* **2020**, *319*, 126368. [CrossRef]
123. Katsuno, S.; Shimamura, T.; Kashiwagi, T.; Izawa, N.; Ukeda, H. Effects of dissolved oxygen on the Maillard reaction during heat treatment of milk. *Int. Dairy J.* **2013**, *33*, 34–37. [CrossRef]
124. Elliott, A.J.; Datta, N.; Amenu, B.; Deeth, H.C. Heat-induced and other chemical changes in commercial UHT milks. *J. Dairy Res.* **2005**, *72*, 442–446. [CrossRef]
125. Corzo, N.; Delgado, T.; Troyano, E.; Olano, A. Ratio of lactulose to furosine as indicator of quality of commercial milks. *J. Food Prot.* **1994**, *57*, 737–739. [CrossRef] [PubMed]
126. Lopez-Fandiño, R.; Corzo, N.; Villamiel, M.; Delgado, T.; Olano, A.; Ramos, M. Assessment of quality of commercial UHT milks by chromatographic and electrophoretic methods. *J. Food Prot.* **1993**, *56*, 263–265. [CrossRef] [PubMed]
127. Van Renterghem, R.; De Block, J. Furosine in consumption milk and milk powders. *Int. Dairy J.* **1996**, *6*, 371–382. [CrossRef]
128. Lee, A.P.; Barbano, D.M.; Drake, M.A. The influence of ultra-pasteurization by indirect heating versus direct steam injection on skim and 2% fat milks. *J. Dairy Sci.* **2017**, *100*, 1688–1701. [CrossRef] [PubMed]
129. Rauh, V.M.; Johansen, L.B.; Bakman, M.; Ipsen, R.; Paulsson, M.; Larsen, L.B.; Hammershøj, M. Protein lactosylation in UHT milk during storage measured by liquid chromatography–mass spectrometry and quantification of furosine. *Int. J. Dairy Technol.* **2015**, *68*, 486–494. [CrossRef]
130. Ding, H.; Han, Z.; Wang, B.; Wang, Y.; Ran, Y.; Zhang, L.; Li, Y.; Lu, C.; Lu, X.; Ma, L. Effect of direct steam injection and instantaneous ultra-high-temperature (DSI-IUHT) sterilization on the physicochemical quality and volatile flavor components of milk. *Molecules* **2023**, *28*, 3543. [CrossRef]
131. Marconi, E.; Messia, M.C.; Amine, A.; Moscone, D.; Vernazza, F.; Stocchi, F.; Palleschi, G. Heat-Treated Milk Differentiation by a Sensitive Lactulose Assay. *Food Chem.* **2004**, *84*, 447–450. [CrossRef]
132. Fox, P.F.; Kelly, A.L. Indigenous enzymes in milk: Overview and historical aspects—Part 2. *Int. Dairy J.* **2006**, *16*, 517–532. [CrossRef]
133. Obaidi, A.; Ahmad, A.H. Role of airway lactoperoxidase in scavenging of hydrogen peroxide damage in asthma. *Ann. Thorac. Med.* **2007**, *2*, 107. [CrossRef]
134. Dickow, J.A.; Nielsen, M.T.; Hammershøj, M. Effect of lenient steam injection (LSI) heat treatment of bovine milk on the activities of some enzymes, the milk fat globule and pH. *Int. J. Dairy Technol.* **2012**, *65*, 191–200. [CrossRef]
135. Griffiths, M.W. Use of milk enzymes as indices of heat treatment. *J. Food Prot.* **1986**, *49*, 696–705. [CrossRef] [PubMed]
136. Villamiel, M.; López-Fandiño, R.; Corzo, N.; Olano, A. Denaturation of β-Lactoglobulin and native enzymes in the plate exchanger and holding tube section during continuous flow pasteurization of milk. *Food Chem.* **1997**, *58*, 49–52. [CrossRef]
137. Lan, X.Y.; Wang, J.Q.; Bu, D.P.; Shen, J.S.; Zheng, N.; Sun, P. Effects of heating temperatures and addition of reconstituted milk on the heat indicators in milk. *J. Food Sci.* **2010**, *75*, C653–C658. [CrossRef] [PubMed]
138. van Asselt, A.J.; Sweere, A.P.J.; Rollema, H.S.; de Jong, P. Extreme high-temperature treatment of milk with respect to plasmin inactivation. *Int. Dairy J.* **2008**, *18*, 531–538. [CrossRef]
139. Leite, J.A.S.; Montoya, C.A.; Loveday, S.M.; Maes, E.; Mullaney, J.A.; McNabb, W.C.; Roy, N.C. Heat-treatments affect protease activities and peptide profiles of ruminants' milk. *Front. Nutr.* **2021**, *8*, 626475. [CrossRef] [PubMed]
140. Sun, Y.; Wang, R.; Li, Q.; Ma, Y. Influence of storage time on protein composition and simulated digestion of UHT milk and centrifugation presterilized UHT milk in vitro. *J. Dairy Sci.* **2023**, *106*, 3109–3122. [CrossRef] [PubMed]
141. Leite, J.A.S.; Montoya, C.A.; Loveday, S.M.; Mullaney, J.A.; Loo, T.S.; McNabb, W.C.; Roy, N.C. The impact of heating and drying on protease activities of ruminant milk before and after in vitro infant digestion. *Food Chem.* **2023**, *429*, 136979. [CrossRef] [PubMed]
142. Fan, K.; Wu, P.; Guo, M.; Wang, Y.; Cao, Y.; Wang, P.; Renb, F.; Luo, J. Destabilization of ultra-instantaneous UHT sterilization milk stored at different temperatures. *J. Dairy Sci.* **2024**, *107*, 5460–5472. [CrossRef]
143. Kiełczewska, K.; Jankowska, A.; Dąbrowska, A.; Wachowska, M.; Ziajka, J. The effect of high pressure treatment on the dispersion of fat globules and the fatty acid profile of caprine milk. *Int. Dairy J.* **2020**, *102*, 104607. [CrossRef]
144. Gao, Y.; Ma, Y.; Pan, L.; Li, W.; Peng, X.; Zhang, M.; Dong, L.; Wang, J.; Gu, R. Comparative analysis of whey proteins in yak milk from different breeds in China using a data-independent acquisition proteomics method. *J. Dairy Sci.* **2023**, *106*, 3791–3806. [CrossRef] [PubMed]
145. Zhang, Y.; Ren, F.; Wang, P.; Liang, Q.; Peng, Y.; Song, L.; Wen, P. The influence of yak casein micelle size on rennet-induced coagulation properties. *J. Sci. Food Agric.* **2021**, *101*, 327–333. [CrossRef] [PubMed]
146. Gammoh, S.; Alu'datt, M.H.; Tranchant, C.C.; Al-U'datt, D.G.; Alhamad, M.N.; Rababah, T.; Kubow, S.; Haddadin, M.S.Y.; Ammari, Z.; Maghaydah, S.; et al. Modification of the functional and bioactive properties of camel milk casein and whey proteins by ultrasonication and fermentation with *Lactobacillus Delbrueckii* Subsp. *Lactis*. *LWT-Food Sci. Technol.* **2020**, *129*, 109501. [CrossRef]
147. Barłowska, J.; Szwajkowska, M.; Litwińczuk, Z.; Król, J. Nutritional value and technological suitability of milk from various animal species used for dairy production. *Compr. Rev. Food Sci. Food Saf.* **2011**, *10*, 291–302. [CrossRef]
148. Heilig, A.; Çelik, A.; Hinrichs, J. Suitability of dahlem cashmere goat milk towards pasteurisation, ultrapasteurisation and UHT-heating with regard to sensory properties and storage stability. *Small Rumin. Res.* **2008**, *78*, 152–161. [CrossRef]

149. Liu, Z.; Suolang, Q.; Wang, J.; Li, L.; Luo, Z.; Shang, P.; Chen, X.D.; Wu, P. Formation of structured clots, gastric emptying and hydrolysis kinetics of yak milk during in vitro dynamic gastrointestinal digestion: Impact of different heat treatments. *Food Res. Int.* **2022**, *162*, 111958. [CrossRef] [PubMed]
150. Farah, Z.; Atkins, D. Heat coagulation of camel milk. *J. Dairy Res.* **1992**, *59*, 229–231. [CrossRef]
151. Farah, Z. Composition and characteristics of camel milk. *J. Dairy Res.* **1993**, *60*, 603–626. [CrossRef] [PubMed]
152. Sagar, S.P.; Mehta, B.M.; Wadhwani, K.N.; Darji, V.B.; Aparnathi, K.D. Evaluation of camel milk for selected processing related parameters and comparisons with cow and buffalo milk. *Int. J. Health Anim. Sci. Food Saf.* **2016**, *3*, 27–37. [CrossRef]
153. Genene, A.; Hansen, E.B.; Eshetu, M.; Hailu, Y.; Ipsen, R. Effect of heat treatment on denaturation of whey protein and resultant rennetability of camel milk. *LWT-Food Sci. Technol.* **2019**, *101*, 404–409. [CrossRef]
154. Zhao, X.; Guo, Y.; Zhang, Y.; Pang, X.; Wang, Y.; Lv, J.; Zhang, S. Effects of different heat treatments on maillard reaction products and volatile substances of camel milk. *Front. Nutr.* **2023**, *10*, 1072261. [CrossRef] [PubMed]
155. Loiseau, G.; Faye, B.; Serikbayeva, A.; Montet, D. Enzymes ability to serve as markers of pasteurized camel milk. In Proceedings of the Conference on New Horizons in Biotechnology, Trivandrum, India, 18–21 April 2001.
156. Tayefi-Nasrabadi, H.; Hoseinpour-fayzi, M.A.; Mohasseli, M. Effect of heat treatment on lactoperoxidase activity in camel milk: A comparison with bovine lactoperoxidase. *Small Rumin. Res.* **2011**, *99*, 187–190. [CrossRef]

International Journal of
Molecular Sciences

MDPI

Review

Does Vitamin C Supplementation Provide a Protective Effect in Periodontal Health? A Systematic Review and Meta-Analysis

Roxana Buzatu [1], Magda Mihaela Luca [2,*] and Bogdan Andrei Bumbu [3]

1 Department of Dental Aesthetics, Faculty of Dental Medicine, "Victor Babes" University of Medicine and Pharmacy Timisoara, Revolutiei Boulevard 9, 300041 Timisoara, Romania; roxana.buzatu@umft.ro
2 Department of Pediatric Dentistry, Faculty of Dental Medicine, "Victor Babes" University of Medicine and Pharmacy Timisoara, Eftimie Murgu Square 2, 300041 Timisoara, Romania
3 Department of Dental Medicine, Faculty of Medicine and Pharmacy, University of Oradea, 410073 Oradea, Romania; bogdanbumbu@uoradea.ro
* Correspondence: luca.magda@umft.ro

Abstract: Recent research has highlighted potential benefits of vitamin C in managing periodontal diseases, yet systematic reviews to consolidate these findings are scarce. This study aims to evaluate the effectiveness of vitamin C supplementation in preventing and treating periodontal diseases and elucidate the biological mechanisms underlying these effects. We conducted a systematic review following PRISMA guidelines, searching three databases up to 13 April 2024, for studies from 2010 onward. Our selection criteria aimed to capture a wide range of studies regarding vitamin C's impact on periodontal health. After rigorous screening, 16 studies were included in the final analysis. Meta-analysis techniques were employed to synthesize data and evaluate the association between vitamin C intake and periodontal disease outcomes. The meta-analysis included 17,853 participants from studies with diverse geographical and demographic settings. Notable findings indicated that higher vitamin C intake was associated with a reduction in periodontal disease risk, with a pooled odds ratio (OR) of 1.52 (95% CI: 1.49–1.55). The individual studies reported ORs ranging from 0.62 (95% CI: 0.38–0.94) indicating significant protective effects, to 1.66 (95% CI: 1.04–2.64), suggesting increased risks associated with inadequate vitamin C levels. The heterogeneity among the studies was high (I^2 = 95.46%), reflecting variability in study design and population characteristics. This systematic review confirms that vitamin C supplementation has a beneficial effect on periodontal health. The significant variability across studies suggests that individual dietary needs and baseline vitamin C levels might influence the effectiveness of supplementation. These findings underscore the importance of personalized nutritional guidance as part of comprehensive periodontal care. Future research should focus on longitudinal studies to better understand the causal relationships and potential confounding factors affecting the link between vitamin C intake and periodontal health.

Keywords: vitamin C; dentistry; periodontal disease; stomatology; oral health; nutritional supplementation

Citation: Buzatu, R.; Luca, M.M.; Bumbu, B.A. Does Vitamin C Supplementation Provide a Protective Effect in Periodontal Health? A Systematic Review and Meta-Analysis. *Int. J. Mol. Sci.* **2024**, *25*, 8598. https://doi.org/10.3390/ijms25168598

Academic Editors: Vassilis Athanasiadis and Theodoros G. Chatzimitakos

Received: 3 July 2024
Revised: 28 July 2024
Accepted: 6 August 2024
Published: 7 August 2024

1. Introduction

Periodontal disease, encompassing a range of inflammatory conditions affecting the supporting structures of the teeth, is a major public health concern globally [1,2]. Despite advancements in dental practices, periodontitis remains highly prevalent, affecting approximately 20–50% of the global population [3,4]. The pathophysiology of periodontal disease involves complex interactions between microbial biofilms and the host's immune response, leading to the destruction of periodontal ligament and alveolar bone [5,6]. The role of nutrition, particularly micronutrients, in the modulation of periodontal health has garnered considerable scientific interest.

Vitamin C, a potent antioxidant, is essential for the synthesis of collagen and the maintenance of connective tissue integrity, functions critical to periodontal structure and health [7–10]. The antioxidant properties of vitamin C contribute to its protective role in periodontal health by neutralizing reactive oxygen species (ROS) generated during inflammatory processes [11]. Excessive ROS can exacerbate periodontal destruction by damaging cellular structures and modifying signaling pathways that regulate inflammatory responses [12]. Moreover, vitamin C influences the immune system; it enhances the function of phagocytes and proliferation of T-lymphocytes, crucial components in the host defense against periodontal pathogens [13].

There is also emerging interest in the synergistic effects of vitamin C with other micronutrients, such as vitamin D and calcium, on periodontal health [14–17]. This synergy could potentially enhance the therapeutic outcomes of supplementation regimens, offering a multifaceted approach to managing periodontal disease. Nevertheless, the clinical application of such findings necessitates rigorous randomized controlled trials to establish clear guidelines and recommendations.

However, the oral cavity's complexity as a system, with its unique environmental conditions and microbial interactions, poses significant challenges in isolating the effects of vitamin C. While it enhances the function of phagocytes and promotes T-lymphocyte proliferation—key defenses against periodontal pathogens—the application of vitamin C as an anti-inflammatory cofactor in clinical settings remains underexplored and not fully substantiated by clear, causal evidence [7–13]. This highlights the need for more targeted research to verify vitamin C's role within this intricate biological system, especially as a potential therapeutic agent in the management of periodontal diseases.

To address the gaps in existing research, this systematic review poses specific research questions: Does regular vitamin C supplementation reduce the risk and severity of periodontal diseases? What are the mechanisms through which vitamin C exerts its protective effects on periodontal health? Furthermore, the study hypothesizes that adequate vitamin C intake, as part of a balanced diet or supplementation regimen, significantly reduces the incidence and severity of periodontal diseases by enhancing immune function and reducing oxidative stress.

Based on the current understanding of vitamin C's role in periodontal health and the gaps identified in existing research, this systematic review aims to comprehensively assess the evidence on the effectiveness of vitamin C in the prevention and treatment of periodontal diseases. Our hypothesis is that adequate intake of vitamin C significantly reduces the risk and severity of periodontal diseases. The objectives of this study are to evaluate the impact of vitamin C supplementation on clinical periodontal outcomes across different populations and to clarify the underlying biological mechanisms through which vitamin C influences periodontal health.

2. Materials and Methods

2.1. Eligibility Criteria and Information Sources

For this systematic review, we established specific inclusion and exclusion criteria to ensure the rigorous selection of studies: (1) Studies must involve human participants diagnosed with periodontal diseases; (2) Research must explicitly examine the impact of vitamin C intake—through diet or supplementation—on periodontal health. This encompasses studies assessing clinical outcomes such as gingival inflammation, bleeding on probing, pocket depth reduction, and alveolar bone preservation; (3) The review included a broad array of study designs, such as randomized controlled trials, observational studies, cohort studies, case-control studies, and cross-sectional studies; (4) Regarding the outcome measures, the current review included studies that utilize validated instruments or clearly defined clinical parameters to assess periodontal health outcomes that consist of periodontal disease indices, biochemical markers of inflammation, and radiographic evidence of alveolar bone health; (5) Only peer-reviewed articles published in English were considered.

The exclusion criteria encompassed: (1) Non-human studies; (2) Studies that examine the impact of general multivitamin supplements or broad dietary patterns without specific focus on vitamin C were excluded; (3) Studies that do not provide clear, quantifiable outcomes related to periodontal health, or lack sufficient detail for a comprehensive analysis, were also excluded; (4) To maintain the credibility and reliability of the data included in the review, grey literature, including non-peer-reviewed articles, preprints, conference proceedings, general reviews, commentaries, editorials, systematic reviews and meta-analyses were also excluded. The exclusion of grey literature, including general reviews and commentaries, from this systematic review ensures the inclusion of only peer-reviewed studies, upholding the highest standards of scientific accuracy and reducing bias. These excluded sources often lack rigorous peer review and can introduce secondary interpretations not grounded in primary data, potentially skewing the review's findings.

This study employed a comprehensive search strategy across three electronic databases, including PubMed, Scopus, and Web of Science. The literature search targeted publications up to 13 April 2024, and no older than 2010, capturing the most recent and relevant studies on the topic.

2.2. Search Strategy

The search strategy utilized a comprehensive array of keywords and phrases pertinent to the study's objectives, focusing on the impact of vitamin C on periodontal health. Key search terms included: "vitamin C", "ascorbic acid", "periodontal health", "periodontitis", "gingivitis", "periodontal disease", "dental health", "gingival health", "oral health", "collagen synthesis", "antioxidant properties", "immune response in periodontal disease", "inflammatory response", and "clinical outcomes in periodontal treatment".

To ensure a comprehensive and efficient literature retrieval, Boolean operators (AND, OR, NOT) were employed to effectively combine and refine search terms. The search string included the following: (("vitamin C" OR "ascorbic acid") AND ("periodontal health" OR "periodontitis" OR "gingivitis") AND ("collagen synthesis" OR "antioxidant properties") AND ("immune response" OR "inflammatory response") AND ("clinical outcomes" OR "treatment efficacy" OR "disease progression")). This strategy was designed to capture the most relevant studies assessing the therapeutic impacts of vitamin C on various aspects of periodontal health.

2.3. Data Collection and Selection Process

This systematic review evaluates the effectiveness of vitamin C supplementation on periodontal health, focusing on individuals diagnosed with periodontal diseases (Population). The intervention in question is vitamin C supplementation, administered either through diet or direct supplementation (Intervention). This study compares the outcomes of individuals receiving vitamin C supplementation against those with no supplementation or inadequate vitamin C levels (Comparison). The primary outcomes assessed include clinical measures of periodontal health such as gingival inflammation, bleeding on probing, pocket depth reduction, and alveolar bone preservation (Outcome).

The current study followed the Preferred Reporting Items for Systematic Reviews and Meta-Analyses (PRISMA) guidelines [18]. Initially, all retrieved records were independently screened by two reviewers to determine their eligibility based on the predefined inclusion and exclusion criteria. Discrepancies between reviewers were resolved through discussion or, if necessary, consultation with a third reviewer. The review protocol, including the detailed selection methodology, has been registered and is openly accessible on the Open Science Framework (OSF) with the registration code osf.io/by274, ensuring transparency of our research process and findings.

The data collection process for this systematic review began with the removal of duplicate entries, followed by a rigorous screening of abstracts by two independent reviewers to assess each study's relevance based on predefined inclusion and exclusion criteria. Discrepancies between reviewers were resolved through discussion or, if necessary, consultation with a third reviewer to achieve consensus. The initial database search yielded a total number of articles, which were evaluated and subsequently identified for inclusion in the final study. This process ensured a comprehensive and unbiased collection of data pertinent to the effects of vitamin C on periodontal health.

2.4. Data Items

The primary outcomes of interest were clinical measures of periodontal health, including gingival inflammation, periodontal pocket depth, clinical attachment loss, and alveolar bone level changes. To provide a detailed analysis of how vitamin C influences periodontal health, we also collected data on secondary outcomes related to systemic markers of inflammation and antioxidant status. Additionally, we gathered information on participant demographics (age, gender, ethnicity, smoking status), study characteristics (country, year, design, sample size), and quality of studies to assess the representativeness and applicability of the findings. Nutritional status and intake levels of other relevant micronutrients were also recorded to evaluate potential confounding factors and synergistic effects with vitamin C.

This systematic review defined periodontal health outcomes according to standardized clinical parameters commonly used in dental research, such as the Community Periodontal Index (CPI) and bleeding on probing. We included studies that provided clear definitions of these measures and whose methodologies adhered to internationally recognized standards. For each included study, we extracted data related to the type and amount of vitamin C consumed, whether through diet or supplementation, and the specific methods used to assess vitamin C intake (e.g., dietary questionnaires, blood serum levels).

2.5. Risk of Bias and Quality Assessment

For the systematic assessment of study quality and determination of risk of bias within the included studies, our review employed a dual approach, integrating both qualitative and quantitative evaluation methods. The quality of observational studies was evaluated using the Newcastle–Ottawa Scale [19], and Grading of Recommendations Assessment, Development, and Evaluation (GRADE) tool. Each study is awarded stars in these categories, cumulating in a score that classifies the study quality as either low, medium, or high.

For randomized controlled trials, the Cochrane Collaboration's tool for assessing risk of bias was utilized [20]. This tool evaluates several domains including random sequence generation, allocation concealment, blinding of participants and personnel, blinding of outcome assessment, incomplete outcome data, selective reporting, and other biases. Each domain is judged as "low risk", "high risk", or "unclear risk" of bias. To ensure the objectivity and reproducibility of our quality assessment process, each study was independently evaluated by two researchers. Discrepancies in quality assessment scores were resolved through discussion, or if necessary, consultation with a third researcher.

2.6. Synthesis Methods

In this systematic review, we synthesized findings from selected studies through a qualitative and quantitative synthesis, focusing on the role of vitamin C in periodontal health. The selection of studies for synthesis was based on their alignment with predefined inclusion criteria, specifically those studies that provided data on vitamin C intake and its impact on clinical periodontal outcomes such as pocket depth, bleeding on probing, and clinical attachment loss. To prepare data for synthesis, we tabulated outcomes related to periodontal health improvements, inflammatory markers, and vitamin C intake levels, while noting any missing data and acknowledging potential impacts on our findings.

Results from individual studies were summarized and presented descriptively, comparing periodontal health outcomes across diverse geographic and clinical settings.

A meta-analysis was performed to evaluate the effectiveness of vitamin C supplementation or dietary intake on specific periodontal health outcomes. Heterogeneity among study results was quantified using the I^2 statistic, which describes the percentage of total variation across studies that is due to heterogeneity rather than chance. A high I^2 value indicates substantial variability among the studies. All analyses were performed using standard statistical software, ensuring that all estimates were accompanied by 95% confidence intervals to assess the precision of the findings.

3. Results

3.1. Study Selection and Study Characteristics

A total of 885 articles were identified according to the initial search (308 in PubMed, 241 in Scopus, and 336 in Web of Science), of which 113 duplicate entries were eliminated, 636 records excluded before screening based on title and abstract, and 120 articles excluded after full read for not matching the inclusion criteria or having no available data (Figure 1). The systematic review included a total of 16 eligible studies in the final analysis, as presented in Table 1.

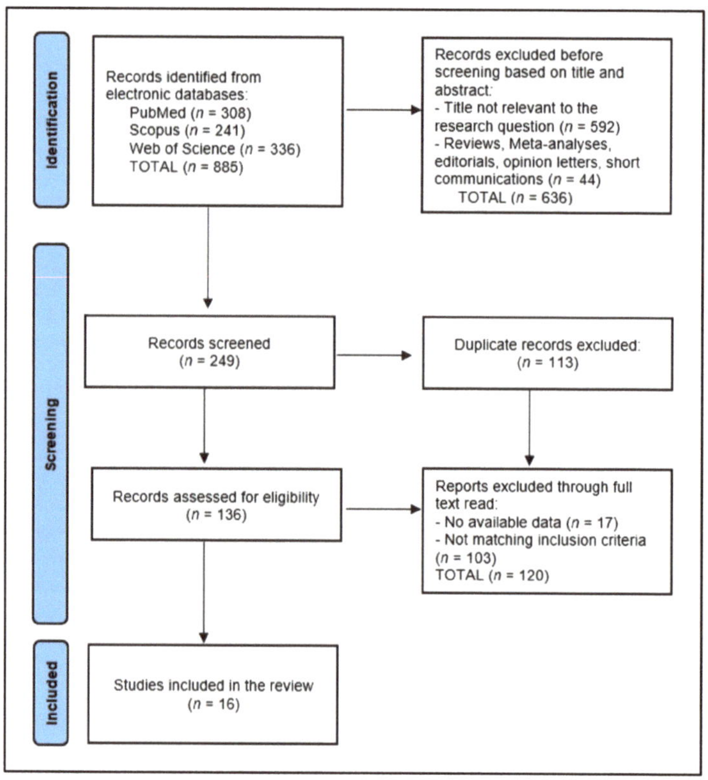

Figure 1. PRISMA Flow Diagram.

Table 1. Study characteristics.

Study and Author	Country	Study Year	Study Design	Study Quality
[21] Hosoda et al.	Japan	2020	Cross-sectional	Medium
[22] Lee et al.	South Korea	2017	Cross-sectional	Medium
[23] Li et al.	China	2022	Cross-sectional	High
[24] Luo et al.	China	2018	Cross-sectional	High
[25] Park et al.	South Korea	2017	Cross-sectional	High
[26] Watson et al.	United Kingdom	2022	Cross-sectional	High
[27] Kuzmanova et al.	The Netherlands	2012	Cross-sectional	Medium
[28] Iwasaki et al.	Japan	2012	Retrospective cohort	Medium
[29] Shimabukuro et al.	Japan	2015	Randomized trial	High
[30] Gokhale et al.	India	2013	Randomized trial	Medium
[31] Sulaiman et al.	Syria	2010	Randomized trial	Medium
[32] Munday et al.	Australia	2020	Cross-sectional	Medium
[33] Assaf et al.	Palestine	2022	Cross-sectional	Medium
[34] Yoshihara et al.	Japan	2022	Cross-sectional	High
[35] Amaliya et al.	The Netherlands	2015	Prospective cohort	High
[36] Güner et al.	Turkey	2023	Cross-sectional	Medium

3.2. Results of Individual Studies

The individual studies included in this meta-analysis examined the impact of vitamin C on periodontal disease across a diverse set of populations and age groups, offering a rich dataset spanning various global regions. The studies included a total of 17,853 participants, with individual study sizes ranging from 21 participants in Kuzmanova et al. [27] to 12,980 participants in Lee et al. [22], reflecting both small-scale community studies and larger, more extensive research efforts. Age distributions varied significantly, with Hosoda et al. [21] focusing on a remarkably young cohort with an average age of 20.4 years, while Iwasaki et al. [28] targeted a much older population, over 75 years, as presented in Table 2. The gender distribution across these studies also varied, although several studies showed a female predominance or nearly equal gender distribution, which is somewhat atypical in periodontal disease studies where male prevalence is often higher.

Table 2. Patients' background characteristics.

Study and Author	Number of Participants	Comparison Group	Age (Category/ Mean/Median)	Gender
[21] Hosoda et al.	49 with PD	71 without PD	20.4 years	Female: 100%
[22] Lee et al.	3812 with PD	7118 without PD	112 (2.9%) 19–29 years; 482 (12.6%) 30–39 years; 881 (23.1%) 40–49 years; 930 (24.4%) 50–59 years; 860 (22.6%) 60–69 years; 547 (14.4%) $\geq$ 70 years	Female: 44.2%
[23] Li et al.	4466 with PD	4493 without PD	56.7 years	Female: 51.2%
[24] Luo et al.	2274 with moderate PD, 676 with severe PD	3465 without PD or mild PD	55.3 years: moderate PD; 54.5 years severe PD	Female: 45.9% moderate PD; 29.6% severe PD
[25] Park et al.	279 with PD	1770 without PD	33.0 years	Female: 47.7%
[26] Watson et al.	1634 with PD	7842 with PD	56.4 years	Female: 55.9%
[27] Kuzmanova et al.	21 with PD	21 without PD	46.9 years	Female: 52.0%
[28] Iwasaki et al.	264 with PD	NR	>75 years	Female: 46.6%
[29] Shimabukuro et al.	150 with vitamin C supplementation (dentifrice)	150 without vitamin C supplementation (dentifrice)	38.1 years	Female: 48.0%
[30] Gokhale et al.	90 with PD and	30 without PD	30–60 years	NR

Table 2. *Cont.*

Study and Author	Number of Participants	Comparison Group	Age (Category/ Mean/Median)	Gender
[31] Sulaiman et al.	30 with PD (15 with vitamin C supplementation, 15 without vitamin C supplementation)	30 without PD	41.0 years	Female: 70.0%
[32] Munday et al.	20 with PD	NR	65.0 years	Female: 50.0%
[33] Assaf et al.	25 with PD	NR	55.0 years	Female: 48.0%
[34] Yoshihara et al.	353 with PD	NR	70.0 years	Female: 45.6%
[35] Amaliya et al.	98 with PD and vitamin C supplementation for 90 days (60 mg/day)	NR	33–43 years	NR
[36] Güner et al.	25 with PD	24 without PD	42.5 years	Female: 52.0%

PD—Periodontal Disease; NR—Not Reported.

3.3. Results of Synthesis

Notably, the studies by Lee et al. [22] and Li et al. [23] featured large sample sizes and reported moderate effect sizes with ORs of 1.16 and 1.13, respectively, suggesting a consistent but moderate protective effect of adequate vitamin C intake against periodontal disease. On the other hand, Park et al. [25] and Luo et al. [24] reported higher odds ratios (1.66 and 1.40), indicating a stronger association in specific populations. Kuzmanova et al. [27] and Iwasaki et al. [28] provided evidence of a significant protective effect of high vitamin C intake with odds ratios substantially below 1 (0.62 and 0.72), highlighting the potential for dietary vitamin C to mitigate periodontal disease risks significantly. Lastly, Yoshihara et al. [34] found that low serum vitamin C levels were associated with an increased risk of periodontal attachment loss (OR = 1.58), reinforcing the role of nutritional status in periodontal health (Table 3).

Table 3. Assessment of periodontal disease risk.

Study and Author	Periodontal Disease Characteristics	Diabetes/Smoking	Vitamin C Assessment	Risk Assessment (OR/HR/RR—95% CI)
[21] Hosoda et al.	Periodontal pocket > 4 mm	NR	48 mg/1000 kcal (PD group) vs. 52 mg/1000 kcal (non-PD group)	NR
[22] Lee et al.	Periodontal pocket > 3.5 mm (CPI scores 3–4)	Diabetes 13.4% (PD group) vs. 6.2% (non-PD group) *	Inadequate dietary vitamin C 48.1% (PD group) vs. 45.2% (non-PD group) *; Quartile of vitamin C in the diet 23.2% ≥ 132.22 mg/day (PD group) vs. 25.9% ≥ 132.22 mg/day (non-PD group) *	Inadequate dietary vitamin C OR = 1.16 (95% CI = 1.04–1.29) * for periodontal disease
[23] Li et al.	Severe/moderate periodontitis defined as ≥2 Interproximal sites with CAL of ≥4 mm or pocket depth of ≥5 mm	Diabetes 21.2% (PD group) vs. 11.2% (non-PD group) *; Smoking 24.3% (PD group) vs. 12.6% (non-PD group) *	Inadequate dietary vitamin C 58.0% (PD group) vs. 55.8% (non-PD group)	Vitamin C > 90 mg/day in men OR = 1.13 (95% CI = 1.03–1.22) * for periodontal disease

Table 3. *Cont.*

Study and Author	Periodontal Disease Characteristics	Diabetes/Smoking	Vitamin C Assessment	Risk Assessment (OR/HR/RR—95% CI)
[24] Luo et al.	Severe/moderate periodontitis defined as ≥2 Interproximal sites with CAL of ≥4 mm or pocket depth of ≥5 mm	Diabetes 15.4% (moderate PD group) vs. 13.6% (severe PD group) vs. 7.0% (non-PD group) *; Smoking 29.8% (moderate PD group) vs. 27.4% (severe PD group) vs. 25.1% (non-PD group) *	Quartile of vitamin C in the diet 25.1% ≥ 112.91 mg/day (moderate PD group) vs. 22.4% ≥ 112.91 mg/day (severe PD group) 25.5% ≥ 112.92 mg/day (non-PD group) *	Inadequate dietary vitamin C OR = 1.40 (95% CI = 1.12–1.74) * for periodontal disease
[25] Park et al.	Periodontal pocket > 3.5 mm (CPI scores 3–4)	Diabetes 1.4% (PD group) vs. 1.3% (non-PD group); Smoking 37.3% (PD group) vs. 24.7% (non-PD group) *	Inadequate dietary vitamin C (<100 mg/day) 63.1% (PD group) vs. 59.8% (non-PD group) *	Inadequate dietary vitamin C OR = 1.66 (95% CI = 1.04–2.64) * for periodontal disease
[26] Watson et al.	Self-reported questionnaire about painful gums, bleeding gums, or loose teeth	NR	NR	Vitamin C >100 mg/day in men OR = 0.81 (95% CI = 0.70–0.96)
[27] Kuzmanova et al.	Radiographic bone loss > 1/3 of the root length at ≥1 tooth per quadrant and ≥20 teeth	Smoking 52% (PD group) vs. 52% (non-PD group)	Vitamin C supplements: 14% (PD group) vs. 10% (non-PD group); Plasma vitamin C: 8.3 mg/L (PD group) vs. 11.3 mg/L (non-PD group) *	Adequate dietary vitamin C OR = 0.62 (95% CI = 0.38–0.94) * for periodontal disease
[28] Iwasaki et al.	CAL of ≥3 mm or pocket depth of ≥3 mm	Diabetes 8.3%; Smoking 47.7%	Median 91.8 mg/day	Adequate dietary vitamin C OR = 0.72 (95% CI = 0.56–0.93) * for periodontal disease
[29] Shimabukuro et al.	GI at 3 months 0.73 (dentifrice group) vs. 0.84 (control group) *; GSI at 3 months 0.21 (dentifrice group) vs. 0.15 (control group)	Smoking 18.7% (dentifrice group) vs. 20.0% (no dentifrice group)	NR	NR
[30] Gokhale et al.	SBI score of ≥2	Diabetes 0.0% (excluded); Smoking: 0.0% (excluded)	450 mg daily ascorbic acid supplementation; Sulcus bleeding index difference 0.56 (vitamin C supplementation) vs. 0.28 (no vitamin C supplementation) *	NR
[31] Sulaiman et al.	GI > 3.5 mm and CAL > 3.5 mm	NR	Plasma TAOC levels 625 mm Teq (without PD) vs. 559 mm Teq (with PD) *	NR
[32] Munday et al.	Radiographic bone loss around teeth in the coronal third, extending to the mid third of the root or beyond	Diabetes 20.0%; Smoking 20.0%	Vitamin C < 40 µmol/L: 30.0% of patients; CRP levels > 4 umol/L 25.0% of patients—significant inverse correlation between vitamin C levels and CRP	NR
[33] Assaf et al.	NR	NR	40% of patients with low vitamin C < 40 µmol/L had stage IV PD *	NR

Table 3. *Cont.*

Study and Author	Periodontal Disease Characteristics	Diabetes/Smoking	Vitamin C Assessment	Risk Assessment (OR/HR/RR—95% CI)
[34] Yoshihara et al.	PAD ≥ 4 mm and CAL ≥ 4 mm	Smoking 100%	Serum vitamin C levels (mean): 7.36 ug/mL	Low serum vitamin C levels OR = 1.58 (95% CI = 1.54–1.61) for CAL ≥ 4 *
[35] Amaliya et al.	NR	Diabetes 6.1%	Serum vitamin C levels (mean): 5.19 mg/L, Optimal vitamin C levels (>8.8 mg/L) in 13.3% of patients	Supplementation with VitC/Ca/F reduced the subgingival load of all studied bacteria
[36] Güner et al.	In all groups, periodontal status was evaluated with plaque index and GI	NR	Serum vitamin C levels (mean): 7.00 mg/L (PD group) vs. 8.83 (non PD group) *; Vitamin C intake 124.6 mg/day (PD group) vs. 176.7 mg/day (non-PD group) *	The plasma ascorbic acid levels and total oxidant status (r = −0.42) and superoxide radical levels (r = −0.53) were inversely correlated in patients with periodontitis

*—statistically significant values; NR—Not Reported; OR—Odds Ratio; RR—Risk Ratio; PD—Periodontal Disease; CPI—Community Periodontal Index; CAL—Clinical Attachment Loss.

The meta-analysis of seven studies examining the relationship between vitamin C intake and periodontal health yielded a pooled odds ratio (OR) of 1.52, with a narrow 95% confidence interval (CI) of 1.49 to 1.55, indicating a significant association between higher vitamin C intake and improved periodontal outcomes. However, the high heterogeneity (I^2 = 95.46%) among the studies suggests considerable variability in how vitamin C affects periodontal health across different populations or under different study conditions (Figure 2).

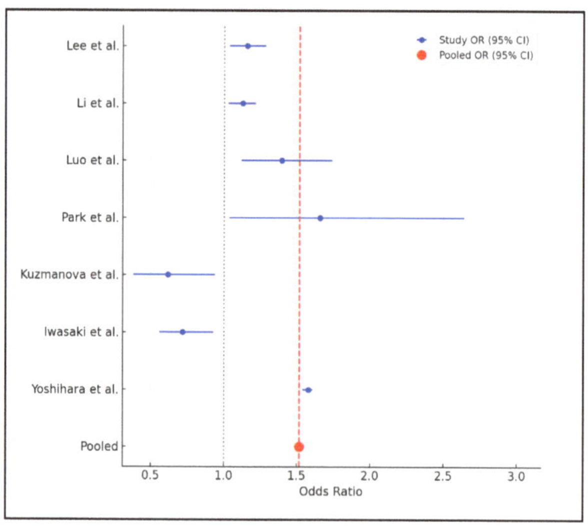

Figure 2. Forest plot of meta-analysis [22–25,27,28,34].

4. Discussion

The studies incorporated into the meta-analysis reflect a significant diversity in methodology and demographics, showcasing the complex interaction between vitamin C intake and periodontal health. Notably, studies with larger cohorts such as those conducted by Lee et al. [22] and Li et al. [23] suggested a consistent yet moderate protective effect of adequate vitamin C intake. These findings were further supported by Park et al. [25] and Luo et al. [24], who reported stronger associations between inadequate vitamin C levels and increased periodontal disease risk. Such results underscore the potential of dietary modification as a part of preventive dental care, particularly in reducing the risk of periodontal diseases.

Conversely, Kuzmanova et al. [27] and Iwasaki et al. [28] demonstrated that higher dietary vitamin C intake could significantly mitigate periodontal disease risks, with odds ratios substantially below one (0.62 and 0.72, respectively). These studies highlight the potential therapeutic benefits of vitamin C in maintaining periodontal health and suggest that supplementation could be particularly effective in populations at higher risk of periodontal diseases. Furthermore, Yoshihara et al. [34] provided compelling evidence linking low serum vitamin C levels to increased risks of periodontal attachment loss, reinforcing the importance of adequate nutritional intake in periodontal disease prevention.

The observed high heterogeneity (I^2 = 95.46%) among the included studies is indeed noteworthy and merits thorough investigation to identify potential sources. This variability could stem from several factors, including differences in study populations, such as variations in baseline nutritional status, demographic characteristics, or the severity of periodontal disease. Intervention disparities, such as the dosage and form of vitamin C supplementation, along with differences in study designs and outcome measures, may also contribute significantly to this heterogeneity.

In recent studies examining the impact of vitamin C on periodontal ligament cells (PDLCs), distinct protective mechanisms against periodontal degradation have been highlighted. The work by Yan et al. [37] elucidated that vitamin C not only promotes the osteogenic differentiation of PDL progenitor cells but does so via specific molecular pathways. Specifically, they demonstrated that vitamin C activates the ERK pathway, which in turn up-regulates PELP1 expression and facilitates the expression of the osteogenesis marker Runx2. This finding suggests a targeted molecular mechanism whereby vitamin C could be utilized in regenerative medicine specifically tailored for periodontal disease treatment, leveraging the PELP1-ERK axis for bone and tissue regeneration. Meanwhile, Wu et al. [38] explored the antioxidative role of vitamin C against oxidative stress in PDLCs induced by hydrogen peroxide, a common byproduct of inflammation in periodontitis. Their study showed that vitamin C significantly mitigated the cytotoxic effects of hydrogen peroxide by reducing apoptosis in PDLCs, as evidenced by decreased activation of apoptosis markers such as caspases-3, caspases-9, and poly (ADP-ribose) polymerase. These findings highlight vitamin C's potential as a therapeutic agent that not only counters oxidative stress but also supports cellular survival and function in the inflammatory environments typical of periodontal diseases. Older studies such as those by Amarasena et al. [39] and Ekuni et al. [40] also found interesting outcomes after vitamin C use in periodontal health. In the study by Amarasena et al. [39], an investigation into the serum vitamin C levels and periodontal health among elderly Japanese citizens revealed a statistically significant but modest inverse relationship between serum vitamin C concentration and clinical attachment loss (CAL). Specifically, the study found that CAL was 4% greater in subjects with lower serum vitamin C levels, even after accounting for variables such as smoking, diabetes, oral hygiene, gender, and the number of teeth present ($r = -0.23$, $p < 0.00005$). Conversely, Ekuni et al. [40] explored the effects of vitamin C on oxidative stress and atherosclerosis in a rat model with ligature-induced periodontitis. They demonstrated that vitamin C supplementation significantly reduced oxidative markers and lipid deposition in the aorta, thus attenuating the progression of atherosclerosis triggered by periodontal disease. Notably, vitamin C intake increased plasma vitamin C levels and the GSH:GSSG ratio by 178% and 123%, respectively, while decreasing serum hexanoyl-lysine (HEL) and aortic levels of nitrotyrosine, HEL, and 8-hydroxydeoxyguanosine by 23%, 87%, 84%, and 38%, respectively.

Other older studies explored the relationship between vitamin C levels and periodontal disease, offering insights into how nutritional status impacts periodontal health. Timmerman et al. [41] found a modest negative correlation ($r = -0.199$, $p < 0.05$) between plasma vitamin C levels and periodontal attachment loss among subjects from a tea estate in Indonesia, revealing that low vitamin C levels explained 3.9% of the variance in periodontal attachment loss. Notably, subjects with vitamin C deficiency experienced significantly more attachment loss compared to those with normal levels, underscoring the potential role of vitamin C in mitigating periodontal breakdown. Conversely, Pussinen et al. [42] demonstrated a significant relationship between low plasma vitamin C concentrations and higher serological markers for *P. gingivalis*, a key pathogen in periodontitis. This correlation remained significant even after adjusting for confounding factors, indicating that low vitamin C levels might enhance the colonization or impact the healing processes related to *P. gingivalis* infections. Collectively, these findings emphasize the importance of adequate vitamin C intake in maintaining periodontal health and potentially moderating the pathogenicity of periodontal infections.

The clinical implications of these findings are profound. Regular monitoring and enhancement of dietary vitamin C intake could be recommended as part of routine dental care, especially for individuals at elevated risk of periodontal diseases. Dental professionals might consider dietary assessments and vitamin C supplementation as adjunctive strategies to conventional periodontal treatments, particularly for patients showing early signs of periodontal diseases or those who are at a higher risk due to systemic health conditions. Overall, these findings advocate for a more integrated approach in dentistry that combines nutritional management with traditional periodontal therapies to optimize patient outcomes.

Although a significant risk was identified with low vitamin C levels or insufficient vitamin C supplementation in periodontal disease, there was a high heterogeneity among studies. This variability limits the study findings, and it could be attributed to differences in study designs, population characteristics, definitions of periodontal health, and methods of assessing vitamin C intake. Notably, studies showing both protective effects (Kuzmanova et al. [27], Iwasaki et al. [28]) and risk associations (Lee et al., Li et al., Luo et al., Park et al., Yoshihara et al. [22–25,34]) highlight the complex role of vitamin C in periodontal health, underscoring the need for personalized dietary recommendations and further research to clarify these relationships.

This study aligns with recent findings that highlight the role of vitamin C in directing gene differentiation pathways at the stem cell level, which could be pivotal for future regenerative therapies. The ability of ascorbic acid to activate specific molecular pathways, such as the ERK pathway observed in periodontal ligament cells, underlines its potential to influence stem cell plasticity and differentiation orientation, enhancing the therapeutic outcomes in tissue engineering and regenerative medicine [43].

The manuscript highlights the potential benefits of vitamin C supplementation for periodontal health. To better translate these findings into clinical practice, we recommend that dental practitioners advise patients on optimal vitamin C intake levels based on current dietary guidelines, which suggest a daily intake of 65 to 90 milligrams for adults, depending on age and sex. For patients with or at risk of periodontal disease, a higher intake, tailored to individual health needs and confirmed by a healthcare provider, may be beneficial. Additionally, we suggest integrating discussions about dietary sources rich in vitamin C, such as citrus fruits and leafy greens, during dental consultations.

The call for further research into the role of vitamin C in periodontal health should be emphasized by suggesting specific study designs that could provide more definitive evidence. We recommend conducting longitudinal studies and randomized controlled trials to ascertain causality and the effectiveness of vitamin C supplementation over time.

Future research should also aim to include diverse populations to explore the variability in response based on genetic, dietary, and environmental factors. Addressing these gaps, such as the impact of different forms of vitamin C (e.g., from natural sources versus supplements) and the interaction with other micronutrients, will enhance our understanding and inform more effective, personalized periodontal treatment protocols.

This study acknowledges certain inherent limitations in its design, such as to limit the search to studies published from 2010 onwards was aimed at ensuring the inclusion of contemporary research with standardized definitions and outcome measures for periodontal disease. While this approach provided a focused review of recent findings, it may have excluded relevant historical data and findings reported in other databases or journals.

5. Conclusions

This systematic review and meta-analysis supports the association between adequate vitamin C intake and a decreased risk of periodontal diseases, with a pooled odds ratio of 1.52 suggesting a positive correlation. However, the considerable heterogeneity observed across the studies, as indicated by an I^2 of 95.46%, necessitates a cautious interpretation of these results. This variability may stem from differences in dietary behavior, baseline nutritional status, and methodological discrepancies between studies. Given these findings, while there is evidence supporting the benefit of vitamin C, the results underscore the importance of considering individual dietary needs when making nutritional recommendations for the prevention and management of periodontal diseases. Further research is essential to more definitively determine the mechanisms by which vitamin C impacts periodontal health and to assess the efficacy of vitamin C supplementation in diverse clinical scenarios and populations.

Author Contributions: Conceptualization, B.A.B.; methodology, B.A.B.; software, B.A.B.; validation, B.A.B.; formal analysis, B.A.B.; investigation, M.M.L.; resources, M.M.L.; data curation, M.M.L.; writing—original draft preparation, M.M.L.; writing—review and editing, R.B.; visualization, R.B.; supervision, R.B.; project administration, R.B. All authors have read and agreed to the published version of the manuscript.

Funding: The article processing charge was paid by the "Victor Babes" University of Medicine and Pharmacy Timisoara.

Institutional Review Board Statement: Not applicable.

Data Availability Statement: Not applicable.

Conflicts of Interest: The authors declare no conflicts of interest.

References

1. Könönen, E.; Gursoy, M.; Gursoy, U.K. Periodontitis: A Multifaceted Disease of Tooth-Supporting Tissues. *J. Clin. Med.* **2019**, *8*, 1135. [CrossRef] [PubMed]
2. Nazir, M.A. Prevalence of periodontal disease, its association with systemic diseases and prevention. *Int. J. Health Sci.* **2017**, *11*, 72–80. [PubMed] [PubMed Central]
3. Kwon, T.; Lamster, I.B.; Levin, L. Current Concepts in the Management of Periodontitis. *Int. Dent. J.* **2021**, *71*, 462–476. [CrossRef] [PubMed] [PubMed Central]
4. Janakiram, C.; Dye, B.A. A public health approach for prevention of periodontal disease. *Periodontol. 2000* **2020**, *84*, 202–214. [CrossRef] [PubMed] [PubMed Central]
5. Hajishengallis, G.; Chavakis, T.; Lambris, J.D. Current understanding of periodontal disease pathogenesis and targets for host-modulation therapy. *Periodontol. 2000* **2020**, *84*, 14–34. [CrossRef] [PubMed] [PubMed Central]
6. Ferrà-Cañellas, M.D.M.; Garcia-Sureda, L. Exploring the Potential of Micro-Immunotherapy in the Treatment of Periodontitis. *Life* **2024**, *14*, 552. [CrossRef] [PubMed] [PubMed Central]
7. Najeeb, S.; Zafar, M.S.; Khurshid, Z.; Zohaib, S.; Almas, K. The Role of Nutrition in Periodontal Health: An Update. *Nutrients* **2016**, *8*, 530. [CrossRef] [PubMed] [PubMed Central]
8. Barnawi, B.M.; Alrashidi, N.S.; Albalawi, A.M.; Alakeel, N.S.; Hamed, J.T.; Barashid, A.A.; Alduraibi, M.S.; Alhussain, G.S.; Alghadeer, J.Y.; Alarifi, N.A.; et al. Nutritional Modulation of Periodontal Diseases: A Narrative Review of Recent Evidence. *Cureus* **2023**, *15*, e50200. [CrossRef] [PubMed] [PubMed Central]

9. DePhillipo, N.N.; Aman, Z.S.; Kennedy, M.I.; Begley, J.P.; Moatshe, G.; LaPrade, R.F. Efficacy of Vitamin C Supplementation on Collagen Synthesis and Oxidative Stress after Musculoskeletal Injuries: A Systematic Review. *Orthop. J. Sports Med.* **2018**, *6*, 2325967118804544. [CrossRef] [PubMed] [PubMed Central]
10. Murererehe, J.; Uwitonze, A.M.; Nikuze, P.; Patel, J.; Razzaque, M.S. Beneficial Effects of Vitamin C in Maintaining Optimal Oral Health. *Front. Nutr.* **2022**, *8*, 805809. [CrossRef] [PubMed] [PubMed Central]
11. Kaźmierczak-Barańska, J.; Boguszewska, K.; Adamus-Grabicka, A.; Karwowski, B.T. Two Faces of Vitamin C—Antioxidative and Pro-Oxidative Agent. *Nutrients* **2020**, *12*, 1501. [CrossRef] [PubMed]
12. Kaur, G.; Kathariya, R.; Bansal, S.; Singh, A.; Shahakar, D. Dietary antioxidants and their indispensable role in periodontal health. *J. Food Drug Anal.* **2016**, *24*, 239–246. [CrossRef] [PubMed] [PubMed Central]
13. Carr, A.C.; Maggini, S. Vitamin C and Immune Function. *Nutrients* **2017**, *9*, 1211. [CrossRef] [PubMed] [PubMed Central]
14. Liang, F.; Zhou, Y.; Zhang, Z.; Zhang, Z.; Shen, J. Association of vitamin D in individuals with periodontitis: An updated systematic review and meta-analysis. *BMC Oral Health* **2023**, *23*, 387. [CrossRef] [PubMed] [PubMed Central]
15. Shah, M.; Poojari, M.; Nadig, P.; Kakkad, D.; Dutta, S.B.; Sinha, S.; Chowdhury, K.; Dagli, N.; Haque, M.; Kumar, S. Vitamin D and Periodontal Health: A Systematic Review. *Cureus* **2023**, *15*, e47773. [CrossRef] [PubMed] [PubMed Central]
16. Shodhan Shetty, A.; Shenoy, R.; Dasson Bajaj, P.; Rao, A.; Ks, A.; Pai, M.; Br, A.; Jodalli, P. Role of nutritional supplements on oral health in adults—A systematic review. *F1000Research* **2023**, *12*, 492. [CrossRef] [PubMed] [PubMed Central]
17. Cagetti, M.G.; Wolf, T.G.; Tennert, C.; Camoni, N.; Lingström, P.; Campus, G. The Role of Vitamins in Oral Health. *A Systematic Review and Meta-Analysis. Int. J. Environ. Res. Public Health* **2020**, *17*, 938. [CrossRef] [PubMed] [PubMed Central]
18. Page, M.J.; McKenzie, J.E.; Bossuyt, P.M.; Boutron, I.; Hoffmann, T.C.; Mulrow, C.D.; Shamseer, L.; Tetzlaff, J.M.; Akl, E.A.; Brennan, S.E.; et al. The PRISMA 2020 statement: An updated guideline for reporting systematic reviews. *Syst. Rev.* **2021**, *10*, 89. [CrossRef] [PubMed] [PubMed Central]
19. Stang, A. Critical evaluation of the Newcastle-Ottawa scale for the assessment of the quality of nonrandomized studies in meta-analyses. *Eur. J. Epidemiol.* **2010**, *25*, 603–605. [CrossRef] [PubMed]
20. Higgins, J.P.; Altman, D.G.; Gøtzsche, P.C.; Jüni, P.; Moher, D.; Oxman, A.D.; Savovic, J.; Schulz, K.F.; Weeks, L.; Sterne, J.A. Cochrane Bias Methods Group; Cochrane Statistical Methods Group. The Cochrane Collaboration's tool for assessing risk of bias in randomised trials. *BMJ* **2011**, *343*, d5928. [CrossRef] [PubMed] [PubMed Central]
21. Hosoda, A.; Komagamine, Y.; Kanazawa, M.; Hama, Y.; Kojo, A.; Minakuchi, S. The Association between Dietary Habits and Periodontal Disease in Young Adult Women. *J. Nutr. Sci. Vitaminol.* **2021**, *67*, 48–56. [CrossRef] [PubMed]
22. Lee, J.H.; Shin, M.S.; Kim, E.J.; Ahn, Y.B.; Kim, H.D. The association of dietary vitamin C intake with periodontitis among Korean adults: Results from KNHANES IV. *PLoS ONE* **2017**, *12*, e0177074. [CrossRef] [PubMed] [PubMed Central]
23. Li, W.; Shang, Q.; Yang, D.; Peng, J.; Zhao, H.; Xu, H.; Chen, Q. Abnormal Micronutrient Intake Is Associated with the Risk of Periodontitis: A Dose-response Association Study Based on NHANES 2009-2014. *Nutrients* **2022**, *14*, 2466. [CrossRef] [PubMed] [PubMed Central]
24. Luo, P.P.; Xu, H.S.; Chen, Y.W.; Wu, S.P. Periodontal disease severity is associated with micronutrient intake. *Aust. Dent. J.* **2018**, *63*, 193–201. [CrossRef] [PubMed]
25. Park, J.A.; Lee, J.H.; Lee, H.J.; Jin, B.H.; Bae, K.H. Association of Some Vitamins and Minerals with Periodontitis in a Nationally Representative Sample of Korean Young Adults. *Biol. Trace Elem. Res.* **2017**, *178*, 171–179. [CrossRef] [PubMed]
26. Watson, S.; Woodside, J.V.; Winning, L.; Wright, D.M.; Srinivasan, M.; McKenna, G. Associations between self-reported periodontal disease and nutrient intakes and nutrient-based dietary patterns in the UK Biobank. *J. Clin. Periodontol.* **2022**, *49*, 428–438. [CrossRef] [PubMed] [PubMed Central]
27. Kuzmanova, D.; Jansen, I.D.; Schoenmaker, T.; Nazmi, K.; Teeuw, W.J.; Bizzarro, S.; Loos, B.G.; van der Velden, U. Vitamin C in plasma and leucocytes in relation to periodontitis. *J. Clin. Periodontol.* **2012**, *39*, 905–912. [CrossRef] [PubMed]
28. Iwasaki, M.; Moynihan, P.; Manz, M.C.; Taylor, G.W.; Yoshihara, A.; Muramatsu, K.; Watanabe, R.; Miyazaki, H. Dietary antioxidants and periodontal disease in community-based older Japanese: A 2-year follow-up study. *Public Health Nutr.* **2013**, *16*, 330–338. [CrossRef] [PubMed] [PubMed Central]
29. Shimabukuro, Y.; Nakayama, Y.; Ogata, Y.; Tamazawa, K.; Shimauchi, H.; Nishida, T.; Ito, K.; Chikazawa, T.; Kataoka, S.; Murakami, S. Effects of an ascorbic acid-derivative dentifrice in patients with gingivitis: A double-masked, randomized, controlled clinical trial. *J. Periodontol.* **2015**, *86*, 27–35. [CrossRef] [PubMed]
30. Gokhale, N.H.; Acharya, A.B.; Patil, V.S.; Trivedi, D.J.; Thakur, S.L. A short-term evaluation of the relationship between plasma ascorbic acid levels and periodontal disease in systemically healthy and type 2 diabetes mellitus subjects. *J. Diet. Suppl.* **2013**, *10*, 93–104. [CrossRef] [PubMed]
31. Abou Sulaiman, A.E.; Shehadeh, R.M. Assessment of total antioxidant capacity and the use of vitamin C in the treatment of non-smokers with chronic periodontitis. *J. Periodontol.* **2010**, *81*, 1547–1554. [CrossRef] [PubMed]
32. Munday, M.-R.; Rodricks, R.; Fitzpatrick, M.; Flood, V.M.; Gunton, J.E. A Pilot Study Examining Vitamin C Levels in Periodontal Patients. *Nutrients* **2020**, *12*, 2255. [CrossRef]
33. Assaf, M.; Rabi, H. Assessment of Vitamin C Levels in Periodontal Patients: A Cross-Sectional Study in Palestine. *J. Pharm. Bioallied Sci.* **2022**, *14* (Suppl. 1), S903–S906. [CrossRef] [PubMed] [PubMed Central]
34. Yoshihara, A.; Nakashima, K.; Suwama, K.; Odajima, A.; Yamaga, T.; Ogawa, H. Interaction between serum vitamin C levels and smoking on the periodontal condition in older adults. *J. Periodontal Res.* **2022**, *57*, 587–593. [CrossRef] [PubMed]

35. Amaliya, A.; Laine, M.L.; Loos, B.G.; Van der Velden, U. Java project on periodontal diseases: Effect of vitamin C/calcium threonate/citrus flavonoids supplementation on periodontal pathogens, CRP and HbA1c. *J. Clin. Periodontol.* **2015**, *42*, 1097–1104. [CrossRef] [PubMed]
36. Emiroğlu, E.; Cakir, T.; Yüksel, M.; Güneş, F.; Cekici, A.; Karakoyun, B. The relationship between vitamin C and oxidative status in periodontitis: A case-control study. *Ann. Clin. Anal. Med.* **2023**, *14*, 630–634. [CrossRef]
37. Yan, Y.; Zeng, W.; Song, S.; Zhang, F.; He, W.; Liang, W.; Niu, Z. Vitamin C induces periodontal ligament progenitor cell differentiation via activation of ERK pathway mediated by PELP1. *Protein Cell* **2013**, *4*, 620–627. [CrossRef]
38. Wu, W.; Yang, N.; Feng, X.; Sun, T.; Shen, P.; Sun, W. Effect of vitamin C administration on hydrogen peroxide-induced cytotoxicity in periodontal ligament cells. *Mol. Med. Rep.* **2015**, *11*, 242–248. [CrossRef] [PubMed]
39. Amarasena, N.; Ogawa, H.; Yoshihara, A.; Hanada, N.; Miyazaki, H. Serum vitamin C-periodontal relationship in community-dwelling elderly Japanese. *J. Clin. Periodontol.* **2005**, *32*, 93–97. [CrossRef] [PubMed]
40. Ekuni, D.; Tomofuji, T.; Sanbe, T.; Irie, K.; Azuma, T.; Maruyama, T.; Tamaki, N.; Murakami, J.; Kokeguchi, S.; Yamamoto, T. Vitamin C intake attenuates the degree of experimental atherosclerosis induced by periodontitis in the rat by decreasing oxidative stress. *Arch. Oral. Biol.* **2009**, *54*, 495–502. [CrossRef] [PubMed]
41. Amaliya; Timmerman, M.F.; Abbas, F.; Loos, B.G.; Van der Weijden, G.A.; Van Winkelhoff, A.J.; Winkel, E.G.; Van der Velden, U. Java project on periodontal diseases: The relationship between vitamin C and the severity of periodontitis. *J. Clin. Periodontol.* **2007**, *34*, 299–304. [CrossRef] [PubMed]
42. Pussinen, P.J.; Laatikainen, T.; Alfthan, G.; Asikainen, S.; Jousilahti, P. Periodontitis is associated with a low concentration of vitamin C in plasma. *Clin. Diagn. Lab. Immunol.* **2003**, *10*, 897–902. [CrossRef] [PubMed] [PubMed Central]
43. Bhandi, S.; Alkahtani, A.; Mashyakhy, M.; Abumelha, A.S.; Albar, N.H.M.; Renugalakshmi, A.; Alkahtany, M.F.; Robaian, A.; Almeslet, A.S.; Patil, V.R.; et al. Effect of Ascorbic Acid on Differentiation, Secretome and Stemness of Stem Cells from Human Exfoliated Deciduous Tooth (SHEDs). *J. Pers. Med.* **2021**, *11*, 589. [CrossRef] [PubMed]

International Journal of *Molecular Sciences*

MDPI

Article

Angiotensin-Converting Enzyme and Renin-Inhibitory Activities of Protein Hydrolysates Produced by Alcalase Hydrolysis of Peanut Protein

Sukanya Poddar and Jianmei Yu *

Food and Nutritional Sciences Program, Department of Family and Consumer Sciences, North Carolina Agricultural and Technical State University, Greensboro, NC 27411, USA
* Correspondence: jyu@ncat.edu

Abstract: Hypertension is a major controllable risk factor associated with cardiovascular disease (CVD) and overall mortality worldwide. Most people with hypertension must take medications that are effective in blood pressure management but cause many side effects. Thus, it is important to explore safer antihypertensive alternatives to regulate blood pressure. In this study, peanut protein concentrate (PPC) was hydrolyzed with 3–5% Alcalase for 3–10 h. The in vitro angiotensin-converting enzyme (ACE) and renin-inhibitory activities of the resulting peanut protein hydrolysate (PPH) samples and their fractions of different molecular weight ranges were determined as two measures of their antihypertensive potentials. The results show that the crude PPH produced at 4% Alcalase for 6 h of hydrolysis had the highest ACE-inhibitory activity with IC_{50} being 5.45 mg/mL. The PPH samples produced with 3–5% Alcalase hydrolysis for 6–8 h also displayed substantial renin-inhibitory activities, which is a great advantage over the animal protein-derived bioactive peptides or hydrolysate. Remarkably higher ACE- and renin-inhibitory activities were observed in fractions smaller than 5 kDa with IC_{50} being 0.85 and 1.78 mg/mL. Hence, the PPH and its small molecular fraction produced under proper Alcalase hydrolysis conditions have great potential to serve as a cost-effective anti-hypertensive ingredient for blood pressure management.

Keywords: peanut protein hydrolysate; Alcalase concentration; hydrolysis time; ACE inhibition; renin inhibition; Immunoglobulin E (IgE)-binding inhibition

Citation: Poddar, S.; Yu, J. Angiotensin-Converting Enzyme and Renin-Inhibitory Activities of Protein Hydrolysates Produced by Alcalase Hydrolysis of Peanut Protein. *Int. J. Mol. Sci.* **2024**, *25*, 7463. https://doi.org/10.3390/ijms25137463

Academic Editors: Vassilis Athanasiadis and Theodoros G. Chatzimitakos

Received: 4 June 2024
Revised: 3 July 2024
Accepted: 5 July 2024
Published: 7 July 2024

1. Introduction

Hypertension, a risk factor for both CVD and overall mortality [1,2], affected 31.1% of the global adult population (1.39 billion people) in 2010 [3]. Per the Centers for Disease Control (CDC), approximately 500,000 annual deaths in the United States can be attributed to high blood pressure [4]. Given its prevalence, lifestyle changes, dietary interventions, and pharmaceutical treatments are widely employed to address this condition [5]. The US CDC's estimate suggests that the economic impact of hypertension in the United States ranges from USD 131 billion to USD 198 billion each year, encompassing healthcare expenses and productivity losses [4]. Renin and ACE play pivotal roles in the renin–angiotensin–aldosterone system (RAAS) pathway. Renin transforms angiotensinogen into angiotensin I (Ang I), which was further hydrolyzed by ACE to yield angiotensin II (Ang II), a potent vasoconstrictor. Ang II leads to vasoconstriction and prompts the release of aldosterone, elevating sodium levels and blood pressure [6]. Given this, ACE and renin inhibition stands as one of the strategies in hypertension treatment. Presently used synthetic drugs for hypertension management, such as captopril, alacepril, lisinopril, and enalapril, are mostly ACE inhibitors, and aliskiren is the only synthetic renin inhibitor approved by the FDA for managing hypertension [7]. The use of them causes undesirable side effects such as hypotension, cough, hyperkalemia, headache, dizziness, fatigue, nausea, and renal impairment, although they are effective [8]. Consequently, the exploration and development of safe and cost-effective antihypertensive compounds from food sources or the development of food therapy has garnered significant attention [9].

Various ACE-inhibitory peptides have been identified within diverse food protein sources like milk, fish, soybean, amaranth, and rapeseed [10–14]. Despite the limited understanding of the structure–function relationship of bioactive peptides, many of them share common attributes. Typically comprising 2 to 20 amino acids, these peptides tend to be rich in hydrophobic amino acids [15]. Numerous bioactive peptides derived from food have showcased their ability to counteract hypertension. These peptides often function by inhibiting ACE and decreasing renin activity [16,17]. Targeting renin alone within the Renin–Angiotensin System (RAS) does not completely prevent the breakdown of bradykinin catalyzed by ACE, which may result in vasoconstriction [18]. Consequently, a more effective strategy for reducing elevated blood pressure involves the development of natural therapeutic substances capable of producing diverse effects, such as concurrently inhibiting renin and ACE activities [19]. So far, the commercially available food-derived antihypertensive peptide products are mostly made from milk, fish, and algae [20]. No side effects have been reported for the intended use of food-derived bioactive peptides. Recently, peptides with renin-inhibiting activities have been found through the enzymatic breakdown of plant proteins produced from flaxseed and pea, opening new paths for research and potential therapeutic applications [21]. However, the study of antihypertensive properties of peptides and hydrolysates from peanuts is limited, mostly due to peanut allergy issues.

Peanuts, scientifically known as *Arachis hypogaea*, consist of approximately 50% lipid, primarily monounsaturated fat, 25% protein, and 8% dietary fiber, but minimal amounts of easily digestible carbohydrates like sugars and starches [22]. Due to the high oil content, peanuts are widely used to produce cooking oil in the world, which results in a large quantity of peanut flour/meal with a protein content of around 50%. Thus, peanut flour is a sustainable source of food protein and can be processed into peanut protein isolate (PPI) or PPC [23]. Limited studies show that protease hydrolysis of PPI and peanut flour also results in hydrolysates with certain ACE-inhibitory activities [22,24]. The production of antihypertensive peptides from food involves several steps: protein extraction, controlled hydrolysis, fractionation by ultrafiltration or preparative HPLC, and purification [20,25]. This process is complex, time-consuming, low yield, and high cost. The purpose of this study was to optimize the protease treatment condition to produce PPH with high ACE- and renin-inhibitory activity for possible food therapy of hypertension.

2. Results

2.1. Protein/Peptide and Amino Acid Concentrations of PPH

The PPC yield from peanut flour was 22.02 ± 1.41% (n = 9) under the extraction condition used in this study and the protein concentration of dried PPC was 80.39 ± 1.94% (n = 9). According to the product specification provided by the vendor, the protein content of peanut flour used in this study is 50 ± 2 g/100 g flour. Thus, the protein recovery of the extraction process was approximately 35.4%. The PPH was made by hydrolyzing 10% PPC at pH 8 and 40 °C. The highest protein concentration of 64 mg/mL was detected in the unhydrolyzed PPC solution, which means that only 79.61% of the protein in PPC was solubilized at the PH of enzymatic hydrolysis (pH 8). This is because of the lower protein solubility of peanut protein at pH 8 than at pH 10.

As hydrolysis progresses, the protein concentration decreases due to the formation of peptides and amino acids, as shown by the gradually reduced protein concentration and steadily increased amount of free amino acids (FAA) with increasing hydrolysis time (Figure 1). At the same enzyme concentration, the protein/peptide concentration decreased logarithmically (R^2 = 0.845–0.894) (Figure 1A), and the FAA increased linearly with treatment time ($p < 0.0001$, R^2 = 0.896–0.987) (Figure 1B). Nevertheless, changes in protein concentrations with increasing Alcalase were small although significant in some cases at the same hydrolysis time. The protein concentration of the control sample, which did not undergo Alcalase treatment, was measured at 64.07 mg/mL. Among the treated samples, the highest protein concentration of 33.80 mg/mL was observed in the sample treated with a 3% Alcalase solution for a duration of 3 h. Conversely, the lowest protein concentration of 27.45 mg/mL was identified in the sample subjected to a 10 h hydrolysis at 3% Alcalase. Regarding the amino acid concentration, the control sample exhibited a minimum amino acid concentration of 3.19 mg/mL. Among the treated samples, the sample hydrolyzed for 3 h with 3% Alcalase demonstrated the lowest FAA concentration of 8.97 mg/mL, while the samples subjected to a 10 h treatment with 5% Alcalase peaked a FAA concentration of 13.38 mg/mL.

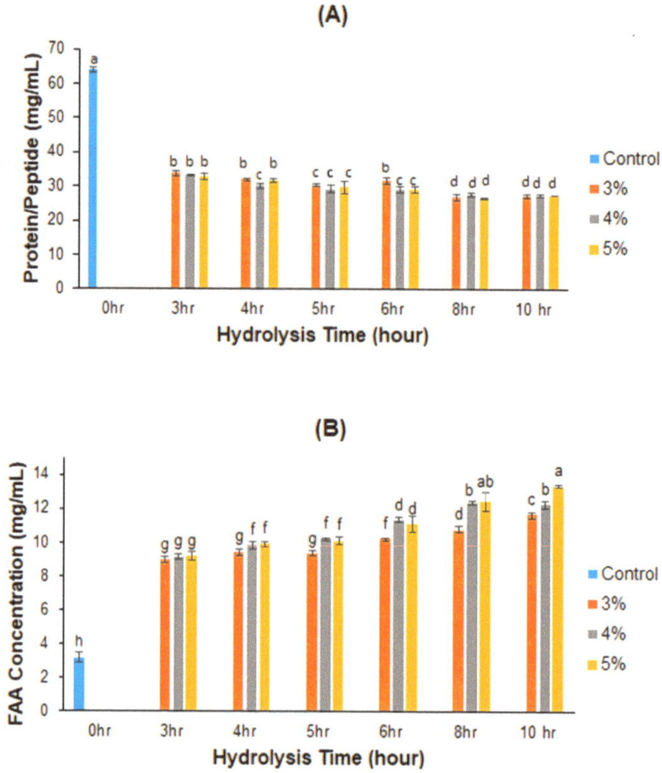

Figure 1. Effects of Alcalase concentration and hydrolysis time on the FAA and protein concentration of PPH. (**A**) Protein concentration, (**B**) FAA concentration (data bars with different labels represent significantly different values at $p < 0.05$).

2.2. *ACE-Inhibitory Activity (%) of Crude PPHs*

Alcalase hydrolysis of PPC resulted in hydrolysates with different degrees of ACE-inhibitory activities depending on hydrolysis time and Alcalase concentration. In this study, the ACE-inhibitory activity of PPC and PPH was assessed across concentrations ranging from 1 to 12 mg/mL. The percentages of ACE inhibition of both PPC and PPH increased with their concentration in a sigmoid pattern, as shown in Figure 2. The untreated PPC samples inhibited 4.4% to 24.5% ACE activity at a concentration of 1–12 mg/mL, while the percentage of ACE inhibition of PPH samples was in the range of 11.1–68.6%, greatly increased due to Alcalase hydrolysis. The IC_{50} values, the PPH concentrations required for 50% ACE inhibition, were determined and presented in Table 1. The smaller IC_{50} value corresponds to higher ACE-inhibitory potential. Notably, the lowest IC_{50} value was obtained for samples treated with different concentrations of Alcalase for 6 h. Specifically, the PPH obtained at 4% Alcalase over a 6 h hydrolysis period exhibited the lowest IC_{50}

value of 5.45 mg/mL, indicating the highest ACE-inhibitory activity among tested samples. In contrast, samples subjected to 8 and 10 h of hydrolysis exhibited IC_{50} values lower than PPH samples treated for 3, 4, and 5 h but higher than those treated for 6 h. Table 1 also shows that the IC_{50} values of most PPH samples did not change significantly with Alcalase concentration at the same treatment time.

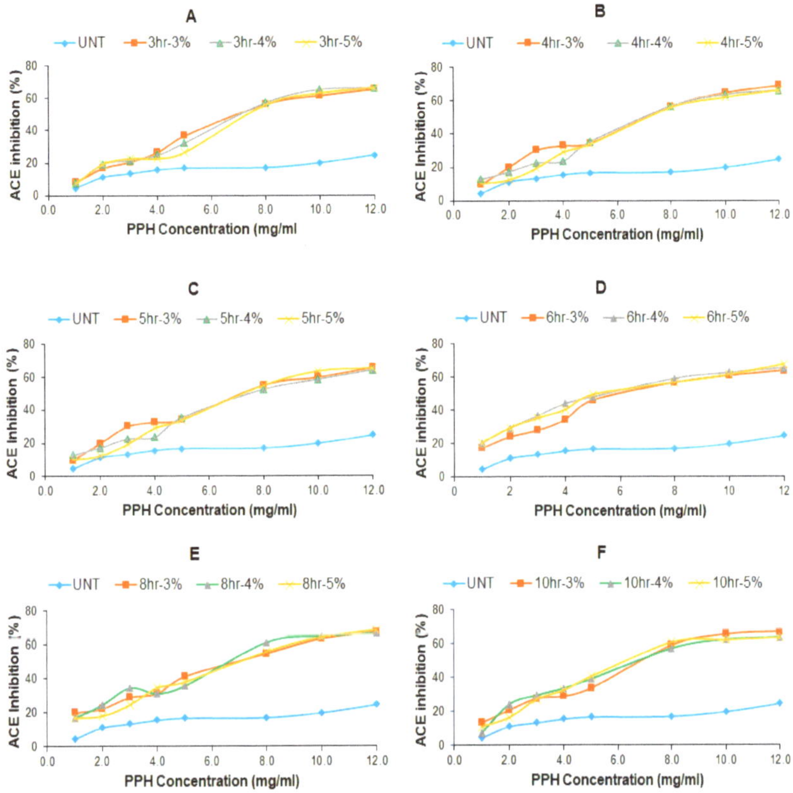

Figure 2. The sigmoid relationship between the ACE Inhibition (%) and concentration of crude PPH obtained at different hydrolysis times and Alcalase concentrations (%). UNT—untreated PPC, (**A**) 3 h, (**B**) 4 h, (**C**) 5 h, (**D**) 6 h, (**E**) 8 h, and (**F**) 10 h.

Table 1. IC_{50} (mg/mL) for ACE inhibition of crude PPHs produced at different Alcalase concentrations and hydrolysis times.

Alcalase Concentration	Hydrolysis Time (h)					
	3 h	4 h	5 h	6 h	8 h	10 h
3%	6.94 ± 0.19 [b]	6.89 ± 0.13 [b]	6.63 ± 0.02 [a]	6.36 ± 0.08 [c]	6.84 ± 0.02 [b]	6.67 ± 0.22 [b]
4%	6.78 ± 0.34 [b]	6.73 ± 0.14 [b]	6.13 ± 0.09 [a]	5.45 ± 0.20 [d]	6.37 ± 0.06 [c]	6.46 ± 0.20 [c]
5%	7.40 ± 0.30 [a]	7.05 ± 0.06 [b]	6.34 ± 0.24 [a]	5.93 ± 0.37 [d]	6.78 ± 0.31 [c]	6.17 ± 0.14 [d]

Note: in the same row, data with different superscripts are significantly different at $p < 0.05$.

2.3. ACE-Inhibitory Activity (%) of Different Fractionations of PPH

This study evaluated the ACE-inhibitory activities of three fractions (F1: >10 kDa, F2: 5–10 kDa, and F3: <5 kDa) obtained from samples treated with different concentrations of Alcalase (3%, 4%, and 5%) for 6, 8, and 10 h because the crude PPH obtained under these hydrolysis conditions showed higher ACE-inhibitory activities. The ACE-inhibitory potential of these fractions was measured at protein/peptide concentrations of 0.5, 1.0, 1.5, and 2 mg/mL. Table 2 shows that the IC_{50} of various fractions of the same PPH sample decreased in the order of F1 > F2 > F3, indicating the increase in ACE-inhibitory activity of PPH with decreasing peptide size. Fraction 3 demonstrated the lowest IC_{50} values (0.85–1.68 mg/mL), which varied with the PPC hydrolysis condition. Among the different treatment conditions, samples treated for 6 h at Alcalase concentrations of 3, 4, and 5% displayed substantially higher ACE inhibition across all fractions ($p < 0.05$). The samples treated for 6 h with a 4% Alcalase exhibited the highest ACE inhibition, as evidenced by the lowest IC_{50} values across all fractions (0.87–3.68 mg/mL).

Table 2. IC_{50} (mg/mL) values for ACE inhibition of different fractions of PPHs produced at different Alcalase concentrations and hydrolysis times.

Hydrolysis Time	Alcalase Concentration	IC_{50} (mg/mL)		
		Fraction 1	Fraction 2	Fraction 3
6 h	3%	5.43 ± 0.49^{c}	3.38 ± 0.30^{b}	0.89 ± 0.02^{e}
	4%	3.68 ± 1.09^{d}	1.57 ± 0.11^{d}	0.87 ± 0.05^{e}
	5%	5.75 ± 1.22^{c}	3.56 ± 0.29^{b}	0.85 ± 0.01^{e}
8 h	3%	7.35 ± 0.20^{a}	5.23 ± 1.07^{c}	1.50 ± 0.01^{bc}
	4%	5.23 ± 1.27^{c}	4.34 ± 0.23^{bc}	1.56 ± 0.06^{bc}
	5%	6.51 ± 0.22^{bc}	4.30 ± 0.66^{bc}	1.68 ± 0.03^{a}
10 h	3%	5.76 ± 0.13^{c}	5.39 ± 0.04^{ac}	1.59 ± 0.05^{ab}
	4%	4.03 ± 0.06^{d}	3.34 ± 0.59^{b}	1.47 ± 0.01^{c}
	5%	5.90 ± 0.09^{c}	4.39 ± 0.02^{bc}	1.36 ± 0.04^{d}

Note: in the same column, data with different superscripts are significantly different at $p < 0.05$.

2.4. *Renin-Inhibitory Activity of PPH*

In this study, the renin-inhibitory potential of crude PPH at concentrations of 5 and 10 mg/mL, alongside fractions smaller than 5 kDa (F3) at a protein/peptide's concentration of 0.5, 1.0, 1.5, and 2.0 mg/mL, were assessed (Figure 3). Both the crude PPH and F3 demonstrated obviously higher renin inhibition compared to the control (PPC) ($p < 0.05$) across all concentrations (Figure 3A,B). Specifically, the crude PPH inhibited the renin activity by 17.08–32.56% at a concentration of 5 mg/mL and 41.25–54.88% at 10 mg/mL (Figure 3A), respectively. At the same concentrations, control samples only showed renin inhibitions of 10.03% and 17.59%, respectively. Figure 3A also demonstrated that (1) at PPH concentration 5 mg/mL, the renin-inhibitory activities of samples were in the order of 6 h hydrolysis > 8 h hydrolysis > 10 h hydrolysis, and (2) at PPH concentration of 10 mg/mL, the renin-inhibitory activities of samples obtained from 6 h and 10 h hydrolysis time were the same, but slightly lower than that of PPH samples obtained from 8 h hydrolysis. Notably, the crude PPH from 6 h hydrolysis with 4% Alcalase and the PPH from 8 h hydrolysis with 5% Alcalase exhibited the highest renin-inhibitory activity at 5 and 10 mg/mL, respectively.

Figure 3B shows that the smaller fraction of F3 of PPH displayed a renin inhibition rate of 16.57–36.80% at a concentration of 0.5 mg/mL, similar to crude PPH at 5 mg/mL but significantly higher than the F3 of control (untreated PPC). As total peptide concentration increased from 0.5 mg/mL to 1.0 mg/mL, the renin inhibition of F3 increased sharply. Further increase in F3 concentration from 1.0 mg/mL to 2 mg/mL resulted in a slight but significant increase in the renin inhibition with the exception of the samples hydrolyzed with 4% Alcalase for 6 h. It is worth mentioning that a significant renin inhibition of 54 to 57% was observed at a concentration of 2 mg/mL in all samples treated for 8 h (Figure 3B).

We also assessed the IC_{50} values of renin inhibition for the F3 samples (Figure 3C). The IC_{50} values ranged from 1.779 to 2.199 mg/mL. For the control sample, a low concentration of F3 (0.5 mg/mL) was used for the renin-inhibition test because the highest protein/peptide concentration in the F3 of the control sample was 0.5 mg/mL. Overall, the F3 samples from PPH treated for 8 h exhibited lower IC_{50} and the lowest IC_{50} was observed for samples treated for 8 h with a 3% Alcalase. The impact of Alcalase concentration on the renin-inhibitory activity of F3 is limited although the 6 h and 10 h hydrolysis with 5% Alcalase resulted in F3 having lower IC_{50} than other F3 samples under the same hydrolysis time.

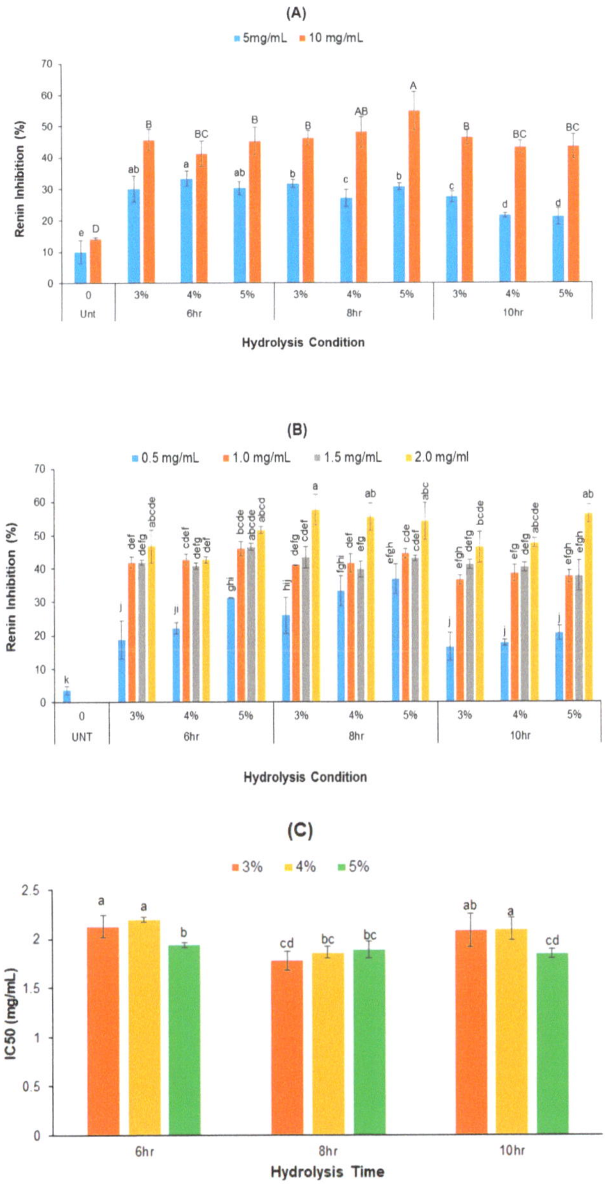

Figure 3. Effects of Alcalase concentration and hydrolysis time on the renin-inhibitory activity (%) of PPH. (**A**) Renin inhibition at the concentration 5 and 10 mg/mL of crude PPH (lowercase letters are for 5mg/mL, and uppercase letters are for 10 mg/mL), (**B**) renin-inhibitory activities of F3 of PPH samples at the different protein/peptides concentration, and (**C**) IC_{50} values of F3 samples (PPH fractions smaller than 5 kDa). (Data bars with different labels represent significantly different values at $p < 0.05$.)

2.5. IgE-Binding Inhibition of PPH

This study also evaluated the *in vitro* allergenicity of PPH produced by extensive hydrolysis of PPC using a competitive inhibitory ELISA. Figure 4 shows that the IgE-binding inhibition of all crude PPH was higher than that of PPC. At all Alcalase concentrations, the percentage of IgE-binding inhibition of PPH increased with hydrolysis time, but no further increase was observed after 6 h of hydrolysis, which suggests the optimal hydrolysis time for reducing IgE binding of PPC at the Alcalase concentration tested in this study is 6 h. The IgE-binding inhibition of PPH samples produced by 6, 8, and 10 h hydrolysis at 100, 1000, and 10,000 μg/mL were 48–52%, 44–50%, and 55–63% higher than those of PPC, respectively, varying slightly with Alcalase concentration.

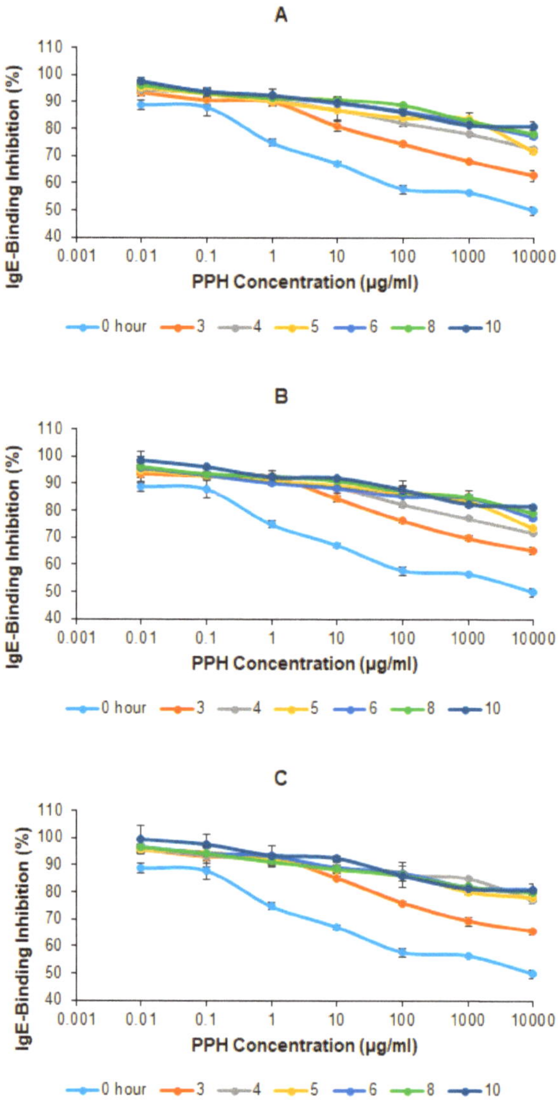

Figure 4. Percentages of IgE-binding inhibition of PPH samples produced at different Alcalase concentrations and hydrolysis time. (**A**) 3% Alcalase, (**B**) 4% Alcalase, and (**C**) 5% Alcalase.

3. Discussion

The use of Alcalase to hydrolyze PPC resulted in a decrease in protein concentration but a consistent increase in the quantity of free amino acids (FAA) with time. The decrease in detectable soluble protein after hydrolysis can be explained by the following two reasons. First, the peptides turned into amino acids, which could not be measured using the BCA method due to the lack of peptide bonds. This is evidenced by the increase in free amino acids with the progress of hydrolysis. Secondly, hydrophobic peptides form aggregates and become insoluble in a water or buffer solution due to their nature [26]. Regarding the increase in the amino acid concentration, enzymatic hydrolysis by protease is a process by which proteins are broken down into peptides and amino acids [27]. The assessment of FAA production is a reliable method for evaluating the progress of the protein hydrolysis process [28]. Alcalase is a protease preparation dominated by subtilisin, which is an endopeptidase with broader substrate specificity than chymotrypsin, although they both have a catalytic triad of Asp, His, and Ser [29]. Depending on the position of Asp, His, and Ser, the end products can be peptides and amino acids. In the present study, the amino acid concentration of PPH increased significantly with both Alcalase concentration and hydrolysis time.

Our investigation found that the hydrolysis of PPC led to a considerable increase in ACE-inhibitory activity at various concentrations and treatments. The results suggest a significant dependence of ACE inhibition on the hydrolysis duration, rather than on the concentration of Alcalase when its concentration was 3% or higher. This is consistent with the findings of prior studies [22,24]. In an early study reported by Jamdar and colleagues [24], the hydrolyzed PPI exhibited 90 to 97% ACE inhibition depending on the degree of hydrolysis, while the unhydrolyzed PPI inhibited ACE by 66%, which was greatly higher than the ACE inhibition of unhydrolyzed plant proteins reported in other studies [22,30] and that of this study. In the study of Giromini et al. (2017), the ACE inhibitions of undigested and gastric-digested plant proteins were 8–36% and 32–81%, respectively [30]. Our previous study demonstrated that the ACE inhibition of untreated peanut protein extract was 3.8–4.7% at a protein concentration of 1 mg/mL, and it increased to 35–37% after 4 h of hydrolysis [22]. The different ACE sources might be the reason for the large difference between Jamdar's study and other studies. In Jamder's study, the ACE was self-prepared from swine lung, while the ACE used in our previous study and the current study was from rabbit lung and was purchased from Sigma-Aldrich.

It has been reported that the ACE-inhibitory activities of peptides are affected by the size of the peptide, and the smaller peptides often have stronger ACE-inhibitory activity [31,32]. The results of this study further confirmed that peptide size directly impacts ACE-inhibitory activity, with smaller peptides showing stronger inhibition as shown by the much smaller IC_{50} of F3 compared to the IC_{50} values of crude PPH, F1, and F2 (Table 1). Smaller peptides are more likely to bind to ACE's active site, leading to enzyme inhibition [33]. The F3 obtained from PPH hydrolyzed for 6 h had IC_{50} value of 0.85–0.89 mg/mL, which is close to the IC_{50} of Captopril (0.5 mg/mL) and the 3 kDa permeate (0.59 mg/mL) of brown seaweed *Laminaria digitate* hydrolyzed by 1% (*v*/*w*) Viscozyme® and Alcalase®, as reported in a recent study [34]. However, the quantity of fraction 3 was limited and its concentration remained relatively low (between 1.5 and 2 mg/mL). Thus, it is resource intensive and time consuming to produce Fraction 3. Thus, it is important to optimize the protease hydrolysis conditions to produce crude PPH with the highest ACE-inhibitory activity. Consequently, the consumption of PPH as a dietary supplement or food could offer blood pressure-lowering benefits for individuals with hypertension, at a relatively higher dose.

The ACE-inhibitory activity of a hydrolysate can be influenced by both the duration of hydrolysis and the concentration of protease employed. Deviating from the optimal hydrolysis time, either by making it longer or shorter, can result in reduced or diminished ACE inhibition. This occurs because shorter hydrolysis times may fail to produce sufficient effective peptides, whereas longer hydrolysis times may break down effective peptides into smaller, less potent peptides or free amino acids [32]. The ideal hydrolysis conditions depend on the specific protein source and the type of protease utilized [33]. In this study, the hydrolysis time of 6 h at an Alcalase concentration of 4% (*v*/*w*) resulted in crude PPH and diffractions with the lowest IC_{50} values, that is, the highest ACE-inhibitory activities. The lower percentages of ACE inhibition of fractions obtained at 8 and 10 h of enzymatic treatment compared to those derived from the 6 h samples is most likely due to the degradation of some ACE-inhibitory peptides into FAA as evidenced by the significantly lower protein/peptide concentration and higher FAA concentration in the PPH (Figure 1).

Renin inhibitors provide precise inhibition within RAS, yielding improved therapeutic profiles by selectively targeting renin-catalyzed hydrolysis of angiotensinogen [35]. Compared to ACE inhibition, there have been limited studies concerning the renin-inhibitory characteristics of peptides derived from food proteins. The slower progress in this domain might be attributed to the greater challenge in renin inhibition assay, coupled with the notably higher expenses associated with renin assay procedures compared to those for ACE [25]. Only a few studies reported the renin-inhibitory abilities of some food protein hydrolysates, including hemp seed, rapeseed, canola seed, macroalga, kidney bean, and African yam seed [36–41]. A common finding of the above studies is that the crude protein hydrolysate displayed higher ACE and renin-inhibitory activity than the smaller weight fractions. In contrast to the above studies, the small molecular fraction of PPH showed higher ACE and renin-inhibitory activity than crude PPH in this study. This might be due to the difference in protein source and protease used for protein hydrolysis. It was reported that the PPH produced by hydrolysis of microfluidized peanut protein using the combined Neutrase and Protamex contained peptides mostly smaller than 1 kDa and displayed 77% of renin-inhibitory activity at 2 mg/mL [42], which is higher than the crude PPH produced by direct Alcalase hydrolysis and comparable to the F3 in this study. The studies summarized in a recent review show that the rank of the renin-inhibitory activities of protein hydrolysates from different protein sources is oil seeds (including canola, rape seeds, hemp seeds, and peanuts) $\geq$ pea > chicken meats > beans > bovine proteins [43]. Among protein hydrolysates of animal origin, fish protein hydrolysates exhibited the highest renin inhibition, while bovine protein hydrolysates showed negative or no renin inhibition compared to unhydrolyzed bovine proteins, although they were reported to have significant ACE-inhibitory activity both in vitro and in vivo [43]. This suggests that the plant-protein hydrolysates may be more effective antihypertensive ingredients, and thus function better than animal-protein hydrolysates in blood pressure control.

Research shows that about 2% of the Western population is affected by peanut allergy, with 7–14% of cases caused by accidental peanut exposure annually, and one-third to one-half may experience anaphylaxis [44]. As a result, studies on using peanut protein to produce dietary supplements and nutraceuticals are limited. Peanut allergy is an IgE-mediated food allergy that occurs when patients develop IgE antibodies against the peanut proteins, followed by exposure to that protein [45]. This study shows that the IgE binding of peanut protein was reduced by 48–63% after 5 h of hydrolysis by Alcalase, which was consistent with the skin-prick test results from a previous study [31] and confirmed what was stated in a recent study that longer than 4 h of hydrolysis duration was needed to further reduce the allergenicity of peanut protein [22]. The significant reduction of IgE binding indicates the formation of non-allergenic peptides during PPC hydrolysis. Due to the remaining allergenicity, allergen labeling is still needed for the products containing antihypertensive crude PPH.

4. Materials and Methods

4.1. Materials

The light roasted defatted peanut flour (containing 12% lipid and 50 ± 2% protein) was purchased from the Golden Peanut Company (Albany, GA, USA). The Novozyme 2.4 L of Alcalase, leucine, ninhydrin reagent, ACE (0.5 U/mL) derived from rabbit lung, an enzyme substrate known as FAPGG peptide, goat anti-human IgE peroxidase conjugate, and o-phenylenediamine were obtained from Sigma-Aldrich (St. Louis, MO, USA). Sodium hydroxide, Bicinchoninic acid (BCA) reagents, Tris base, Tween 20, 30% hydrogen peroxide, and other chemicals were acquired from Fisher Scientific (Waltham, MA, USA). Renin-inhibitory Screen Kits were purchased from Cayman Chemical Company (Ann Arbor, MI, USA).

4.2. Preparation of Peanut Protein Concentrate

PPC was produced using the isoelectric precipitation method, following a modified version of the procedure described by [23]. Briefly, 100 g of defatted light roasted peanut flour was mixed with 900 mL of deionized (DI) water at a 1:10 ratio (w/v). The pH of the mixture was then adjusted to 10.0 using a 2N solution of sodium hydroxide (NaOH). The resulting suspension of peanut flour was agitated at 35 °C and 250 rpm for 1 h in a water bath shaker. Subsequently, the suspension was centrifuged at 3000× g for 20 min to separate the solid and liquid components. The supernatant, the liquid portion, was collected and its pH was adjusted to 4.8–5.0 using a 2N solution of hydrochloric acid (HCl). Following this, the suspension was centrifuged again at 3000× g for 20 min. The supernatant was discarded, and the remaining precipitate was collected and freeze-dried at −5 °C to obtain the final PPC product. The protein content of dry PPC was determined by a nitrogen analysis method using a 2400 CHN Elemental Analyzer (PerkinElmer, Waltham, MA, USA) and a nitrogen-to-protein conversion factor of 5.46 [46].

4.3. Preparation of Peanut Protein Hydrolysate (PPH)

A 3 × 6 two-factor factorial design was used to optimize the PPH production condition. The protease hydrolysis was conducted with 3, 4, and 5% Alcalase for 3, 4, 5, 6, 8, and 10 h at pH 8 and 40 °C. The process involved suspending 1.00 g of PPC in 10 mL of phosphate buffer solution. The pH of the mixture was adjusted to 8.0 using a 1N NaOH solution. Subsequently, the PPC suspension was incubated with different concentrations (3, 4, and 5%) of 2.4 L Alcalase for a duration of 3 to 10 h in a water bath shaker at 40 °C and 200 rpm. The pH of the solution was maintained at 8.0 throughout the enzymatic hydrolysis process by adding 2N NaOH solution as necessary. Control samples were prepared by incubating the peanut flour suspensions without the addition of Alcalase. At specific time points, samples were taken and immediately subjected to enzyme inactivation by placing them in a 90 °C water bath for 10 min. Thereafter, the samples were cooled in ice water and then centrifuged at 3000× g for 20 min. The resulting supernatants (PPH) were collected, divided into small quantities, and stored at −20 °C for further utilization.

4.4. Fractionation of Peanut Protein Hydrolysates

To obtain PPH fractions with different ranges of molecular weights, the PPH was diluted with phosphate buffer at a ratio of 1:7 and subjected to sequential centrifugation using ultrafiltration (UF) centrifugal tubes of molecular weight cut-off (10 kDa and 5 kDa). This process resulted in three fractions with a molecular weight greater than 10 kDa (Fraction 1), within 10 to 5 kDa (Fraction 2), and smaller than 5 kDa (Fraction 3). Each fraction was split into small portions and stored at −20 °C until use.

4.5. Determination of Protein/Peptide and Amino Acid Concentrations in PPC and PPH

The protein/peptide concentrations in the samples were measured by the Bicinchoninic Acid (BCA) method, following the microplate procedure using a Synergy HTX Microplate Reader (Bio Tek, Winooski, VT, USA). Bovine serum albumin (BSA) solution (ThermoFisher Scientific, High Point, NC, USA) was employed as the standard. The protein concentration of each sample was determined in triplicate. FAA concentrations in PPC and PPH samples were determined by the ninhydrin method using a 2% ninhydrin reagent solution purchased from Millipore Sigma and leucine standard solutions according to the manufacturer's instructions [47].

4.6. ACE-Inhibitory Activity Assay

The ACE-inhibitory activity was assessed following a previously described method [9]. In brief, 10 μL of ACE solution (0.5 U/mL, from rabbit lung) and 10 μL of either PPH/fractions of PPH or demineralized water (used as a blank) were added to the wells of a 96-well microplate. The reaction was initiated by adding 150 μL of prewarmed substrate solution containing FAPGG (1 mM in 50 mM Tris-HCl with 0.3 M NaCl, pH 7.5) to each well. The microplate was then incubated at 37 °C, and the absorbance at 340 nm was recorded every minute for a duration of 30 min using an HTX Microplate Reader using kinetic mode. The slope of the linear portion of the absorbance vs. reaction time curve was utilized as a measure for the calculation of ACE-inhibitory activity according to the provided equation:

$$\text{ACE inhibition (\%)} = [1 - (\Delta A_{Inhibitor}/\Delta A_{Control})] \times 100$$

where $\Delta A_{Inhibitor}$ is the slope of the sample with the inhibitor and $\Delta A_{Control}$ is the slope of the control. Each experiment was performed in three replicates. Using non-linear regression analysis of ACE-inhibitory activity (%) versus peptide concentration, the IC_{50} value or the concentration of PPH/fractions of PPH needed to generate 50% ACE inhibition under the conditions specified was calculated.

4.7. Renin Inhibition Assay

The inhibition of human recombinant renin activity was assessed using the Renin Inhibitor Screening Assay Kit, following previously established protocols [35]. Before conducting the assay, the renin buffer was diluted 10 times with Milli-Q water and the renin enzyme solution was diluted 20 times with assay buffer prior to use, and the assay buffer was pre-warmed to 37 °C before the reaction was started. Prior to the reaction, specific amounts of substrate, assay buffer, and Milli-Q water were added to different wells. For the assay setup, the initial activity wells contained 20 μL substrate, 160 μL assay buffer, and 10 μL Milli-Q water. Background wells were prepared with 20 μL substrate, 150 μL assay buffer, and 10 μL Milli-Q water, while inhibitor wells contained 20 μL substrate, 150 μL assay buffer, and 10 μL sample. Subsequently, the reaction was started by adding 10 μL renin to the control and sample wells. Following a 10 s shaking step for thorough mixing, the microplate was then incubated at 37 °C, and the fluorescence intensity (FI) was measured using excitation and emission wavelengths of 340 nm and 490 nm, respectively, every minute for a duration of 30 min using the HTX Microplate Reader (Bio Tek, Winooski, VT, USA) using kinetic mode. The percentage inhibition was computed using the following formula:

$$\text{Renin inhibition (\%)} = [1 - (\Delta FI_{Inhibitor}/\Delta FI_{Control})] \times 100$$

where $\Delta FI_{Inhibitor}$ is the slope of the sample with the inhibitor and $\Delta FI_{Control}$ is the slope of the control. Each experiment was performed in three replicates.

4.8. Assessing the IgE Binding of PPH Using Human Serum

In accordance with previously established protocols [9], a competitive inhibition enzyme-linked immunosorbent assay (ciELISA) was conducted to determine the IgE binding of PPH at concentrations of 0–10,000 μg/mL. A microplate was coated with untreated PPC, which was diluted at a ratio of 1:20 with PBS and incubated at 37 °C for 2 h. Following three washes with PBST (pH 7.4), the plate was subjected to a three-blocking step with 200 μL of 1% BSA-PBST at room temperature, each 5 min. The plate was washed three times after each blocking. After this, 50 μL of PPC or PPH sample was mixed with 50 μL of 1:50 diluted pooled human serum obtained from 7 peanut-allergic patients (Lab Plasma International Inc., Everett, WA, USA) and added to respective wells. Subsequently, the plate was gently shaken for 45 min at room temperature, followed by three washes with PBST. The presence of immunoglobulin E (IgE) antibodies bound to the plate was then detected using a goat anti-human IgE peroxidase conjugate (Sigma-Aldrich, St. Louis, MO, USA) (diluted at 1:1000, 100 μL) and a substrate solution (100 μL) containing o-phenylenediamine

(0.5 mg/mL) and 0.03% hydrogen peroxide in 0.1 M citrate buffer (pH 5.5). The enzymatic reaction was halted with 2 N sulfuric acid (50 μL), and the absorbance was measured at 490 nm using an HTX Microplate Reader. The absorbance of the sample containing both IgE and peanut sample was denoted as B, while B_0 represented the absorbance of a control comprising IgE alone. Notably, the higher the allergenicity of the sample, the lower the absorbance, and conversely, as previously described [48]. The degree of IgE binding inhibition within a sample was quantified as $B \times 100/B_0$.

4.9. Statistical Analysis

Changes in protein and FAA concentrations during hydrolysis were analyzed using linear regression and post-ANOVA Tukey tests at a significance level of 5%. Additionally, IC_{50} values for the ACE and renin-inhibitory activities of crude PPH and the fractions of PPH were determined across various hydrolysis conditions using a non-linear regression curve. To evaluate distinctions among IC_{50} values of different samples, post-ANOVA Tukey tests were conducted at a significance level of 5%. Regression analysis and Tukey tests were conducted using GraphPad Prism version 8.0.2 (263) (GraphPad Software, Boston, MA, USA).

5. Conclusions

This study revealed that the hydrolysis of PPC with Alcalase under optimal conditions could result in crude PPH with moderate capacity to inhibit the activities of ACE and renin, two critical enzymes in blood pressure regulation. This is an important advantage over the protein hydrolysates/peptides derived from animal proteins, except for fish proteins. Since renin catalyzes the rate-determining step in the renin–angiotensin–aldosterone system, the protein hydrolysates with both renin and ACE-inhibitory activities and should work better in blood pressure attenuation. The optimal Alcalase concentration and hydrolysis time for high ACE-inhibitory PPH were 4% and 6 h, respectively, different from that for high renin-inhibitory PPH (5% Alcalase and 8 h hydrolysis time). Within the spectrum of PPH fractions, F3 (<5 kDa) consistently showcased higher ACE-and renin-inhibitory activities regardless of the Alcalase concentration or hydrolysis time, but its availability was limited because of the low yield from fractionation and low concentration (1.5–2 mg/mL). Thus, the production of F3 could be both costly and time-consuming. Consequently, crude PPH holds promise as a potential dietary supplement or functional food for blood pressure regulation among individuals with hypertension. In addition, the crude PPH showed significantly reduced IgE binding, which implicates lower allergenicity, compared to untreated PPC. However, hydrolysis of plant protein has been reported to generate bitter peptides [49,50]; thus, it is important to evaluate the sensory properties of PPH and explore practical methods to eliminate or mask the bitterness without reducing its antihypertensive potential need to be studied. To establish the therapeutic effects of PPH- or PPH-enriched foods on blood pressure control, further investigations encompassing in vivo hypertension studies and human clinical trials are needed.

Author Contributions: Conceptualization, J.Y.; methodology, J.Y. and S.P.; formal analysis, S.P.; investigation, S.P. and J.Y.; resources, J.Y.; data curation, S.P.; writing—original draft preparation, S.P.; writing—review and editing, J.Y.; supervision, J.Y.; project administration, J.Y.; funding acquisition, J.Y. All authors have read and agreed to the published version of the manuscript.

Funding: This research was funded by the USDA National Institute of Food and Agriculture, Capacity Building program, project number: 2022-38821-37357.

Institutional Review Board Statement: Not applicable.

Informed Consent Statement: Not applicable.

Data Availability Statement: Data will be made available on request.

Acknowledgments: We acknowledge the administrative support from the Department of Family and Consumer Sciences and the Agricultural Research Station at the North Carolina A&T State University.

Conflicts of Interest: The authors declare no conflicts of interest.

References

1. Stanaway, J.D.; Afshin, A.; Gakidou, E.; Lim, S.S.; Abate, D.; Abate, K.H.; Abbafati, C.; Abbasi, N.; Abbastabar, H.; Abd-Allah, F.; et al. Global, regional, and national comparative risk assessment of 84 behavioural, environmental and occupational, and metabolic risks or clusters of risks for 195 countries and territories, 1990–2017: A systematic analysis for the Global Burden of Disease Study 2017. *Lancet* **2018**, *392*, 1923–1994. [CrossRef]
2. Roth, G.A.; Abate, D.; Abate, K.H.; Abay, S.M.; Abbafati, C.; Abbasi, N.; Abbastabar, H.; Abd-Allah, F.; Abdela, J.; Abdelalim, A.; et al. Global, regional, and national age-sex-specific mortality for 282 causes of death in 195 countries and territories, 1980–2017: A systematic analysis for the Global Burden of Disease Study 2017. *Lancet* **2018**, *392*, 1736–1788. [CrossRef] [PubMed]
3. Mills, K.T.; Bundy, J.D.; Kelly, T.N.; Reed, J.E.; Kearney, P.M.; Reynolds, K.; Chen, J.; He, J. Global Disparities of Hypertension Prevalence and Control. *Circulation* **2016**, *134*, 441–450. [CrossRef] [PubMed]
4. CDC. *Health Topics—High Blood Pressure—POLARIS*; Centers for Disease Control and Prevention: Tlanta, GA, USA, 2021. Available online: https://www.cdc.gov/policy/polaris/healthtopics/highbloodpressure/index.html (accessed on 12 April 2024).
5. Mills, K.T.; Stefanescu, A.; He, J. The Global Epidemiology of Hypertension. *Nat. Rev. Nephrol.* **2020**, *16*, 223–237. [CrossRef] [PubMed]
6. Ma, K.; Gao, W.; Xu, H.; Liang, W.; Ma, G. Role and Mechanism of the Renin-Angiotensin-Aldosterone System in the Onset and Development of Cardiorenal Syndrome. *J. Renin-Angiotensin-Aldosterone Syst.* **2022**, *2022*, 3239057. [CrossRef] [PubMed]
7. Gradman, A.H.; Kad, R. Renin inhibition in hypertension. *J. Am. Coll. Cardiol.* **2008**, *51*, 519–528. [CrossRef]
8. Sidorenkov, G.; Navis, G. Safety of ACE inhibitor therapies in patients with chronic kidney disease. *Expert Opin. Drug Saf.* **2014**, *13*, 1383–1395. [CrossRef]
9. Li, H.; Yu, J.; Ahmedna, M.; Goktepe, I. Reduction of major peanut allergens Ara h 1 and Ara h 2, in roasted peanuts by ultrasound assisted enzymatic treatment. *Food Chem.* **2013**, *141*, 762–768. [CrossRef] [PubMed]
10. López-Fandiño, R.; Otte, J.; van Camp, J. Physiological, chemical and technological aspects of milk-protein-derived peptides with antihypertensive and ACE-inhibitory activity. *Int. Dairy J.* **2006**, *16*, 1277–1293. [CrossRef]
11. Abachi, S.; Bazinet, L.; Beaulieu, L. Antihypertensive and Angiotensin-I-Converting Enzyme (ACE)-Inhibitory Peptides from Fish as Potential Cardioprotective Compounds. *Mar. Drugs* **2019**, *17*, 613. [CrossRef]
12. Daliri EB, M.; Ofosu, F.K.; Chelliah, R.; Park, M.H.; Kim, J.H.; Oh, D.H. Development of a soy protein hydrolysate with an antihypertensive effect. *Int. J. Mol. Sci.* **2019**, *20*, 1496. [CrossRef] [PubMed]
13. Nardo, A.E.; Suárez, S.; Quiroga, A.V.; Añón, M.C. Amaranth as a Source of Antihypertensive Peptides. *Front. Plant Sci.* **2020**, *11*, 578631. [CrossRef] [PubMed]
14. Marczak, E.D.; Usui, H.; Fujita, H.; Yang, Y.; Yokoo, M.; Lipkowski, A.W.; Yoshikawa, M. New antihypertensive peptides isolated from rapeseed. *Peptides* **2003**, *24*, 791–798. [CrossRef] [PubMed]
15. Karami, Z.; Akbari-adergani, B. Bioactive food derived peptides: A review on correlation between structure of bioactive peptides and their functional properties. *J. Food Sci. Technol.* **2019**, *56*, 535–547. [CrossRef] [PubMed]
16. Bleakley, S.; Hayes, M.; O' Shea, N.; Gallagher, E.; Lafarga, T. Predicted release and analysis of novel ACE-I, renin, and DPP-IV inhibitory peptides from common oat (*Avena sativa*) protein hydrolysates using in Silico analysis. *Foods* **2017**, *6*, 108. [CrossRef] [PubMed]
17. Ciau-Solís, N.A.; Acevedo-Fernández, J.J.; Betancur-Ancona, D. In vitro renin–angiotensin system inhibition and in vivo antihypertensive activity of peptide fractions from lima bean (*Phaseolus lunatus* L.). *J. Sci. Food Agric.* **2018**, *98*, 781–786. [CrossRef]
18. Segall, L.; Covic, A.; Goldsmith DJ, A. Direct renin inhibitors: The dawn of a new era, or just a variation on a theme? *Nephrol. Dial. Transplant.* **2007**, *22*, 2435–2439. [CrossRef]
19. Udenigwe, C.C.; Lin, Y.-S.; Hou, W.-C.; Aluko, R.E. Kinetics of the inhibition of renin and angiotensin I-converting enzyme by flaxseed protein hydrolysate fractions. *J. Funct. Foods* **2009**, *1*, 199–207. [CrossRef]
20. Miralles, B.; Amigo, L.; Recio, I. Critical Review and Perspectives on Food-Derived Antihypertensive Peptides. *J. Agric. Food Chem.* **2018**, *66*, 9384–9390. [CrossRef]
21. Li, H.; Aluko, R.E. Identification and Inhibitory Properties of Multifunctional Peptides from Pea Protein Hydrolysate. *J. Agric. Food Chem.* **2010**, *58*, 11471–11476. [CrossRef]
22. Yu, J.; Mikiashvili, N.; Bonku, R.; Smith, I.N. Allergenicity, antioxidant activity and ACE-inhibitory activity of protease hydrolyzed peanut flour. *Food Chem.* **2021**, *360*, 129992. [CrossRef] [PubMed]
23. Yu, J.; Ahmedna, M.; Goktepe, I. Peanut protein concentrate: Production and functional properties as affected by processing. *Food Chem.* **2007**, *103*, 121–129. [CrossRef]
24. Jamdar, S.N.; Rajalakshmi, V.; Pednekar, M.D.; Juan, F.; Yardi, V.; Sharma, A. Influence of degree of hydrolysis on functional properties, antioxidant activity and ACE inhibitory activity of peanut protein hydrolysate. *Food Chem.* **2010**, *121*, 178–184. [CrossRef]
25. Aluko, R.E. Antihypertensive peptides from food proteins. In *Annual Review of Food Science and Technology*; Annual Reviews Inc.: San Mateo, CA USA, 2015; Volume 6, pp. 235–262. [CrossRef]
26. Zhang, Y.; Zhou, F.; Zhao, M.; Ning, Z.; Sun-Waterhouse, D.; Sun, B. Soy peptide aggregates formed during hydrolysis reduced protein extraction without decreasing their nutritional value. *Food Funct.* **2017**, *8*, 4384–4395. [CrossRef]

27. López-Otín, C.; Bond, J.S. Proteases: Multifunctional Enzymes in Life and Disease. *J. Biol. Chem.* **2008**, *283*, 30433–30437. [CrossRef]
28. Aguirre, L.; Garro, M.S.; Savoy de Giori, G. Enzymatic hydrolysis of soybean protein using lactic acid bacteria. *Food Chem.* **2008**, *111*, 976–982. [CrossRef]
29. Azrin NA, M.; Ali MS, M.; Rahman RN ZR, A.; Oslan, S.N.; Noor, N.D.M. Versatility of subtilisin: A review on structure, characteristics, and applications. *Biotechnol. Appl. Biochem.* **2022**, *69*, 2599–2616. [CrossRef]
30. Giromini, C.; Fekete, Á.A.; Givens, D.I.; Baldi, A.; Lovegrove, J.A. Short-communication: A comparison of the in vitro angiotensin-1-converting enzyme inhibitory capacity of dairy and plant protein supplements. *Nutrients* **2017**, *9*, 1352. [CrossRef]
31. Li, Y.; Yu, J. Research Progress in Structure-Activity Relationship of Bioactive Peptides. *J. Med. Food* **2015**, *18*, 147–156. [CrossRef] [PubMed]
32. Puspitojati, E.; Indrati, R.; Cahyanto, M.N.; Marsono, Y. Effect of fermentation time on the molecular weight distribution of ACE inhibitory peptide from jack bean tempe. *IOP Conf. Series. Earth Environ. Sci.* **2023**, *1177*, 012026. [CrossRef]
33. Manzanares, P.; Gandía, M.; Garrigues, S.; Marcos, J.F. Improving Health-Promoting Effects of Food-Derived Bioactive Peptides through Rational Design and Oral Delivery Strategies. *Nutrients* **2019**, *11*, 2545. [CrossRef]
34. Purcell, D.; Packer, M.A.; Hayes, M. Angiotensin-I-converting enzyme inhibitory activity of protein hydrolysates generated from the macroalga *Laminaria digitata* (Hudson) JV Lamouroux 1813. *Foods* **2022**, *11*, 1792. [CrossRef] [PubMed]
35. Stanton, A. Review: Potential of renin inhibition in cardiovascular disease. *J. Renin-Angiotensin-Aldosterone Syst.* **2003**, *4*, 6–10. [CrossRef]
36. Girgih, A.T.; Udenigwe, C.C.; Li, H.; Adebiyi, A.P.; Aluko, R.E. Kinetics of enzyme inhibition and antihypertensive effects of hemp seed (*Cannabis sativa* L.) protein hydrolysates. *J. Am. Oil Chem. Soc.* **2011**, *88*, 1767–1774. [CrossRef]
37. He, R.; Alashi, A.; Malomo, S.A.; Girgih, A.T.; Chao, D.; Ju, X.; Aluko, R.E. Antihypertensive and free radical scavenging properties of enzymatic rapeseed protein hydrolysates. *Food Chem.* **2013**, *141*, 153–159. [CrossRef] [PubMed]
38. Alashi, A.M.; Blanchard, C.L.; Mailer, R.J.; Agboola, S.O.; Mawson, A.J.; He, R.; Malomo, S.A.; Girgih, A.T.; Aluko, R.E. Blood pressure lowering effects of Australian canola protein hydrolysates in spontaneously hypertensive rats. *Food Res. Int.* **2014**, *55*, 281–287. [CrossRef]
39. Fitzgerald, C.; Mora-Soler, L.; Gallagher, E.; O'Connor, P.; Prieto, J.; Soler-Vila, A.; Hayes, M. Isolation and Characterization of Bioactive Pro-Peptides with In Vitro Renin Inhibitory Activities from the Macroalga *Palmaria palmata*. *J. Agric. Food Chem.* **2012**, *60*, 7421–7427. [CrossRef] [PubMed]
40. Mundi, S.; Aluko, R.E. Inhibitory properties of kidney bean protein hydrolysate and its membrane fractions against renin, angiotensin converting enzyme, and free radicals. *Austin J. Nutr. Food Sci* **2014**, *2*, 1008–1019.
41. Ajibola, C.F.; Fashakin, J.B.; Fagbemi, T.N.; Aluko, R.E. Renin and angiotensin converting enzyme inhibition with antioxidant properties of African yam bean protein hydrolysate and reverse phase HPLC separated peptide fractions. *Food Res. Int.* **2013**, *52*, 437–444. [CrossRef]
42. Gong, K.; Deng, L.; Shi, A.; Liu, H.; Liu, L.; Hu, H.; Adhikari, B.; Wang, Q. High-pressure microfluidisation pretreatment disaggregate peanut protein isolates to prepare antihypertensive peptide fractions. *Int. J. Food Sci. Technol.* **2017**, *52*, 1760–1769. [CrossRef]
43. Aluko, R.E. Food protein-derived renin-inhibitory peptides: In vitro and in vivo properties. *J. Food Biochem.* **2019**, *43*, e12648. [CrossRef] [PubMed]
44. Lieberman, J.A.; Gupta, R.; Knibb, R.C.; Haselkorn, T.; Tilles, S.; Mack, D.P.; Pouessel, G. The Global Burden of Illness of Peanut Allergy: A Comprehensive Literature Review. *Allergy* **2020**, *76*, 1367–1384. [CrossRef] [PubMed]
45. Anvari, S.; Miller, J.; Yeh, C.-Y.; Davis, C.M. IgE-Mediated Food Allergy. *Clin. Rev. Allergy Immunol.* **2018**, *57*, 244–260. [CrossRef] [PubMed]
46. Mariotti, F.; Tomé, D.; Mirand, P.P. Converting nitrogen into protein—Beyond 6.25 and Jones' factors. *Crit. Rev. Food Sci. Nutr.* **2008**, *48*, 177–184. [CrossRef]
47. Moore, S. Amino acid analysis: Aqueous dimethyl sulfoxide as solvent for the ninhydrin reaction. *J. Biol. Chem.* **1968**, *243*, 6281–6283. [CrossRef]
48. Chung, S.-Y.; Yang, W.; Krishnamurthy, K. Effects of pulsed UV-light on peanut allergens in extracts and liquid peanut butter. *J. Food Sci.* **2008**, *73*, C400–C404. [CrossRef] [PubMed]
49. Iwaniak, A.; Minkiewicz, P.; Darewicz, M.; Hrynkiewicz, M. Food protein-originating peptides as tastants—Physiological, technological, sensory, and bioinformatic approaches. *Food Res. Int.* **2016**, *89*, 27–38. [CrossRef]
50. Görgüç, A.; Gençdağ, E.; Yılmaz, F.M. Bioactive peptides derived from plant origin by-products: Biological activities and techno-functional utilizations in food developments—A review. *Food Res. Int.* **2020**, *136*, 109504. [CrossRef] [PubMed]

International Journal of *Molecular Sciences*

MDPI

Review

Narrative Review of the Current and Future Perspectives of Phycobiliproteins' Applications in the Food Industry: From Natural Colors to Alternative Proteins

Simeon Minić [1,*,†], Nikola Gligorijević [2,†], Luka Veličković [1] and Milan Nikolić [1]

1 Department of Biochemistry and Center of Excellence for Molecular Food Sciences, Faculty of Chemistry, University of Belgrade, Studentski trg 12–16, 11000 Belgrade, Serbia
2 Department of Chemistry, Institute of Chemistry, Technology, and Metallurgy, National Institute of the Republic of Serbia, University of Belgrade, Studentski trg 12–16, 11000 Belgrade, Serbia
* Correspondence: sminic@chem.bg.ac.rs; Tel.: +381-11-3282111
† These authors contributed equally to this work.

Abstract: Vivid-colored phycobiliproteins (PBPs) have emerging potential as food colors and alternative proteins in the food industry. However, enhancing their application potential requires increasing stability, cost-effective purification processes, and consumer acceptance. This narrative review aimed to highlight information regarding the critical aspects of PBP research that is needed to improve their food industry potential, such as stability, food fortification, development of new PBP-based food products, and cost-effective production. The main results of the literature review show that polysaccharide and protein-based encapsulations significantly improve PBPs' stability. Additionally, while many studies have investigated the ability of PBPs to enhance the techno-functional properties, like viscosity, emulsifying and stabilizing activity, texture, rheology, etc., of widely used food products, highly concentrated PBP food products are still rare. Therefore, much effort should be invested in improving the stability, yield, and sensory characteristics of the PBP-fortified food due to the resulting unpleasant sensory characteristics. Considering that most studies focus on the C-phycocyanin from Spirulina, future studies should concentrate on less explored PBPs from red macroalgae due to their much higher production potential, a critical factor for positioning PBPs as alternative proteins.

Keywords: phycobiliproteins; food industry; natural colors; alternative proteins; C-phycocyanin; spirulina

Citation: Minić, S.; Gligorijević, N.; Veličković, L.; Nikolić, M. Narrative Review of the Current and Future Perspectives of Phycobiliproteins' Applications in the Food Industry: From Natural Colors to Alternative Proteins. *Int. J. Mol. Sci.* **2024**, *25*, 7187. https://doi.org/10.3390/ijms25137187

Academic Editors: Vassilis Athanasiadis and Theodoros G. Chatzimitakos

Received: 13 June 2024
Revised: 22 June 2024
Accepted: 25 June 2024
Published: 29 June 2024

1. Introduction

Phycobiliproteins (PBPs), pigment proteins from cyanobacteria and algae, are water-soluble multimeric proteins which form light-harvesting complexes called phycobilisomes. So far, there are sixty-one phycobiliprotein structures deposited in the Protein Data Bank [1], and all of them are divided into three groups: allophycocyanins, phycocyanins (including phycoerythrocyanin), and phycoerythrins [2]. PBPs are differently colored due to the presence of covalently bound chromophores called phycobilins. Phycobilins are tetrapyrrole molecules attached to the specific Cys residues of apoproteins via a thioether bond [1]. There are four types of phycobilins: phycocyanobilin (PCB), phycoerythrobilin (PEB), phycourobilin (PUB), and phycoviolobilin/crypto-violobilin (CVB) [2]. PBPs are hexameric proteins in their native and functional state with $(\alpha\beta)_6$ structure. Allophycocyanin is the exception, where the functional unit is trimer $(\alpha\beta)_3$ [3]. Each $\alpha\beta$ monomer is composed of two subunits, α and β. Molecular masses of these subunits range from 16 to 17 kDa for α and 18 to 19 kDa for β. The number of bound chromophores per $\alpha\beta$ monomer is associated with their division into three mentioned classes, with allophycocyanin having two chromophores, phycocyanins and phycoerythrocyanins having three, and phycoerythrins having five or six [4,5]. Within hexamers, two trimers are associated with face-to-face orientation, and one trimer is rotated by 30° relative to the other [6].

The presence of covalently bound phycobilins in PBPs provides them with a high potential for application in several areas, including the food and confectionery industry and biomedical applications [7]. Intensive colors allow these proteins to be used as natural pigments in the confectionery industry. Considering their high nutritional values [8], these protein sources are increasingly considered alternative food options. Phycobilins are also fluorescent, and this property is essential for their application as fluorescent biomarkers. Many recent review papers have covered these aspects [1,9–13]. Multiple biological activities of PBPs have also been investigated, and it is reported that they have antioxidant, anticancer, anti-inflammatory, antidiabetic, antibacterial, antiobesity, and neuroprotective properties [11]. Their bioactivity is also ascribed to covalently bound phycobilins, and, through digestion, bioactive chromopeptides can be released [14,15].

The primary sources of PBPs, cyanobacteria and red algae, are rich in protein content, with cyanobacteria Arthrospira platensis (commercially named Spirulina) having 55–70% protein in their dry weight [16], while red seaweeds range from 2.7 to 47% in dry weight, depending on the species [17]. PBPs are present in high abundance. Blue-colored PBPs can be obtained from cyanobacteria, like Spirulina, comprising up to 8–13% of dry biomass [18]. The sources of red and purple PBPs, like: Gracilaria spp. and *Pyropia*/*Porphyra* spp. (known as Nori), *Rhodella* spp., *Bangia* spp., and *Porphyridium* spp. yield fewer PBPs (e.g., [9]). On the other hand, the annual production of these species is about 50,000 tons for Spirulina [19] and much more for red algae, with China being the leading producer [20]. Considering that Spirulina gained a lot of attention due to its high nutritional value and many health benefits [20], most of the research regarding the applications of PBPs in food fortification or as food colors are conducted on Spirulina biomass or its main PBP, C-phycocyanin (C-PC). Much less research is accomplished on red algae and their PBPs, with new research emerging in the past several years. It should be noted that the high production of red algae (more than five million tons per year) gives these species massive potential for broader food applications, and they represent a good alternative food source [19].

Regarding human consumption, highly purified PBPs are not used, considering the high costs of final products due to the purification process. The purity requirements of PBPs for food usage are easily achievable and do not require expensive techniques [12], lowering the price of PBP-based food products. Biomasses containing PBPs are often used in dry powder, tablets, leaves, capsules, or liquid form [20]. These products can be considered as food supplements and not as alternative food. Fortification with PBPs or biomasses containing PBPs into other edible food products could have several benefits. First, it would increase the nutritional and bioactive values of other foods. Second, it would make seaweed and, thus, PBPs more acceptable to consumers.

The critical aspect that should be considered in the field of alternative proteins is their allergenic potential. While the allergenicity of plant and insect proteins has been demonstrated in many studies [21], the allergenic potential of PBPs is still not explored, with only a few case studies indicating the allergenic potential of C-phycocyanin from Spirulina [22,23]. On the other hand, several studies suggested anti-allergenic properties of C-phycocyanin from Spirulina [24] and R-phycocyanin from red macroalgae *Porphyra haitanensis* [25]. Although further studies are required to explore PBPs' allergenicity more comprehensively, current findings do not point to the high allergenic potential of PBPs, giving them a comparative advantage over insect and plant proteins.

This narrative review article's uniqueness is that it looks at the potential use of PBPs as alternative proteins. Fortifying existing food is one possible approach to consuming larger amounts of PBPs; however, this paper aims to show potential methods for obtaining new attractive food based exclusively on PBPs. To achieve this, it is necessary to solve several significant problems mentioned in the paper. Having PBPs used as food colorants is one thing compared to using them as alternative proteins in the diet. The first goal can be achieved with a few milligrams of PBPs per liter/kilogram [26]; the other requires larger amounts, bringing several obstacles that must be overcome. Those obstacles, regarding biomass cultivation, protein yield, purification with odor and taste issues, and stabilization of PBPs, are critically discussed in this review, together with the economic aspect for broader usage. The lower stability of PBPs needs to be overcome if PBP-based food products are to be made, especially using thermal treatment, and these stability issues and means to improve them are also discussed.

Additionally, a detailed overview of current inventions in PBP-based food products, with technology to make them and fortification of the commonly used foods with PBPs, is presented and is the main focus of this review. Overall, the presented review paper should provide the reader with a solid literature overview of the current state-of-the-art application of PBPs as alternative food/protein sources (Figure 1). Ultimately, we proposed the future trends in the PBP science and industry domains that are required for positioning them as promising alternative proteins.

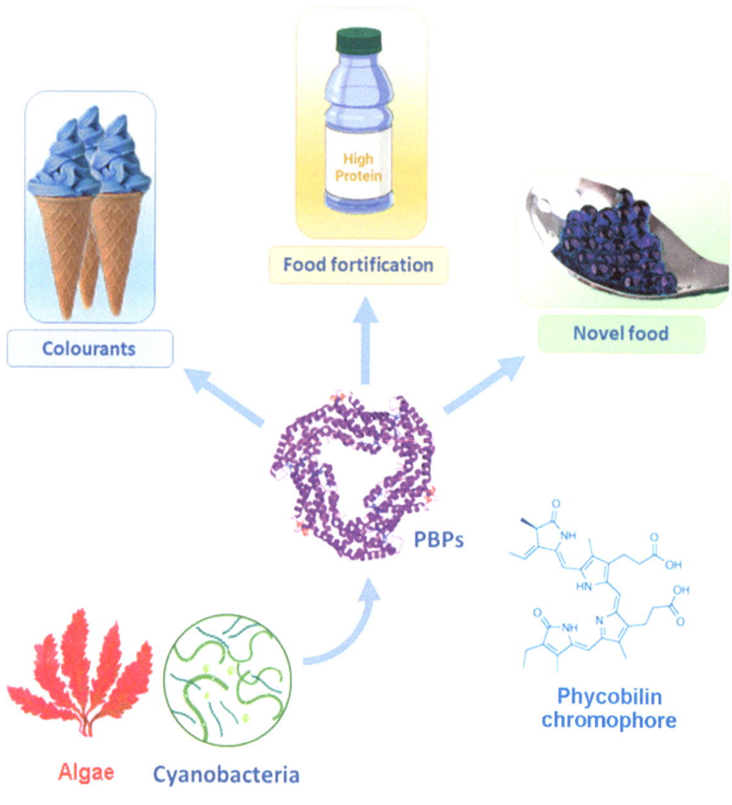

Figure 1. Overview of the applications of phycobiliproteins in the food industry.

2. Methods

The presented narrative review was realized by conducting a literature search, papers abstract, and full-text analysis, followed by discussions based on the results. The literature search was performed as described previously [27]. Relevant studies were obtained using the following databases: Google Scholar, Scopus, Web of Science, and PubMed. Papers were searched by combining several keywords, including phycobiliproteins, Spirulina, Nori, food fortification, stabilization, red algae, encapsulation, colors, etc. For this review to reflect recent, state-of-the-art progress in phycobiliprotein research in food application, most of the references included are no older than 5 years.

3. Phycobiliproteins as Food Colors

Color represents an essential characteristic of food and is a critical factor that makes food products more appealing to customers [28]. Considering that the general public is more interested in what they eat, natural food additives, including colors, are gaining more consumer and food industry interest [29]. Among colors, blue still represents a real challenge, and blue food is usually labeled by consumers as artificial [30]. The industry still uses artificial blue colors since natural ones lack physical and chemical stability, coloring power, and easy scale-up production [28]. The characteristic vivid coloration of PBPs makes them attractive in the food industry as natural food colors [31]. Their colors range from blue (C-PC), turquoise-blue (allophycocyanin), purple (R-phycocyanin), and red (R-phycoerythrin). Extract from Arthrospira platensis, rich in C-PC, is the only FDA-approved phycobiliprotein-based natural color [32], and it is also approved in the EU (EFSA regulation numbers 1333/2008 and 231/2012). It is approved for coloring confections, frostings, ice cream and frozen desserts, dessert coatings and toppings, beverage mixes and powders, yogurts, custards, puddings, cottage cheese, gelatin, breadcrumbs, and ready-to-eat cereals and as a color of hard-boiled eggshells.

As with other natural colors, PBPs have certain limitations for broader usage. Based on stability studies of C-PC, for example, optimal storage conditions for this protein are in the dark with a temperature below 45 °C and at pH 5.5–6 [33]. This protein, on its own, without stabilizers, is sensitive to light, high temperature, and pH, meaning it cannot withstand many food processing techniques. Also, it cannot be efficiently used as a color for acidic drinks, as it is unstable under such conditions [34]. On the other hand, some results point out that acidic drinks, with pH around 3, can be colored with C-PC (concentration up to 0.11 mg/mL) and that the color was stable for 11 days when kept in the fridge [35]. C-phycocyanin was found to be a more versatile natural blue color than indigo and gardenia blue, as it gave a more acceptable light blue color in jelly gum and soft cover candy, as evidenced in exploratory studies [36].

R-phycoerythrin has a broader pH stability range, from 3 to 10 [37], but it is also sensitive to thermal treatment above 40 °C [38]. However, it was shown that high-pressure treatment of R-phycoerythrin solution at 450 MPa is much less detrimental for protein color than thermal treatment [39]. R-phycocyanin is stable in a pH range from 4 to 8 and has a melting point of about 53 °C [40]. These stability characteristics of PBPs are only guidelines, and specific characteristics will depend on the algal/cyanobacterial source. For instance, PBPs isolated from thermophilic organisms are much more stable at higher temperatures and have higher potential as food colors [41,42]. However, cultivating such organisms on a large scale is not achievable since they require much more energy. One important feature is the influence of PBPs purity on their stability. The thermal stability of food-grade C-PC was greater than that of reagent-grade C-PC [43].

For PBPs to be utilized as food colors to their full potential, stability issues must be overcome. Several approaches for stabilizing PBPs are published and covered in several recently published review papers, indicating the topicality of the problem [33,44,45]. They can be divided into (1) those involving stabilizing proteins by the addition of various additives (Table 1) and approaches using different processing methods to strengthen PBP stability, such as encapsulation, cross-linking, and high-pressure processing (Table 2). PBP stabilization is achieved by the addition of edible oils [46]; sugars, including glucose, sucrose, sorbitol, trehalose, fructose, etc. (e.g., [43,47–50]); proteins [51]; and other natural preservatives, like ε-polylysine [52] or cysteine [53]. Chemical modification of PBPs is also a possibility [54].

Table 1. Effects of additives on stabilities of phycobiliproteins.

Protein (Source)	The Experimental Approach	The Results	Reference
C-phycocyanin (*Arthrospira platensis*)	Color stabilization of protein via λ-carrageenan (λC) in liquid formulations	Complexation with λC increased the protein color stability at pH < pI, especially at a pH of 3.0, even when heated to 90 °C.	[55]
C-Phycocyanin (*Arthrospira platensis*)	Stabilization effects of protein complexation with λ-carrageenan on its intrinsic blue color	The electrostatic complexation stabilized protein color in the acidic pH range (2.5–6.0) and against a heat treatment at 70 °C.	[56]
C-Phycocyanin (*Spirulina platensis*)	Stabilization of protein in aqueous solutions via sucrose and trehalose (20 and 40%, *w*/*w*)	The stabilizing effect of saccharides on the thermal discoloration of protein with sucrose performed better than trehalose.	[43]
C-Phycocyanin (Hawaiian Spirulina)	Investigation of the potential of twelve food-derived antioxidants to bind and stabilize the protein	Complexation of protein with quercetin and coenzyme Q_{10} improved its thermal stability (higher melting point).	[57]
C-phycocyanin (*Arthrospira platensis*)	Enhancement of protein productivity and stability using organic acids (citric, acetic, succinic, fumaric, and oxalic acid)	Organic acids, primarily citric acid (7.5%), act as preservatives to stabilize protein (promoting the half-live) at high temperatures.	[58]
C-phycocyanin (*Arthrospira platensis*)	Improvement of protein stability by adding saccharides (glucose, mannose, galactose, and maltose) and sugar alcohols (mannitol and maltitol)	Sugars effectively improved the protein's thermal stability in correlation with the additive concentration and inhibited its oxidative degradation.	[59]
C-phycocyanin (*Spirulina platensis*)	Increasing protein stability via 0.5% cysteine addition during enzyme-assisted extraction	Cysteine increased the thermal stability of protein extracted with (endopeptidase) Collupulin.	[53]
C-phycocyanin (Spirulina)	Improvement of protein stability by forming soluble complexes with poly-saccharides (κ-/ι-/λ-carrageenans, xanthan gum, high-methoxyl pectin, and guar gum)	Improved protein's colloidal and color stabilities against acidic pH (standard beverage processing) and heating conditions.	[60]
R-phycoerythrin (*Porphyra haitanensis*)	Stabilization of protein by self-assembly with oligochitosan (at a 1:20 reaction ratio)	The thermal (40–80 °C), natural light, and ultraviolet light irradiation (254 nm) protein stabilities were all improved.	[61]
Phycobiliproteins (*Oscillatoria* sp. BTA-170)	Stabilization of C-PC, A-PC, and PE in the presence of different monosaccharides (glucose, fructose, glucose, and lactose)	Glucose was the most critical monosaccharide that stabilizes the degradation of proteins at 65 °C and higher temperatures.	[62]
Phycobiliproteins (*Spirulina platensis*)	More efficient extraction of protein using NaCl as an extraction enhancer	Protein stability was improved by adding NaCl, which had unaffected antioxidant activity and a secondary structure.	[63]
C-Phycocyanin	Improving the protein color stability with epigallocatechin gallate (EGCG)	EGCG binding protected protein against color fading under light conditions.	[64]
R-phycocyanin (*Cyanidioschyzon merolae*)	Preservation of thermotolerant protein during storage with salts	The stabilizing effect of $CaCl_2$ and $MgCl_2$ (0.1 M) towards protein during seven days.	[65]
C-Phycocyanin	Improving the protein color stability in acidified conditions with whey protein isolate (WPI)	A low WPI concentration (0.05–0.1%) at pH 3.0 improved the protein's color stability under light exposure.	[66]
C-phycocyanin (Spirulina)	Improvement of protein stability in acidified conditions using whey proteins (α-lactalbumin, β-lactoglobulin, BSA, immunoglobulins, and glycomacropeptides)	Native whey protein (10%) efficiently improves protein colloidal stability and prevents aggregation at pH 3.0.	[51]

Table 2. Processing methods for phycobiliprotein stabilizations.

Protein (Source)	Method and Conditions Used	Result	Reference
R-Phycoerythrin (*Gracilaria gracilis*)	Protein incorporation into the gelatin-based films	Improved the protein photochemical stability in the solid state for eight months.	[67]
C-Phycocyanin (*Arthrospira platensis*)	Preparation of pectin–phycocyanin complexes with different mixing ratios	Improved the colloidal stability of the protein wholly entrapped by the polysaccharide molecules at acidic pH after heating at 85 °C.	[68]
C-phycocyanin (*Spirulina platensis*)	Double encapsulation of protein using aqueous two-phase systems (PEG 4000/Potassium phosphate and PEG 6000/Dextran) by spray drying	Prolonged shelf life with the additional benefit of enhancing the purity of protein compared with conventional (maltodextrin) encapsulation.	[69]
Phycobiliproteins (*Spirulina platensis*)	Proteins were treated with high hydrostatic pressure (HPP) (600 MPa; 300 s) in the presence of sucrose, trehalose, and glucose (20 and 40%, *w*/*w*)	Sugars exerted baroprotective, concentration-dependent action on proteins' (color) stability with preserved antioxidant activity.	[70]
C-Phycocyanin (*Arthrospira platensis* EGEMACC 38)	Spray-dried microencapsulation of protein using various combinations and ratios of wall materials (maltodextrin, gum arabic, whey protein isolate, and sodium caseinate)	The highest blueness index was observed in protein powder encapsulated with maltodextrin and whey protein isolate.	[71]
C-phycocyanin (Spirulina)	Modification of protein with 20 kDa methoxy polyethylene glycol polymers	The conjugates exhibited higher blue color intensity, improved thermodynamic stability, and a gain in pH stability and antioxidant activities.	[72]
Phycocyanin and B-Phycoerythrin	Intercalation of proteins into montmorillonite and laponite laminar nanoclays	Proteins' optical and thermal properties were significantly improved.	[73]
C-Phycocyanin (*Spirulina platensis*)	The protein was modified using formaldehyde crosslinking	Increases photostability of modified protein only upon yellow light irradiation.	[74]
C-phycocyanin (*Arthrospira platensis* IFRPD 1182)	Freeze-dried maltodextrin and gum Arabic (fractions from 0 to 100%) were used as protein microencapsulation wall materials	Increased thermal stability of encapsulated protein, with high antioxidant properties.	[75]
Phycobiliprotein (*Palmaria palmata*)	Phycobiliprotein within liposome (soy lecithin) stabilized using polyethylene glycol adsorbed cellulose nanocrystals.	The encapsulated protein was stable below 60 °C, above pH of 5.0, and against illumination.	[76]
C-Phycocyanin (Spirulina)	High-pressure processing treatment of the protein–whey protein and protein–carrageenan complexes at acidic pH	Protein's complexations improved the color and, therefore, its storage stability under light exposure.	[77]
R-Phycoerythrin (*Porphyra haitanensis*)	Preparation of the various oligochitosan-modified protein complexes (OMPC) via the transglutaminase-catalyzed glycosylation reaction	Emulsifying stability, thermal stability, photostability, and pH stability of the OMPC were all significantly improved.	[54]

While many encapsulating strategies for the stabilization of PBPs exist [33,44,45], not all of them are acceptable for usage in the food industry, with some being more applicable in photovoltaic or pharmaceutical applications [67,78]. Encapsulation can provide higher storage, temperature, and pH stability because PBPs are immobilized into specific matrixes. Production of microcapsules of small size is fine if final products are to be used as carriers of active PBP components, for applications in pharmacy, for preservation and enrichment of other foods, or for use as food supplements. Higher concentrations and amounts of such products should be used as alternative protein sources. Utilizing biomass and/or raw extracts rich in PBPs in this regard is sometimes a more accessible and cheaper approach, but it can have certain limitations. Polysaccharide hydrogels are inexpensive to fabricate and are shown to be very efficient in improving the stability of incorporated PBPs. Another alternative would be to use protein-based gels. While they are more expensive, final products are more nutritional since they contain higher protein content. Using appropriate gelling conditions and/or cross-linkers is also very important as some may have adverse effects if eaten in larger amounts, like using sodium tripolyphosphate for cross-linking of water-soluble chitosan [79]. Considering that polysaccharide hydrogels are non-digestible and have no particular taste, additional enrichment in this direction is necessary for the product to be accepted by consumers. Incorporating PBPs in food products like cakes, ice creams, milk, bread, and cookies overcomes this problem since these products already have a rich and pleasant taste. In addition, by incorporating PBPs, they act as immobilizing matrixes, improving the stability of these proteins. The following section will summarize an overview of food products containing PBPs as alternative protein sources.

4. Phycobiliproteins as Alternative Proteins and Food Fortifiers

In the last decades, most research has focused on the various aspects of PBP bioactivities and the improvement of their color stability in the conditions described above. In terms of these studies, PBPs have been usually studied at lower concentrations (generally less than 0.2%). Although PBPs may contribute to the color and health-promoting activities of fortified food at these levels, their concentration is insufficient to influence the food structure and its nutritional properties significantly. On the other hand, there is a significant abundance of PBPs in algae; for example, the PBPs in red macroalgae *Porphyra* spp. contribute up to 3% [80], while the content of PBPs in cyanobacteria Spirulina could be more than 10% of the dry mass [81]. Therefore, there is a substantial potential for cyanobacteria and algae to be explored and positioned as promising sources of alternative proteins, with the PBPs as the key contributors. In this context, fortifying the food products with high concentrations of PBPs and their influence on the structure, techno-functional, and nutritional properties of foods could be more significant. In recent years, studies have focused on how PBPs (mainly C-PC from Spirulina) could change the structure and properties of the existing foods and develop new PBP-based products. Currently, products based on Spirulina are approved for use in the EU (Novel Foods Regulation (EC) 287/1997 (NFR 1997)) since its usage was widely present in EU territory before 1997, while approval is necessary for products from red microalgae novel food.

4.1. Phycobiliprotein-Based Food Products

The development of PBP-based new food products is a new field in food science, and only several studies have been dealing with this topic in recent years. The PBP-based food may have significant potential since PBPs are the major proteins of cyanobacteria (e.g., Spirulina) and red algae (e.g., *Porphyra*), a source with the desirable amino acid profile, especially in terms of essential amino acid content, which is nearly equivalent to the egg albumin [17,82].

One promising approach is developing composite gels obtained from PBPs and algal-derived polysaccharides, which could enable the sustainable production of new foods. It was shown that the addition of polysaccharides such as κ-carrageenan and guar gum (from 0.1 to 0.4%) enhances the structure and influences the water-holding capacity of thermally induced C-PC gels (16%) through modulation of the protein secondary structure content as well as hydrophobic interactions and disulfide bonds [83].

Alginate, the well-known algal polysaccharide, has a significant application in the food industry due to its high propensity to make gels without organic solvents and at room temperature. Moreover, encapsulating proteins within an alginate gel network could significantly improve their stability, which is highly relevant regarding low PBP stability. Indeed, several studies revealed that the thermal stability of C-PC from Spirulina and R-phycocyanin from Porphyra sp. (Figure 2) could be significantly improved via PBP encapsulation within alginate beads [40,84]. Further, it was demonstrated that increased alginate concentration substantially enhances the encapsulation efficiency of C-PC. Alginate beads could not only improve the C-PC stability; they may also influence its bioavailability since the composite C-PC–alginate gel is resistant in the simulated gastric fluid, while rapid release occurs in the small intestine fluid medium [85,86]. Encapsulating R-phycoerythrin within alginate beads also improved its bioavailability at intestine-stage digestion [87]. On the other hand, C-PC, even at low concentrations (0.2%), can influence the properties of the alginate gels in terms of PC-containing beads containing more water structures with weak hydrogen bonds.

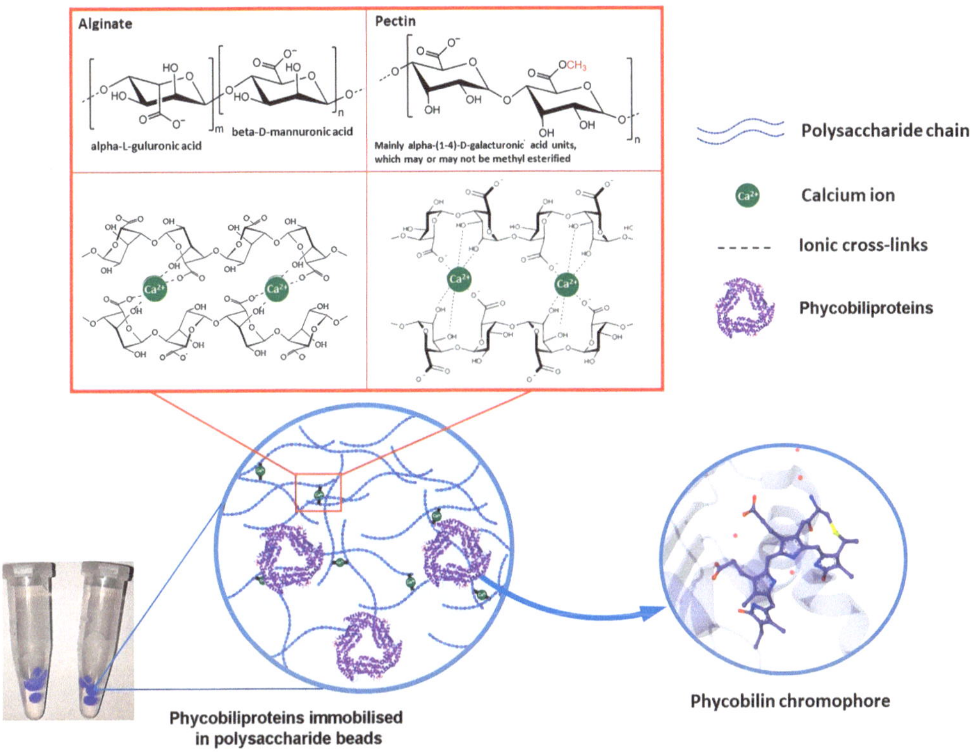

Figure 2. Encapsulation of phycobiliproteins into selected polysaccharide matrices [40].

Meanwhile, pregelatinized corn starch acts as a filler within alginate gels and absorbs surrounding water, which could improve the thermal stability of encapsulated C-PC. Pregelatinized corn starch enhances the encapsulation efficiency of C-PC within alginate beads [85]. Cross-linked starches from different botanical sources can encapsulate C-PC between amorphous chains. Among the five starches (potato, banana, corn, cassava, and breadfruit), potato starch exhibits the highest water uptake capacity and C-PC encapsulation efficiency [88].

Besides polysaccharide hydrogels, PBPs could also be encapsulated within protein-based gels. In terms of this, Spirulina extracts, containing a high amount of C-PC, improve the rheological and mechanical properties of soy protein isolate hydrogel (SPI) by enhancing its strength, hardness, and storage modulus. Furthermore, adding Spirulina extracts increases the content of β-sheet, which is why the hardness and compactness of the SPI hydrogel structure are improved [89]. Therefore, as the major component of Spirulina extracts, the C-PC could improve food texture. Another approach utilizes the interactions between C-PC and gelatin to form the self-assembly complex proteins. Due to the excellent rheological properties, this complex could stabilize high internal phase emulsions (HIPEs) used for the 3D printing of novel foods. The electrostatic interactions and hydrogen bonds between C-PC and gelatin enabled the compact structure, promoted the interfacial adsorption behavior at the oil–water interface, enhanced emulsion stability, and reduced the creaming index of HIPEs, followed by excellent extrudability. The 3D printing resolution and surface quality strongly depend on the C-PC content, and a concentration of 3.4% provides the best results regarding the compact intermolecular network system, the solid interfacial membrane, the improved mechanical strength, and the outstanding thixotropic properties [90]. Water-in-oil-in-water (W/O/W) double emulsions, composed of proteins and polysaccharides, were exploited to simultaneously and efficiently encapsulate hydrophobic astaxanthin and hydrophilic C-PC. The gelatin solution was used as an internal aqueous phase and soybean oil containing polyglycerol polyricinoleates was the oil phase, while sodium caseinate solution was utilized as an external water phase. The addition of gellan gum made the emulsion stable during the 30 days, and it can control the release of nutrients in the simulated digestion. Like alginate beads, W/O/W double emulsions were stable in a highly acidic stomach environment, with a strong protective effect on C-PC. In contrast, a weakly alkaline intestinal environment induced the release of nutrients [91]. Complexation with polysaccharides could significantly influence the digestibility and bioavailability of PBPs. In terms of this, the C-PC–chitosan complex stabilized emulsion showed better digestibility and oxidation stability than the free C-PC emulsion [92]. Moreover, besides alginate beads, alginate-based W/O/W double emulsion substantially improves the bioaccessibility of C-PC during intestinal digestion [93]. Besides the complexation with polysaccharides, lipid-based nano-carriers, such as Nano-phytosomes, are also used to improve C-PC's gastrointestinal stability and bioavailability [94].

Recently, several studies used electrohydrodynamic processes such as electrospinning and electrospraying to encapsulate C-PC. These two processes represent facile, cost-effective, and flexible approaches that utilize electrically charged jets of polymer solution to produce fibers and particles at different scales. The significant difference between electrospinning and electrospraying is that the polymer solution concentration is much higher during the electrospinning process [95]. Incorporating C-PC from Spirulina into zein fibers via electrospinning resulted in packaging material that could protect walnut kernels from lipid oxidation. Furthermore, this approach enabled the high encapsulation efficiency of C-PC, and C-PC-loaded zein fibers were more active against pathogenic bacteria [96]. Electrospinning-induced co-encapsulation of C-PC and probiotics within the polyvinyl alcohol (PVA) nanofibers increases the antioxidant activities of the products. Moreover, the presence of C-PC within fibers improved the survivability of the probiotics [97]. C-Phycocyanin encapsulation in the PVA via electrospraying produced nanofibers with similar size and antioxidant activity properties. Thermogravimetric analysis in the same study demonstrated that encapsulation substantially increased the melting point of C-PC (from 59 °C for free protein to 216 °C for the encapsulated one) [98].

Phycobiliproteins not only promise bioactive and alternative proteins but also have significant potential to be used to produce alternative proteins in meat cultivation. Cultured meat is a promising solution to reduce land and water usage and limit pollution, but it is still costly. The critical challenge in cultivated meat science is to identify/develop fetal bovine serum (FBS) alternatives as growth supplements. Phycobiliprotein production is eco-friendly and cruelty-free; a relatively high concentration of PBPs in algae extracts could have significant potential to be positioned as FBS alternatives. The ability of C-PC-containing extract of cyanobacteria Anabaena sp. to partially replace FBS for the cultivation of muscle cell lines such as mouse embryonic myoblast line C2C12 and quail muscle clone 7 (QM7) cells was demonstrated. The same authors fabricated a 3D cell-dense structure by culturing QM7 cells in the same extract [99]. Another strategy comprises the incorporation of C-PC from Spirulina in chitosan/cellulose-based porous nanofilm, which enabled the controlled delivery of the protein in cell media for improving myoblast proliferation in a serum-reduced environment during long-term cultivation [100]. A similar approach was utilized to incorporate C-PC and growth factors in the edible fish gelatin microsphere. This was followed by their controlled release in a cell medium during myoblast cell cultivation at reduced serum conditions [101]. Although there has been some progress in the research of PBPs as potential FBS replacement components, a substantial workload in the future is required to develop the PBP-based media for the complete replacement of the FBS for meat cultivation.

4.2. Fortification of the Commonly Used Foods with Phycobiliproteins

The research and industry activities could be more pronounced in developing novel food products based on PBPs as alternative proteins. On the other hand, numerous studies focus on fortifying commonly used food products with purified PBPs or PBPs containing algae extracts/biomass (Figure 3). Moreover, several food products (milk, ice cream, yogurt, juices, cookies, cheese, and candies) fortified with Spirulina extracts have appeared in the market in the last few years [102]. Table 3 summarizes the most significant findings about the effects of PBPs on fortified food in terms of texture, rheological properties, sensory acceptability, acidity, antioxidant activities, color stability, etc. It is important to emphasize that PBPs could have dual roles when added to commonly used foods: as food additives and the nutritive role as food ingredients. Although these two roles of PBPs are highly interconnected, PBPs at lower concentrations approach their "additive" role, while at higher concentrations, the nutrient aspect of PBPs becomes more pronounced.

As seen in Table 3, most food products are fortified with Spirulina biomass or with the C-PC as the primary protein in Spirulina. Only a few studies are exploring the potential of red algae PBPs for coloring and food fortification. Considering the substantial annual production of red macroalgae *Gracilaria* spp. and *Poprhyra* spp. (around six million tons per year), fortifying food products with the PBPs from this source is promising [19]. It could be sustainable due to its much higher annual productivity than Spirulina, whose annual production is around tens of thousands of tons yearly.

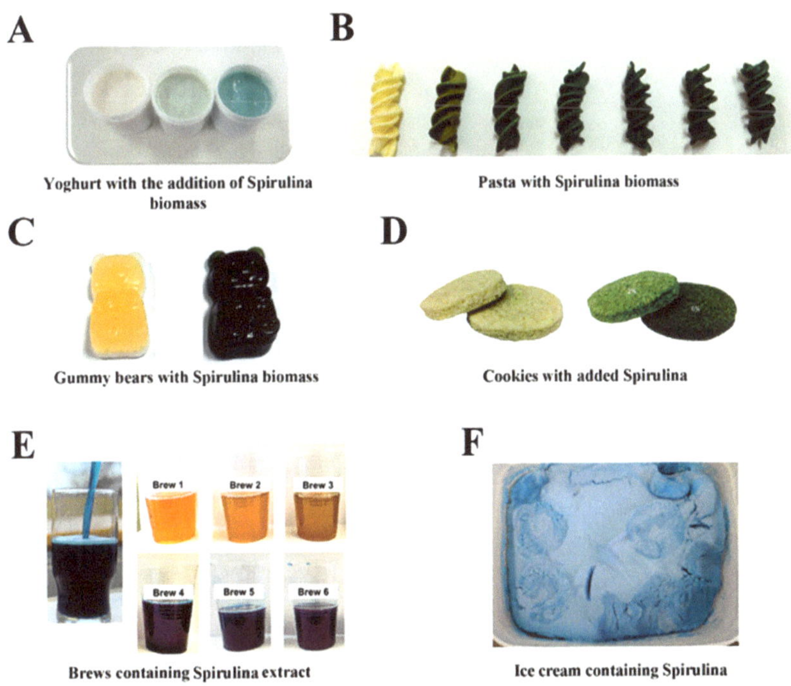

Figure 3. Selected examples of food fortification with the addition of Spirulina biomass or crude extract: (**A**) yogurt [103]; (**B**) fusilli pasta [104]; (**C**) gummy bears [105]; (**D**) wheat flour cookies [106]; (**E**) manufactured beers [107]; and (**F**) ice cream [108].

The development of PBP-fortified food comprises the selection of optimal concentration to improve food texture, rheological properties, nutritional value, antioxidant activity, and color, but without hampering other sensory attributes and consumer acceptance. Adding PBPs or algae biomass improves fortified food products' texture, rheological properties, and antioxidant capacity. Fortified products have higher protein content and improved emulsifying properties (Table 3).

Due to the fishy flavor of algae and cyanobacteria, one of the significant challenges to positioning PBPs from algae as food fortifiers is improving this sensory characteristic and, therefore, consumer acceptance [109]. Lower concentrations of C-PC or Spirulina biomass (up to 1%; w/w) have higher consumer acceptance than their higher content, with some exceptions (Table 3). Several strategies were employed to remove or reduce an unpleasant odor from algal biomass, such as adding sweeteners or aromas, which was covered in a recent review paper [110]. Another approach to avoid Spirulina's unpleasant characteristic flavor is properly combining spices in the enriched food [111]. The type of Spirulina-enriched food also determines consumer acceptance; it has been shown that consumers from France, Germany, and the Netherlands have higher preferences for enriched pasta compared to enriched sushi and jerky [112]. Adding flavors to the Spirulina-enriched pasta, such as lemon-basil, improves consumer acceptance [113]. The response surface methodology was successfully implemented to optimize Spirulina-enriched soy yogurt's ingredient formulation, processing parameters, and sensory parameters [114]. By optimizing the high moisture extrusion technique, it was possible to substitute soy partly with Spirulina biomass, producing firm and fibrous soy-based meat alternative products with a decent algae flavor [112]. This technique also shows promise for Spirulina extracts since mixing with lupin proteins can create meat analogs with improved physicochemical and nutritional properties [115]. Microencapsulation of Spirulina biomass in maltodextrin and Arabic gum could mask the taste of seaweed and seaweed aroma in the ice cream [116], while microencapsulation of Spirulina in alginate spheres resulted in higher consumer acceptability [117].

Table 3. Overview of using natural sources of vivid-colored phycobiliproteins in the food industry to fortify the commonly used food products.

Protein Sample	Concentration (w/w)	Food Type	Major Food Product Characteristics	Reference
Phycoerythrin-rich water extract (*Porphyridium cruentum*)	0.00015–0.00029%	Commercial beverages (e.g., gin and wine)	The pink color was stable during 11 days of storage; well accepted by a semi-trained panelist	[26]
C-PC from Spirulina (*Arthrospira platensis*)	0.025%	Ice cream	Stable color during the 6 months; increased antioxidant activity after in vitro simulated digestion	[108]
PBPs from cyanobacteria *Nostoc* sp.	0.03–0.14%	Skim milk	Satisfactory sensory characteristics	[118]
Encapsulated R-phycoerythrin (*Kappaphycus alvarezii*)	0.1%	Ice cream	Better rheology; pink color intensity increased during 90 days of storage; enhanced antioxidant activity	[119]
C-PC from Spirulina (*Arthrospira platensis*)	0.1–0.2%	Ice cream	Smoother and softer texture; sugar (25%) and fat (50%) content reduction; no significant influence on consumer acceptance	[120]
Porphyridium cruentum spray-dried biomass	0.1–0.3%	Ice cream	Increased consistency index	[121]

Table 3. *Cont.*

Protein Sample	Concentration (*w*/*w*)	Food Type	Major Food Product Characteristics	Reference
C-Phycocyanin extract	0.18–0.32%	Soft beverage	Improved product's antioxidant activity; good sensorial attributes	[122]
Spirulina (*Arthrospira platensis*) biomass	0.25%	Craft beer	Increased total polyphenols, tannins, and antioxidant power; cytoprotective properties towards the oxidative damage	[123]
Spirulina (*Arthrospira platensis*) powder	0.25–1%	Yogurt	Better water holding capacity and lower whey syneresis (28 days of storage); improved antioxidant activity; lower firmness but better elasticity; acceleration of the end of fermentation; acceptable sensory characteristics only at 0.25%	[124]
C-Phycocyanin	0.3%	Biscuit	High oxidative stability during 30 days of storage; satisfactory all main sensory characteristics (e.g., odor, flavor, texture, and overall acceptability)	[125]
C-PC from Spirulina (*Arthrospira platensis*)	0.3–0.4%	Ice cream	Emulsifying and stabilizing activity; lower consumer acceptance	[126]
Spirulina (*Arthrospira platensis*) powder	0.4–1.2%	Bread	Higher moisture content; lower hardness; highest consumer acceptance with 0.8%; higher antioxidant activity	[127]
Spirulina (*Arthrospira platensis*) microencapsulated in alginate	0.5%	Yogurt	Improved viscosity stability during storage; better consumer acceptance	[103]
Spirulina (*Arthrospira platensis*) powder	0.5–1.5%	Feta-type cheese	A higher number of lactic acid bacteria after 60 days of storage; softer texture and sensory characteristics (at 0.5 and 1%)	[128]
C-PC from Spirulina (*Arthrospira platensis*)	0.5–2%	Cow's milk	Increased solid non-fat content; enhanced antioxidant activity; improved sensory characteristics	[129]
Spirulina maxima biomass	0.5–2%	Pasta	The color was relatively stable after cooking, with increased firmness. Higher consumer acceptance scores	[130]
Spirulina (*Arthrospira platensis*) powder	0.5–3%	Processed cheese	Decrease in adhesiveness, cohesiveness, springiness, chewiness; increase in hardness and gumminess; deterioration in the overall sensory acceptability	[131]
Spirulina (*Arthrospira platensis*) powder	0.63–2.5%	Fresh noodles	Increased antioxidant capacity; increased hardness, cohesiveness, springiness, gumminess, and chewiness; the highest consumer acceptance with 1.25%	[132]
Spirulina (*Arthrospira platensis*) powder	1%	Yogurt	Higher antioxidant activity and number of lactic acid bacteria; higher water holding capacity and viscosity; lower syneresis; decreased consumer acceptance	[133]
Spirulina (*Arthrospira platensis*) biomass and wheat germ	1% (both)	Pear–cantaloupe-based beverage	Increased antioxidant capacity, total phenol, and flavonoid content; good organoleptic score	[134]
Spirulina (*Arthrospira platensis*) powder	1.5–3.5%	Yogurt spread	Increased viscosity and spreadability; lower consumer acceptance with a higher Spirulina concentration	[135]
Spirulina (*Arthrospira platensis*) powder	1.6%	Baguette bread	Decreased hardness and gumminess; lower sensory score	[136]

Table 3. *Cont.*

Protein Sample	Concentration (*w*/*w*)	Food Type	Major Food Product Characteristics	Reference
Spirulina (*Arthrospira platensis*) powder	1–15%	Gluten-free fresh pasta	Higher antioxidant activity, without affecting product cooking and texture quality properties; a favorable sensory evaluation	[137]
Spirulina (*Arthrospira platensis*) biomass	1–2%	Cookie	Harder and darker product with increased protein content; questionable sensory quality	[138]
Spirulina (*Arthrospira platensis*) extract	1–5%	Chinese-style pork-sausage	Small changes in pH; inhibition of lipid oxidation; 2.5 and 5% retarded the decrease in sensory acceptability (storage at 4 °C)	[139]
Spirulina (*Arthrospira platensis*) biomass	2 or 6%	Cookie	Color and texture stability over 8 weeks; higher protein and total phenolic content and in vitro antioxidant capacity; without in vitro digestibility changes	[106]
Spirulina (*Arthrospira platensis*) biomass	2%	3D-printed cookie dough	All formulations were suitable for extrusion and microbiologically stable; stable texture after 30 days of storage; improved antioxidant properties and color stability after extract encapsulation in alginate microbeads	[140]
Spirulina (*Arthrospira platensis*) powder	2.5–10%	Pasta	Increased rheological parameters, color, and cooking quality; decreased dough stability; sensory acceptable up to 5%	[141]
Nano-liposomes containing PBPs from *Gracilaria gracilis*	2.5–5%	Carp burger	Lower oxidative spoilage and microbial deterioration; no significant loss of overall consumer acceptability (18 days of refrigerated storage)	[142]
Spirulina (*Arthrospira platensis* F&M-C256) biomass	2–10%	Sourdough bread	Higher antioxidant activity; highest consumer acceptance with 2%	[143]
Spirulina (*Arthrospira platensis*) biomass	2–15%	Pasta	Lower firmness, cut force, and consistency; higher stickiness; highest consumer acceptance for 12.5%	[104]
C-PC from Spirulina (*Arthrospira platensis*)	2–8%	Yogurt	Decreased syneresis; increased firmness and viscosity; higher pH and color stability; no pathogen growth during 21 days of storage; overall acceptability not affected at 4%	[144]
Microencapsulated Spirulina (*Arthrospira platensis*) in alginate	3%	Pasta	Protection of antioxidant potential; higher firmness; acceptable sensory characteristics	[117]
Spirulina (*Arthrospira platensis*) powder	4–6.5%	Dry noodles	Lower cooking loss; higher elongation and tensile strength; highest consumer acceptance with 6%	[145]
Spirulina (*Arthrospira platensis*) powder	5%	Beer	Slightly altered fermentation parameters; typical beer-like product character; odor and taste alteration compromising the consumer acceptance	[107]
Microencapsulated *Spirulina* sp. LEB-18 in maltodextrin and soy lecithin	5–8.75%	Chocolate milk	Increased antioxidant activity; good suspension stability and low hygroscopicity; questionable consumer acceptance.	[146]

4.3. Challenges for Proper Phycobiliproteins Utilization as Alternative Proteins

Several challenges must be addressed and overcome to enable a broader utilization of PBPs as natural colorants and alternative food proteins, mainly if the entire biomass or raw extracts are to be used. These can be divided into issues considering biomass cultivation, safety concerns, and bioavailability of PBPs, as well as their quantity and purity. Connected with these are the economic aspects of the high prices of PBPs.

4.3.1. Cultivation of Cyanobacteria/Algae as Sources of Phycobiliproteins

Some considerations regarding the cultivation risks of cyanobacteria/algae also need to be addressed. They include contamination risks and adverse effects on biodiversity.

Edible seaweed may contain several contaminants, including heavy metals, elevated iodine content, anti-nutritional factors, radioactive isotopes, ammonium, dioxins, and pesticides. However, the safety of seaweed largely depends on the cultivation environment [17,147]. Heavy metal contamination represents the highest risk factor. Biomasses of cyanobacteria and microalgae are known to assimilate toxic heavy metals during their growth due to the presence of alginic acid, proteins, and peptidoglycans present in the extracellular matrix [17,148]. Commercially available seaweed samples were shown to contain As, Cd, and Pb [149]. Cyanobacterial species like *Microcystis aeruginosa* can result in biomass having toxic microcystins [150], which is a significant issue if larger quantities of PBPs are to be used as alternative protein sources. Therefore, it is imperative to implement proper quality control measures that prevent the production of toxic biomass for human nutrition.

Another concern that should be addressed is the potential allergenic properties of cyanobacteria and algae as sources of phycobiliproteins. However, there are limited findings regarding the allergenic properties of PBP-producing cyanobacteria, and only a few case studies about Spirulina and C-PC allergenicity were reported and classified as anaphylaxis [151]. Furthermore, studies addressing other health issues of PBPs in animals and humans are quite scarce. Until now, it was demonstrated that the C-PC from *Galdieria sulphuraria* [152] and Spirulina [153] did not induce any toxic effects in rats and humans, respectively.

Growing particular algae species in large quantities can negatively affect marine ecosystems [154], as they can compete with other species. There is a risk for extreme proliferation and invasion of non-native species when conditions for their overgrowth are met. For example, in the case with China in 2008, a massive green tide occurred, covering about 600 km^2 along the coast of Qingdao [155], which is one of the reasons why algae cultivation needs to be strictly controlled. While macroalgae may enhance biodiversity [156], their subsequent harvest, if not properly managed, may damage this habitation and thus negatively affect other species living there [157]. Marine organisms that live below the level of seaweed and are dependent on sunlight can also be negatively affected by mass production of algae due to lower irradiance [154].

4.3.2. Isolation and Purification of Phycobiliproteins from Cyanobacterial/Algal Biomass

Once biomass that passes safety regulations for human consumption is obtained, the next obstacle is getting PBPs in sufficient quantity and quality. The purity index of PBPs is expressed as the ratio of absorbance at wavelengths of maximal adsorption in the visible part of spectra and absorbance at 280 nm. In the case of C-PC, for example, food-grade protein has an $A_{620\ nm}/A_{280\ nm}$ ratio of 0.7, reagent grade has a ratio of 3.9, and analytical grade has a ratio over 4 [158]. So far, several review papers and book chapters have already covered many published papers about the isolation and purification of PBPs from many algal sources [1,9,33,159,160]. After obtaining protein extracts by simple solvent extraction from biomass or additional treatments, including sonication, nitrogen freezing, lyophilization, or addition of enzyme(s), the next step is the purification of desired PBPs. While standard chromatography techniques give highly purified PBPs, the yield is usually meager. Additionally, they are expensive and complicated to implement for broader usage of PBPs as alternative food proteins. Therefore, other cheaper methods that can be easily scaled up to obtain sufficient quantity and quality PBPs are required. Promising methods are aqueous two-phase separations [161,162], membrane chromatography/technology [163,164], expanded-bed adsorption chromatography [165], and ultrafiltration [166,167]. All of them can yield more and provide sufficient purity of PBPs for food applications.

4.3.3. Economic Aspects of Phycobiliproteins Production

One of the significant obstacles in positioning PBPs as competitive alternative proteins is their high price, mainly due to the costly cultivation of PBP-containing cyanobacteria and algae. The price of Spirulina is usually in the range of EUR 30–70 per kg in Europe [168], but in the Asian market, the Spirulina prices could be much lower (EUR 5 per kg or higher). The price of food-grade C-PC also exhibits significant variations on the market. Generally, it is around EUR 100 per kg [169], which is still costly to establish as an alternative protein. The second significant challenge, closely related to the high C-PC price, is the relatively low production of Spirulina, which is around 50 thousand tons per year according to the report of the Food and Agriculture Organization of the United Nations (FAO) for the 2019 year. This amount may satisfy the market in the context of C-PC as food colorants, but translating PBPs to alternative proteins requires much higher productivity. Therefore, the increase in the Spirulina cultivation fields and the reduction in the price of cultivation are prerequisites for positioning C-PC as a competitive alternative protein. The cheap alternative culture media for growing Spirulina using industrial and processing wastes could be a promising approach to decrease the cultivation costs [170]. Cultivating *Spirulina maxima* in wastewater from the demineralization of cheese whey would lower production costs by 50% [171]. While utilizing the concept of the circulatory economy could significantly reduce the price of the final products, some challenges, such as the uncertainty of nutrient content and toxicity, should be addressed regarding Spirulina cultivation in wastewater [172]. The second, circulatory economy-based approach utilizes the CO_2 from flue gas of biomass power plants and media recycling to reduce the cost of nutrients for *Spirulina maxima* cultivation by up to 42% [173]. Moreover, the exploitation of the biorefinery approach, which comprises the use of residual biomass after C-PC and high-value metabolites extraction as a potential source for the production of low-value, high-volume biofuels, could also be a promising approach for Spirulina and C-PC price reduction [174].

The most economically critical red macroalgae, such as *Porphyra* spp. and *Gracilaria* spp., have substantially higher annual production than Spirulina, and over 5 million tons of these algae are produced annually (FAO report). Although they have nearly one order of magnitude lower PBP content, their high annual production can be a source of alternative proteins [17]. Despite the high yearly production, the price of *Porphyra* spp. is still high, ranging from around EUR 100 per kg in the European market. Meanwhile, in the Asian market, the price starts at around EUR 5 per kg. The *Gracilaria* spp. has a similar price on the European market [168], while the prices on the Asian market may decrease to EUR 1 per kg for high-quantity purchases. However, these prices are still higher than those of the plant protein source [17]. On the other hand, the vivid and strong bioactive effects of PBPs could receive higher consumer acceptance compared to other sources of alternative proteins. There are no food-grade PBPs from red algae on the market; they can only be purchased as expensive research reagents for fluorescent labeling.

Contrary to C-PC from Spirulina, PBPs from red macroalgae are more challenging to extract, representing the obstacles in delivering them to the food market as colorants and alternative proteins [17]. Therefore, future studies are required to improve the extraction yields of PBPs (and other proteins) in cost-effective and food-compatible manners from red macroalgae and position them in the food market. Moreover, significant efforts could be made to improve the flavor characteristics of red algae extracts.

5. Conclusions and Future Perspectives

To date, significant efforts and progress have been made in the context of PBP purification and characterization. C-PC is currently the best-studied PBP, a natural and bioactive colorant for various foods and beverages. However, enhancement of its application potential in terms of availability on the market and competitiveness requires additional steps: (1) increase in Spirulina's annual productivity, (2) improvement of C-PC stability using approaches compatible with food safety, and (3) minimizing or completely removing the unpleasant flavor characteristic arising from Spirulina. Considering the high abundance of

C-PC in Spirulina, future studies should also focus on developing new, C-PC-based and protein-rich foods. These products would significantly strengthen the potential of C-PC as an alternative protein. Moreover, the vivid blue color of C-PC and its substantial bioactive properties could give it a significant advantage in the alternative protein market compared to other sources of alternative proteins.

Phycobiliproteins from other sources are much less explored compared to C-PC from Spirulina. Considering the annual productivity of red macroalgae (such as *Porphyra* and *Gracilaria*) is more than two orders of magnitude higher than Spirulina's yearly yield, there is a vast potential for these algae to be valorized as the source of PBPs for food application. However, establishing the PBPs from red macroalgae as food colorants and alternative proteins in the first place requires substantial research efforts to improve their extraction yield and characterization and enhance their stability. The following steps should also include the improvement of their sensory characteristics.

Based on the presented review, there are several obstacles for the food industry to overcome if phycobiliproteins are to be used more effectively and significantly. The main goals of the food industry regarding a particular product are customer satisfaction, safety, providing product information, and the maintenance of commercial viability. Developing high-quality (in terms of both nutritional and sensory characteristics, but also in terms of their safety) PBP food products, either as novel foods or as fortification in existing food products, would significantly improve consumer acceptance and satisfaction. This achievement will trigger the high demand for PBPs in the food market, which could be a motivation for the higher productivity of cyanobacteria and algae as sources of PBPs. Implementing the concept of a circulatory economy in the production of cyanobacteria and algae could enhance their productivity and decrease the cultivation price, creating a stronger position for PBPs as alternative proteins in the competitive food market by maintaining the commercial viability of PBP-based products.

Author Contributions: Conceptualization, S.M. and N.G.; methodology, S.M. and N.G.; software, L.V.; validation, M.N. and N.G.; formal analysis, S.M., N.G. and M.N.; data curation, S.M., N.G. and M.N.; writing—original draft preparation, S.M., N.G. and L.V.; writing—review and editing, M.N.; visualization, L.V. and N.G.; supervision, M.N.; project administration, S.M.; funding acquisition, S.M. All authors have read and agreed to the published version of the manuscript.

Funding: This study was supported by funding from (1) The Alliance of International Science Organizations (ANSO), Project No. ANSO-CR-PP-2021-01, (2) the Good Food Institute (GFI), and the Field Catalyst Grant Program 2022 (Alg2Meat project), and (3) the Ministry of Science, Technological Development and Innovation of the Republic of Serbia, contract numbers: 451-03-66/2024-03/200168 and 451-03-66/2024-03/200026.

Conflicts of Interest: The authors declare no conflicts of interest.

References

1. Dagnino-Leone, J.; Figueroa, C.P.; Castañeda, M.L.; Youlton, A.D.; Vallejos-Almirall, A.; Agurto-Muñoz, A.; Pavón Pérez, J.; Agurto-Muñoz, C. Phycobiliproteins: Structural aspects, functional characteristics, and biotechnological perspectives. *Comput. Struct. Biotechnol. J.* **2022**, *20*, 1506–1527. [CrossRef]
2. Stadnichuk, I.N.; Tropin, I.V. Phycobiliproteins: Structure, functions and biotechnological applications. *Appl. Biochem. Microbiol.* **2017**, *53*, 1–10. [CrossRef]
3. Brejc, K.; Ficner, R.; Huber, R.; Steinbacher, S. Isolation, crystallization, crystal structure analysis and refinement of allophycocyanin from the cyanobacterium *Spirulina platensis* at 2.3 Å resolution. *J. Mol. Biol.* **1995**, *249*, 424–440. [CrossRef] [PubMed]
4. Glazer, A.N. Light guides. *J. Biol. Chem.* **1989**, *264*, 1–4. [CrossRef] [PubMed]
5. Scheer, H.; Zhao, K.-H. Biliprotein maturation: The chromophore attachment. *Mol. Microbiol.* **2008**, *68*, 263–276. [CrossRef] [PubMed]
6. Adir, N.; Dobrovetsky, Y.; Lerner, N. Structure of c-phycocyanin from the thermophilic cyanobacterium *Synechococcus vulcanus* at 2.5 Å: Structural implications for thermal stability in phycobilisome assembly 1 1Edited by R. Huber. *J. Mol. Biol.* **2001**, *313*, 71–81. [CrossRef]
7. Kannaujiya, V.K.; Kumar, D.; Pathak, J.; Sinha, R.P. Phycobiliproteins and Their Commercial Significance. In *Cyanobacteria*; Elsevier: Amsterdam, The Netherlands, 2019; pp. 207–216.

8. Ma, J.; Hu, J.; Sha, X.; Meng, D.; Yang, R. Phycobiliproteins, the pigment-protein complex form of natural food colorants and bioactive ingredients. *Crit. Rev. Food Sci. Nutr.* **2024**, *64*, 2999–3017. [CrossRef] [PubMed]
9. Kovaleski, G.; Kholany, M.; Dias, L.M.S.; Correia, S.F.H.; Ferreira, R.A.S.; Coutinho, J.A.P.; Ventura, S.P.M. Extraction and purification of phycobiliproteins from algae and their applications. *Front. Chem.* **2022**, *10*, 1065355. [CrossRef]
10. Li, W.; Su, H.-N.; Pu, Y.; Chen, J.; Liu, L.-N.; Liu, Q.; Qin, S. Phycobiliproteins: Molecular structure, production, applications, and prospects. *Biotechnol. Adv.* **2019**, *37*, 340–353. [CrossRef]
11. Pagels, F.; Guedes, A.C.; Amaro, H.M.; Kijjoa, A.; Vasconcelos, V. Phycobiliproteins from cyanobacteria: Chemistry and biotechnological applications. *Biotechnol. Adv.* **2019**, *37*, 422–443. [CrossRef]
12. Tounsi, L.; Ben Hlima, H.; Hentati, F.; Hentati, O.; Derbel, H.; Michaud, P.; Abdelkafi, S. Microalgae: A Promising Source of Bioactive Phycobiliproteins. *Mar. Drugs* **2023**, *21*, 440. [CrossRef]
13. Puzorjov, A.; McCormick, A.J. Phycobiliproteins from extreme environments and their potential applications. *J. Exp. Bot.* **2020**, *71*, 3827–3842. [CrossRef] [PubMed]
14. Minic, S.L.; Stanic-Vucinic, D.; Mihailovic, J.; Krstic, M.; Nikolic, M.R.; Cirkovic Velickovic, T. Digestion by pepsin releases biologically active chromopeptides from C-phycocyanin, a blue-colored biliprotein of microalga *Spirulina*. *J. Proteomics* **2016**, *147*, 132–139. [CrossRef] [PubMed]
15. Liu, R.-Z.; Li, W.-J.; Zhang, J.-J.; Liu, Z.-Y.; Li, Y.; Liu, C.; Qin, S. The Inhibitory Effect of Phycocyanin Peptide on Pulmonary Fibrosis In Vitro. *Mar. Drugs* **2022**, *20*, 696. [CrossRef]
16. Fernandes, R.; Campos, J.; Serra, M.; Fidalgo, J.; Almeida, H.; Casas, A.; Toubarro, D.; Barros, A.I.R.N.A. Exploring the Benefits of Phycocyanin: From Spirulina Cultivation to Its Widespread Applications. *Pharmaceuticals* **2023**, *16*, 592. [CrossRef] [PubMed]
17. Rawiwan, P.; Peng, Y.; Paramayuda, I.G.P.B.; Quek, S.Y. Red seaweed: A promising alternative protein source for global food sustainability. *Trends Food Sci. Technol.* **2022**, *123*, 37–56. [CrossRef]
18. Taufikurahman, T.; Ilhamsyah, D.P.A.; Rosanti, S.; Ardiansyah, M.A. Preliminary Design of Phycocyanin Production from *Spirulina platensis* Using Anaerobically Digested Dairy Manure Wastewater. *IOP Conf. Ser. Earth Environ. Sci.* **2020**, *520*, 012007. [CrossRef]
19. FAO Global Seaweeds and Microalgae Production, 1950–2019. Available online: https://www.fao.org/3/cb4579en/cb4579en.pdf (accessed on 12 January 2024).
20. Zhang, L.; Liao, W.; Huang, Y.; Wen, Y.; Chu, Y.; Zhao, C. Global seaweed farming and processing in the past 20 years. *Food Prod. Process. Nutr.* **2022**, *4*, 23. [CrossRef]
21. Li, P.; Sheng, L.; Ye, Y.; Wang, J.; Geng, S.; Ning, D.; Sun, X. Allergenicity of alternative proteins: Research hotspots, new findings, evaluation strategies, regulatory status, and future trends: A bibliometric analysis. *Crit. Rev. Food Sci. Nutr.* **2024**, 1–12. [CrossRef]
22. Petrus, M.; Culerrier, R.; Campistron, M.; Barre, A.; Rougé, P. First case report of anaphylaxis to spirulin: Identification of phycocyanin as responsible allergen. *Allergy* **2010**, *65*, 924–925. [CrossRef]
23. Le, T.-M.; Knulst, A.C.; Röckmann, H. Anaphylaxis to Spirulina confirmed by skin prick test with ingredients of Spirulina tablets. *Food Chem. Toxicol.* **2014**, *74*, 309–310. [CrossRef] [PubMed]
24. Nemoto-Kawamura, C.; Hirahashi, T.; Nagai, T.; Yamada, H.; Katoh, T.; Hayashi, O. Phycocyanin Enhances Secretary IgA Antibody Response and Suppresses Allergic IgE Antibody Response in Mice Immunized with Antigen-Entrapped Biodegradable Microparticles. *J. Nutr. Sci. Vitaminol.* **2004**, *50*, 129–136. [CrossRef] [PubMed]
25. Liu, Q.; Wang, Y.; Cao, M.; Pan, T.; Yang, Y.; Mao, H.; Sun, L.; Liu, G. Anti-allergic activity of R-phycocyanin from *Porphyra haitanensis* in antigen-sensitized mice and mast cells. *Int. Immunopharmacol.* **2015**, *25*, 465–473. [CrossRef] [PubMed]
26. Carmona, R.; Murillo, M.C.; Lafarga, T.; Bermejo, R. Assessment of the potential of microalgae-derived phycoerythrin as a natural colorant in beverages. *J. Appl. Phycol.* **2022**, *34*, 3025–3034. [CrossRef]
27. Lins, M.; Zandonadi, R.P.; Raposo, A.; Ginani, V.C. Food waste on foodservice: An overview through the perspective of sustainable dimensions. *Foods* **2021**, *10*, 1175. [CrossRef] [PubMed]
28. Landim Neves, M.I.; Silva, E.K.; Meireles, M.A.A. Natural blue food colorants: Consumer acceptance, current alternatives, trends, challenges, and future strategies. *Trends Food Sci. Technol.* **2021**, *112*, 163–173. [CrossRef]
29. Carocho, M.; Morales, P.; Ferreira, I.C.F.R. Natural food additives: Quo vadis? *Trends Food Sci. Technol.* **2015**, *45*, 284–295. [CrossRef]
30. Spence, C. Background colour & its impact on food perception & behaviour. *Food Qual. Prefer.* **2018**, *68*, 156–166. [CrossRef]
31. Santiago-Santos, M.C.; Ponce-Noyola, T.; Olvera-Ramírez, R.; Ortega-López, J.; Cañizares-Villanueva, R.O. Extraction and purification of phycocyanin from *Calothrix* sp. *Process Biochem.* **2004**, *39*, 2047–2052. [CrossRef]
32. FDA Summary of Color Additives for Use in the United States in Foods, Drugs, Cosmetics, and Medical Devices. Available online: https://www.fda.gov/industry/color-additive-inventories/summary-color-additives-use-united-states-foods-drugs-cosmetics-and-medical-devices#table1A (accessed on 12 November 2023).
33. Adjali, A.; Clarot, I.; Chen, Z.; Marchioni, E.; Boudier, A. Physicochemical degradation of phycocyanin and means to improve its stability: A short review. *J. Pharm. Anal.* **2022**, *12*, 406–414. [CrossRef]
34. Patil, G.; Raghavarao, K.S.M.S. Aqueous two phase extraction for purification of C-phycocyanin. *Biochem. Eng. J.* **2007**, *34*, 156–164. [CrossRef]
35. García, A.B.; Longo, E.; Bermejo, R. The application of a phycocyanin extract obtained from *Arthrospira platensis* as a blue natural colorant in beverages. *J. Appl. Phycol.* **2021**, *33*, 3059–3070. [CrossRef]

36. Jespersen, L.; Strømdahl, L.D.; Olsen, K.; Skibsted, L.H. Heat and light stability of three natural blue colorants for use in confectionery and beverages. *Eur. Food Res. Technol.* **2005**, *220*, 261–266. [CrossRef]
37. Liu, L.-N.; Su, H.-N.; Yan, S.-G.; Shao, S.-M.; Xie, B.-B.; Chen, X.-L.; Zhang, X.-Y.; Zhou, B.-C.; Zhang, Y.-Z. Probing the pH sensitivity of R-phycoerythrin: Investigations of active conformational and functional variation. *Biochim. Biophys. Acta-Bioenerg.* **2009**, *1787*, 939–946. [CrossRef] [PubMed]
38. Munier, M.; Jubeau, S.; Wijaya, A.; Morançais, M.; Dumay, J.; Marchal, L.; Jaouen, P.; Fleurence, J. Physicochemical factors affecting the stability of two pigments: R-phycoerythrin of *Grateloupia turuturu* and B-phycoerythrin of *Porphyridium cruentum*. *Food Chem.* **2014**, *150*, 400–407. [CrossRef] [PubMed]
39. Simovic, A.; Combet, S.; Cirkovic Velickovic, T.; Nikolic, M.; Minic, S. Probing the stability of the food colourant R-phycoerythrin from dried Nori flakes. *Food Chem.* **2022**, *374*, 131780. [CrossRef] [PubMed]
40. Veličković, L.; Simović, A.; Gligorijević, N.; Thureau, A.; Obradović, M.; Vasović, T.; Sotiroudis, G.; Zoumpanioti, M.; Brûlet, A.; Ćirković Veličković, T.; et al. Exploring and strengthening the potential of R-phycocyanin from Nori flakes as a food colourant. *Food Chem.* **2023**, *426*, 136669. [CrossRef] [PubMed]
41. Edwards, M.R.; Hauer, C.; Stack, R.F.; Eisele, L.E.; MacColl, R. Thermophilic C-phycocyanin: Effect of temperature, monomer stability, and structure. *Biochim. Biophys. Acta-Bioenerg.* **1997**, *1321*, 157–164. [CrossRef]
42. Ferraro, G.; Imbimbo, P.; Marseglia, A.; Illiano, A.; Fontanarosa, C.; Amoresano, A.; Olivieri, G.; Pollio, A.; Monti, D.M.; Merlino, A. A thermophilic C-phycocyanin with unprecedented biophysical and biochemical properties. *Int. J. Biol. Macromol.* **2020**, *150*, 38–51. [CrossRef]
43. Faieta, M.; Neri, L.; Sacchetti, G.; Di Michele, A.; Pittia, P. Role of saccharides on thermal stability of phycocyanin in aqueous solutions. *Food Res. Int.* **2020**, *132*, 109093. [CrossRef]
44. Nowruzi, B.; Konur, O.; Anvar, S.A.A. The Stability of the Phycobiliproteins in the Adverse Environmental Conditions Relevant to the Food Storage. *Food Bioprocess Technol.* **2022**, *15*, 2646–2663. [CrossRef]
45. Yuan, B.; Li, Z.; Shan, H.; Dashnyam, B.; Xu, X.; McClements, D.J.; Zhang, B.; Tan, M.; Wang, Z.; Cao, C. A review of recent strategies to improve the physical stability of phycocyanin. *Curr. Res. Food Sci.* **2022**, *5*, 2329–2337. [CrossRef] [PubMed]
46. Niccolai, A.; Venturi, M.; Galli, V.; Pini, N.; Rodolfi, L.; Biondi, N.; Granchi, L.; Tredici, M.R. Vegetable oils protect phycocyanin from thermal degradation during cooking of spirulina-based "crostini". *LWT* **2021**, *138*, 110776. [CrossRef]
47. Chaiklahan, R.; Chirasuwan, N.; Bunnag, B. Stability of phycocyanin extracted from Spirulina sp.: Influence of temperature, pH and preservatives. *Process Biochem.* **2012**, *47*, 659–664. [CrossRef]
48. Antelo, F.S.; Costa, J.A.V.; Kalil, S.J. Thermal degradation kinetics of the phycocyanin from *Spirulina platensis*. *Biochem. Eng. J.* **2008**, *41*, 43–47. [CrossRef]
49. Braga, A.R.C.; Figueira, F.D.S.; Silveira, J.T.D.; Morais, M.G.D.; Costa, J.A.V.; Kalil, S.J. Improvement of Thermal Stability of C-Phycocyanin by Nanofiber and Preservative Agents. *J. Food Process. Preserv.* **2016**, *40*, 1264–1269. [CrossRef]
50. Martelli, G.; Folli, C.; Visai, L.; Daglia, M.; Ferrari, D. Thermal stability improvement of blue colorant C-Phycocyanin from *Spirulina platensis* for food industry applications. *Process Biochem.* **2014**, *49*, 154–159. [CrossRef]
51. Zhang, Z.; Li, Y.; Abbaspourrad, A. Improvement of the colloidal stability of phycocyanin in acidified conditions using whey protein-phycocyanin interactions. *Food Hydrocoll.* **2020**, *105*, 105747. [CrossRef]
52. Bhattacharya, S.; Bhayani, K.; Ghosh, T.; Bajaj, S.; Trivedi, N.; Mishra, S. Stability of Phycobiliproteins Using Natural Preservative ε- Polylysine (ε-PL). *Ferment. Technol.* **2018**, *7*, 1000149. [CrossRef]
53. Lee, C.-W.; Bae, G.Y.; Bae, S.-H.; Suh, H.J.; Jo, K. Increased thermal stability of phycocyanin extracted from *Spirulina platensis* by cysteine addition during enzyme extraction. *Food Sci. Technol.* **2022**, *42*, e15021. [CrossRef]
54. Zhang, Y.; Zhang, L.; Hu, J.; Wang, Z.; Meng, D.; Li, H.; Zhou, Z.; Yang, R. The structural characterization and color stabilization of the pigment protein-phycoerythrin glycosylated with oligochitosan. *Food Hydrocoll.* **2023**, *136*, 108241. [CrossRef]
55. Buecker, S.; Grossmann, L.; Loeffler, M.; Leeb, E.; Weiss, J. Influence of storage temperature on the stability of heat treated phycocyanin-λ-carrageenan complexes in liquid formulations. *Green Chem.* **2022**, *24*, 4174–4185. [CrossRef]
56. Buecker, S.; Grossmann, L.; Loeffler, M.; Leeb, E.; Weiss, J. Thermal and acidic denaturation of phycocyanin from *Arthrospira platensis*: Effects of complexation with λ-carrageenan on blue color stability. *Food Chem.* **2022**, *380*, 132157. [CrossRef] [PubMed]
57. Gligorijević, N.; Jovanović, Z.; Cvijetić, I.; Šunderić, M.; Veličković, L.; Katrlík, J.; Holazová, A.; Nikolić, M.; Minić, S. Investigation of the Potential of Selected Food-Derived Antioxidants to Bind and Stabilise the Bioactive Blue Protein C-Phycocyanin from Cyanobacteria Spirulina. *Int. J. Mol. Sci.* **2023**, *25*, 229. [CrossRef] [PubMed]
58. Gomaa, M.; Ali, S.A.; Hifney, A.F. Enhancement of phycocyanin productivity and thermostability from *Arthrospira platensis* using organic acids. *Microb. Cell Fact.* **2023**, *22*, 248. [CrossRef]
59. Huo, Y.; Hou, X.; Yu, Y.; Wen, X.; Ding, Y.; Li, Y.; Wang, Z. Improving the Thermal and Oxidative Stability of Food-Grade Phycocyanin from *Arthrospira platensis* by Addition of Saccharides and Sugar Alcohols. *Foods* **2022**, *11*, 1752. [CrossRef]
60. Li, Y.; Zhang, Z.; Abbaspourrad, A. Improved thermal stability of phycocyanin under acidic conditions by forming soluble complexes with polysaccharides. *Food Hydrocoll.* **2021**, *119*, 106852. [CrossRef]
61. Meng, D.; Zhang, L.; Wang, Q.; Zhang, Y.; Sun, Y.; Zhang, H.; Wang, Z.; Zhou, Z.; Yang, R. Self-Assembly of Phycoerythrin with Oligochitosan by Electrostatic Interaction for Stabilization of Phycoerythrin. *J. Agric. Food Chem.* **2021**, *69*, 12818–12827. [CrossRef] [PubMed]

62. Sharma, R.; Nath, P.C.; Vanitha, K.; Tiwari, O.N.; Bandyopadhyay, T.K.; Bhunia, B. Effects of different monosaccharides on thermal stability of phycobiliproteins from Oscillatoria sp. (BTA-170): Analysis of kinetics, thermodynamics, colour and antioxidant properties. *Food Biosci.* **2021**, *44*, 101354. [CrossRef]
63. Wang, F.; Yu, X.; Cui, Y.; Xu, L.; Huo, S.; Ding, Z.; Hu, Q.; Xie, W.; Xiao, H.; Zhang, D. Efficient extraction of phycobiliproteins from dry biomass of *Spirulina platensis* using sodium chloride as extraction enhancer. *Food Chem.* **2023**, *406*, 135005. [CrossRef]
64. Yang, R.; Ma, T.; Shi, L.; Wang, Q.; Zhang, L.; Zhang, F.; Wang, Z.; Zhou, Z. The formation of phycocyanin-EGCG complex for improving the color protection stability exposing to light. *Food Chem.* **2022**, *370*, 130985. [CrossRef]
65. Yoshida, C.; Murakami, M.; Niwa, A.; Takeya, M.; Osanai, T. Efficient extraction and preservation of thermotolerant phycocyanins from red alga *Cyanidioschyzon merolae*. *J. Biosci. Bioeng.* **2021**, *131*, 161–167. [CrossRef] [PubMed]
66. Zhang, S.; Zhang, Z.; Dadmohammadi, Y.; Li, Y.; Jaiswal, A.; Abbaspourrad, A. Whey protein improves the stability of C-phycocyanin in acidified conditions during light storage. *Food Chem.* **2021**, *344*, 128642. [CrossRef] [PubMed]
67. Bharmoria, P.; Correia, S.F.H.; Martins, M.; Hernández-Rodríguez, M.A.; Ventura, S.P.M.; Ferreira, R.A.S.; Carlos, L.D.; Coutinho, J.A.P. Protein Cohabitation: Improving the Photochemical Stability of R-Phycoerythrin in the Solid State. *J. Phys. Chem. Lett.* **2020**, *11*, 6249–6255. [CrossRef] [PubMed]
68. Buecker, S.; Gibis, M.; Bartmann, L.; Bussler, S.; Weiss, J. Improving the colloidal stability of pectin–phycocyanin complexes by increasing the mixing ratio. *J. Food Sci.* **2024**, *89*, 1086–1097. [CrossRef]
69. Chandralekha, A.; Prashanth, H.S.; Tavanandi, H.; Raghavarao, K.S.M.S. A novel method for double encapsulation of C-phycocyanin using aqueous two phase systems for extension of shelf life. *J. Food Sci. Technol.* **2021**, *58*, 1750–1763. [CrossRef] [PubMed]
70. Faieta, M.; Neri, L.; Di Michele, A.; Di Mattia, C.D.; Pittia, P. High hydrostatic pressure treatment of *Arthrospira (Spirulina) platensis* extracts and the baroprotective effect of sugars on phycobiliproteins. *Innov. Food Sci. Emerg. Technol.* **2021**, *70*, 102693. [CrossRef]
71. İlter, I.; Koç, M.; Demirel, Z.; Conk Dalay, M.; Kaymak Ertekin, F. Improving the stability of phycocyanin by spray dried microencapsulation. *J. Food Process. Preserv.* **2021**, *45*, e15646. [CrossRef]
72. Li, Y.; Zhang, Z.; Abbaspourrad, A. Improved pH stability, heat stability, and functionality of phycocyanin after PEGylation. *Int. J. Biol. Macromol.* **2022**, *222*, 1758–1767. [CrossRef]
73. Micó-Vicent, B.; Perales Romero, E.; Bermejo, R.; Jordán-Núñez, J.; Viqueira, V.; Pérez, J. Using Laminar Nanoclays for Phycocyanin and Phycoerythrin Stabilization as New Natural Hybrid Pigments from Microalgae Extraction. *Appl. Sci.* **2021**, *11*, 11992. [CrossRef]
74. Munawaroh, H.S.H.; Gumilar, G.G.; Alifia, C.R.; Marthania, M.; Stellasary, B.; Yuliani, G.; Wulandari, A.P.; Kurniawan, I.; Hidayat, R.; Ningrum, A.; et al. Photostabilization of phycocyanin from *Spirulina platensis* modified by formaldehyde. *Process Biochem.* **2020**, *94*, 297–304. [CrossRef]
75. Pan-utai, W.; Iamtham, S. Enhanced Microencapsulation of C-Phycocyanin from Arthrospira by Freeze-Drying with Different Wall Materials. *Food Technol. Biotechnol.* **2020**, *58*, 423–432. [CrossRef] [PubMed]
76. Patel, A.S.; Lakshmibalasubramaniam, S.; Nayak, B.; Tripp, C.; Kar, A.; Sappati, P.K. Improved stability of phycobiliprotein within liposome stabilized by polyethylene glycol adsorbed cellulose nanocrystals. *Int. J. Biol. Macromol.* **2020**, *163*, 209–218. [CrossRef] [PubMed]
77. Zhang, Z.; Cho, S.; Dadmohammadi, Y.; Li, Y.; Abbaspourrad, A. Improvement of the storage stability of C-phycocyanin in beverages by high-pressure processing. *Food Hydrocoll.* **2021**, *110*, 106055. [CrossRef]
78. Azari, A.; Ghaboos, S.H.H.; Moghadam, V.E.; Jafari, S.M. Influence of chitosan coating on the physicochemical and antioxidant properties of phycocyanin-loaded nanoliposomes. *Algal Res.* **2023**, *72*, 103120. [CrossRef]
79. Gustiningtyas, A.; Setyaningsih, I.; Hardiningtyas, S.D.; Susila, A.A.R. Improvement stability of phycocyanin from *Spirulina platensis* encapsulated by water soluble chitosan nanoparticles. *IOP Conf. Ser. Earth Environ. Sci.* **2020**, *414*, 012005. [CrossRef]
80. Cao, J.; Wang, J.; Wang, S.; Xu, X. Porphyra Species: A Mini-Review of Its Pharmacological and Nutritional Properties. *J. Med. Food* **2016**, *19*, 111–119. [CrossRef] [PubMed]
81. Silveira, S.T.; Burkert, J.F.M.; Costa, J.A.V.; Burkert, C.A.V.; Kalil, S.J. Optimization of phycocyanin extraction from *Spirulina platensis* using factorial design. *Bioresour. Technol.* **2007**, *98*, 1629–1634. [CrossRef]
82. Tibbettes, S.M.; MacPherson, M.J.; Park, K.C.; Melanson, R.J.; Patelakis, S.J.J. Composition and apparent digestibility coefficients of essential nutrients and energy of cyanobacterium meal produced from Spirulina (*Arthrospira platensis*) for freshwater-phase Atlantic salmon (*Salmo salar* L.) pre-smolts. *Algal Res.* **2023**, *70*, 103017. [CrossRef]
83. Lei, Y.; Zhao, X.; Li, D.; Wang, L.; Wang, Y. Effects of κ-Carrageenan and Guar Gum on the Rheological Properties and Microstructure of Phycocyanin Gel. *Foods* **2022**, *11*, 734. [CrossRef]
84. Qiao, B.-W.; Liu, X.-T.; Wang, C.-X.; Song, S.; Ai, C.-Q.; Fu, Y.-H. Preparation, Characterization, and Antioxidant Properties of Phycocyanin Complexes Based on Sodium Alginate and Lysozyme. *Front. Nutr.* **2022**, *9*, 890942. [CrossRef] [PubMed]
85. Alavi, N.; Golmakani, M.-T.; Hosseini, S.M.H. Fabrication and characterization of phycocyanin-alginate-pregelatinized corn starch composite gel beads: Effects of carriers on kinetic stability of phycocyanin. *Int. J. Biol. Macromol.* **2022**, *218*, 665–678. [CrossRef] [PubMed]
86. Pradeep, H.N.; Nayak, C.A. Enhanced stability of C-phycocyanin colorant by extrusion encapsulation. *J. Food Sci. Technol.* **2019**, *56*, 4526–4534. [CrossRef] [PubMed]

87. Castro-Varela, P.; Rubilar, M.; Martínez-Férez, A.; Fuentes-Ríos, D.; López-Romero, J.M.; Alarcón, C.; Abdala-Díaz, R.; Figueroa, F.L. R-phycoerythrin alginate/shellac beads by external gelation: Process optimization and the effects of gastrointestinal digestion for nutraceutical applications. *Algal Res.* **2024**, *79*, 103473. [CrossRef]
88. Lemos, P.V.F.; Opretzka, L.C.F.; Almeida, L.S.; Cardoso, L.G.; da Silva, J.B.A.; de Souza, C.O.; Villarreal, C.F.; Druzian, J.I. Preparation and characterization of C-phycocyanin coated with STMP/STPP cross-linked starches from different botanical sources. *Int. J. Biol. Macromol.* **2020**, *159*, 739–750. [CrossRef] [PubMed]
89. Wang, M.; Yin, Z.; Sun, W.; Zhong, Q.; Zhang, Y.; Zeng, M. Microalgae play a structuring role in food: Effect of *Spirulina platensis* on the rheological, gelling characteristics, and mechanical properties of soy protein isolate hydrogel. *Food Hydrocoll.* **2023**, *136*, 108244. [CrossRef]
90. Wang, H.; Ouyang, Z.; Hu, L.; Cheng, Y.; Zhu, J.; Ma, L.; Zhang, Y. Self-assembly of gelatin and phycocyanin for stabilizing thixotropic emulsions and its effect on 3D printing. *Food Chem.* **2022**, *397*, 133725. [CrossRef] [PubMed]
91. Yu, H.; Wang, H.; Su, W.; Song, Y.; Zaky, A.A.; Abd El-Aty, A.M.; Tan, M. Co-delivery of hydrophobic astaxanthin and hydrophilic phycocyanin by a pH-sensitive water-in-oil-in-water double emulsion-filled gellan gum hydrogel. *Food Hydrocoll.* **2022**, *131*, 107810. [CrossRef]
92. Zhong, Y.; Sun, S.; Dai, T.; Zhang, H.; Wu, J.; Gong, E.S. Phycocyanin-chitosan complex stabilized emulsion: Preparation, characteristics, digestibility, and stability. *Int. J. Biol. Macromol.* **2024**, *260*, 129253. [CrossRef]
93. Teixé-Roig, J.; Oms-Oliu, G.; Ballesté-Muñoz, S.; Odriozola-Serrano, I.; Martín-Belloso, O. Encapsulation and controlled release of phycocyanin during the in vitro digestion using polysaccharide-added double emulsions (W1/O/W2). *Food Struct.* **2022**, *31*, 100249. [CrossRef]
94. Sahin, O.I.; Dundar, A.N.; Ozdemir, S.; Uzuner, K.; Parlak, M.E.; Dagdelen, A.F.; Saricaoglu, F.T. Nanophytosomes as a protection system to improve the gastrointestinal stability and bioavailability of phycocyanin. *Food Biosci.* **2022**, *50*, 102052. [CrossRef]
95. Anu Bhushani, J.; Anandharamakrishnan, C. Electrospinning and electrospraying techniques: Potential food based applications. *Trends Food Sci. Technol.* **2014**, *38*, 21–33. [CrossRef]
96. Golmakani, M.-T.; Kiani, F.; Hajjari, M.M.; Sharif, N.; Fazaeli, M.; Hosseini, S.M.H. Electrospun zein incorporating phycocyanin and Spirulina extract: Fabrication, characterization, and potential application. *LWT* **2023**, *188*, 115408. [CrossRef]
97. Zhang, Z.; Su, W.; Li, Y.; Zhang, S.; Liang, H.; Ji, C.; Lin, X. High-speed electrospinning of phycocyanin and probiotics complex nanofibrous with higher probiotic activity and antioxidation. *Food Res. Int.* **2023**, *167*, 112715. [CrossRef] [PubMed]
98. Schmatz, D.A.; da Silveira Mastrantonio, D.J.; Vieira Costa, J.A.; de Morais, M.G. Encapsulation of phycocyanin by electrospraying: A promising approach for the protection of sensitive compounds. *Food Bioprod. Process.* **2020**, *119*, 206–215. [CrossRef]
99. Ghosh, J.; Haraguchi, Y.; Asahi, T.; Nakao, Y.; Shimizu, T. Muscle cell proliferation using water-soluble extract from nitrogen-fixing cyanobacteria *Anabaena* sp. PCC 7120 for sustainable cultured meat production. *Biochem. Biophys. Res. Commun.* **2023**, *682*, 316–324. [CrossRef] [PubMed]
100. Park, S.; Jung, S.; Heo, J.; Koh, W.-G.; Lee, S.; Hong, J. Chitosan/Cellulose-Based Porous Nanofilm Delivering C-Phycocyanin: A Novel Platform for the Production of Cost-Effective Cultured Meat. *ACS Appl. Mater. Interfaces* **2021**, *13*, 32193–32204. [CrossRef] [PubMed]
101. Park, S.; Jung, S.; Choi, M.; Lee, M.; Choi, B.; Koh, W.-G.; Lee, S.; Hong, J. Gelatin MAGIC powder as nutrient-delivering 3D spacer for growing cell sheets into cost-effective cultured meat. *Biomaterials* **2021**, *278*, 121155. [CrossRef] [PubMed]
102. Lafarga, T. Effect of microalgal biomass incorporation into foods: Nutritional and sensorial attributes of the end products. *Algal Res.* **2019**, *41*, 101566. [CrossRef]
103. Nourmohammadi, N.; Soleimanian-Zad, S.; Shekarchizadeh, H. Effect of Spirulina (*Arthrospira platensis*) microencapsulated in alginate and whey protein concentrate addition on physicochemical and organoleptic properties of functional stirred yogurt. *J. Sci. Food Agric.* **2020**, *100*, 5260–5268. [CrossRef]
104. Koli, D.; Rudra, S.; Bhowmik, A.; Pabbi, S. Nutritional, Functional, Textural and Sensory Evaluation of Spirulina Enriched Green Pasta: A Potential Dietary and Health Supplement. *Foods* **2022**, *11*, 979. [CrossRef] [PubMed]
105. Bartkiene, E.; Tolpeznikaite, E.; Klupsaite, D.; Starkute, V.; Bartkevics, V.; Skrastina, A.; Pavlenko, R.; Mockus, E.; Lele, V.; Batkeviciute, G.; et al. Bio-Converted Spirulina for Nutraceutical Chewing Candy Formulations Rich in L-Glutamic and Gamma-Aminobutyric Acids. *Microorganisms* **2023**, *11*, 441. [CrossRef] [PubMed]
106. Batista, A.P.; Niccolai, A.; Fradinho, P.; Fragoso, S.; Bursic, I.; Rodolfi, L.; Biondi, N.; Tredici, M.R.; Sousa, I.; Raymundo, A. Microalgae biomass as an alternative ingredient in cookies: Sensory, physical and chemical properties, antioxidant activity and in vitro digestibility. *Algal Res.* **2017**, *26*, 161–171. [CrossRef]
107. Beisler, N.; Sandmann, M. Integration of *Arthrospira platensis* (spirulina) into the brewing process to develop new beers with unique sensory properties. *Front. Sustain. Food Syst.* **2022**, *6*, 918772. [CrossRef]
108. Campos Assumpção de Amarante, M.; Cavalcante Braga, A.R.; Sala, L.; Juliano Kalil, S. Colour stability and antioxidant activity of C-phycocyanin-added ice creams after in vitro digestion. *Food Res. Int.* **2020**, *137*, 109602. [CrossRef] [PubMed]
109. Stanic-Vucinic, D.; Minic, S.; Nikolic, M.R.; Velickovic, T.C. Spirulina Phycobiliproteins as Food Components and Complements. In *Microalgal Biotechnology*; Jacob-Lopes, E., Zepka, L.Q., Queiroz, M.I., Eds.; InTech: Rijeka, Croatia, 2018.
110. Colonia, B.S.O.; de Melo Pereira, G.V.; de Carvalho, J.C.; Karp, S.G.; Rodrigues, C.; Soccol, V.T.; Fanka, L.S.; Soccol, C.R. Deodorization of algae biomass to overcome off-flavors and odor issues for developing new food products: Innovations, trends, and applications. *Food Chem. Adv.* **2023**, *2*, 100270. [CrossRef]

111. Almeida, L.M.R.; da Silva Cruz, L.F.; Machado, B.A.S.; Nunes, I.L.; Costa, J.A.V.; de Souza Ferreira, E.; Lemos, P.V.F.; Druzian, J.I.; de Souza, C.O. Effect of the addition of *Spirulina* sp. biomass on the development and characterization of functional food. *Algal Res.* **2021**, *58*, 102387. [CrossRef]
112. Grahl, S.; Strack, M.; Weinrich, R.; Mörlein, D. Consumer-Oriented Product Development: The Conceptualization of Novel Food Products Based on Spirulina (*Arthrospira platensis*) and Resulting Consumer Expectations. *J. Food Qual.* **2018**, *2018*, 1919482. [CrossRef]
113. Grahl, S.; Strack, M.; Mensching, A.; Mörlein, D. Alternative protein sources in Western diets: Food product development and consumer acceptance of spirulina-filled pasta. *Food Qual. Prefer.* **2020**, *84*, 103933. [CrossRef]
114. Sengupta, S.; Bhowal, J. Optimization of ingredient and processing parameter for the production of *Spirulina platensis* incorporated soy yogurt using response surface methodology. *J. Microbiol. Biotechnol. Food Sci.* **2017**, *6*, 1081–1085. [CrossRef]
115. Palanisamy, M.; Töpfl, S.; Berger, R.G.; Hertel, C. Physico-chemical and nutritional properties of meat analogues based on *Spirulina*/lupin protein mixtures. *Eur. Food Res. Technol.* **2019**, *245*, 1889–1898. [CrossRef]
116. Tiepo, C.B.V.; Gottardo, F.M.; Mortari, L.M.; Bertol, C.D.; Reinehr, C.O.; Colla, L.M. Addition of *Spirulina platensis* in handmade ice cream: Phisicochemical and sensory effects/Adição de *Spirulina platensis* em sorvete caseiro: Efeitos físico-químicos e sensoriais. *Brazilian J. Dev.* **2021**, *7*, 88106–88123. [CrossRef]
117. Zen, C.K.; Tiepo, C.B.V.; da Silva, R.V.; Reinehr, C.O.; Gutkoski, L.C.; Oro, T.; Colla, L.M. Development of functional pasta with microencapsulated *Spirulina*: Technological and sensorial effects. *J. Sci. Food Agric.* **2020**, *100*, 2018–2026. [CrossRef]
118. Galetović, A.; Seura, F.; Gallardo, V.; Graves, R.; Cortés, J.; Valdivia, C.; Núñez, J.; Tapia, C.; Neira, I.; Sanzana, S.; et al. Use of Phycobiliproteins from Atacama Cyanobacteria as Food Colorants in a Dairy Beverage Prototype. *Foods* **2020**, *9*, 244. [CrossRef] [PubMed]
119. Ganesan, A.R.; Shanmugam, M. Isolation of phycoerythrin from *Kappaphycus alvarezii*: A potential natural colourant in ice cream. *J. Appl. Phycol.* **2020**, *32*, 4221–4233. [CrossRef]
120. da Silva Faresin, L.; Devos, R.J.B.; Reinehr, C.O.; Colla, L.M. Development of ice cream with reduction of sugar and fat by the addition of inulin, *Spirulina platensis* or phycocyanin. *Int. J. Gastron. Food Sci.* **2022**, *27*, 100445. [CrossRef]
121. Durmaz, Y.; Kilicli, M.; Toker, O.S.; Konar, N.; Palabiyik, I.; Tamtürk, F. Using spray-dried microalgae in ice cream formulation as a natural colorant: Effect on physicochemical and functional properties. *Algal Res.* **2020**, *47*, 101811. [CrossRef]
122. Benchikh, Y.; Filali, A.; Rebai, S. Modeling and optimizing the phycocyanins extraction from *Arthrospira platensis* (*Spirulina*) algae and preliminary supplementation assays in soft beverage as natural colorants and antioxidants. *J. Food Process. Preserv.* **2021**, *45*, e15170. [CrossRef]
123. Taiti, C.; Stefano, G.; Percaccio, E.; Di Giacomo, S.; Iannone, M.; Marianelli, A.; Di Sotto, A.; Garzoli, S. Addition of Spirulina to Craft Beer: Evaluation of the Effects on Volatile Flavor Profile and Cytoprotective Properties. *Antioxidants* **2023**, *12*, 1021. [CrossRef]
124. Barkallah, M.; Dammak, M.; Louati, I.; Hentati, F.; Hadrich, B.; Mechichi, T.; Ayadi, M.A.; Fendri, I.; Attia, H.; Abdelkafi, S. Effect of *Spirulina platensis* fortification on physicochemical, textural, antioxidant and sensory properties of yogurt during fermentation and storage. *LWT* **2017**, *84*, 323–330. [CrossRef]
125. El Baky, H.H.A.; El Baroty, G.S.; Ibrahem, E.A. Functional characters evaluation of biscuits sublimated with pure phycocyanin isolated from Spirulina and Spirulina biomass. *Nutr. Hosp.* **2015**, *32*, 231–241. [CrossRef]
126. Rodrigues, E.F.; Vendruscolo, L.P.; Bonfante, K.; Reinehr, C.O.; Colla, E.; Colla, L.M. Phycocyanin as substitute for texture ingredients in ice creams. *Br. Food J.* **2019**, *122*, 693–707. [CrossRef]
127. Lee, J.-Y.; Kang, S.-H.; Kim, M.-R. Changes in the Quality Characteristics and Antioxidant Activities of Spirulina Added Bread during Storage. *Korean J. Food Preserv.* **2011**, *18*, 111–118. [CrossRef]
128. Golmakani, M.-T.; Soleimanian-Zad, S.; Alavi, N.; Nazari, E.; Eskandari, M.H. Effect of Spirulina (*Arthrospira platensis*) powder on probiotic bacteriologically acidified feta-type cheese. *J. Appl. Phycol.* **2019**, *31*, 1085–1094. [CrossRef]
129. Nefasa, A.N.; Wulandari, E.C.; Christwardana, M.; Hadiyanto, H. Quality of Macronutrient of Cow's Milk with Addition of Soybean Oil and Phycocyanin Extract as Functional Food. *Food Sci. Technol.* **2020**, *3*, 17. [CrossRef]
130. Fradique, M.; Batista, A.P.; Nunes, M.C.; Gouveia, L.; Bandarra, N.M.; Raymundo, A. Incorporation of *Chlorella vulgaris* and *Spirulina maxima* biomass in pasta products. Part 1: Preparation and evaluation. *J. Sci. Food Agric.* **2010**, *90*, 1656–1664. [CrossRef] [PubMed]
131. Garbowska, M.; Berthold-Pluta, A.; Stasiak-Różańska, L.; Kalisz, S.; Pluta, A. The Impact of White Mulberry, Green Barley, Chia Seeds, and Spirulina on Physicochemical Characteristics, Texture, and Sensory Quality of Processed Cheeses. *Foods* **2023**, *12*, 2862. [CrossRef] [PubMed]
132. Park, G.; Cho, H.; Kim, K.; Kweon, M. Quality Characteristics and Antioxidant Activity of Fresh Noodles Formulated with Flour-Bran Blends Varied by Particle Size and Blend Ratio of Purple-Colored Wheat Bran. *Processes* **2022**, *10*, 584. [CrossRef]
133. Atallah, A.A.; Morsy, O.M.; Gemiel, D.G. Characterization of functional low-fat yogurt enriched with whey protein concentrate, Ca-caseinate and spirulina. *Int. J. Food Prop.* **2020**, *23*, 1678–1691. [CrossRef]
134. Hassanzadeh, H.; Ghanbarzadeh, B.; Galali, Y.; Bagheri, H. The physicochemical properties of the spirulina-wheat germ-enriched high-protein functional beverage based on pear-cantaloupe juice. *Food Sci. Nutr.* **2022**, *10*, 3651–3661. [CrossRef]
135. Nazir, F.; Saeed, M.; Abbas, A.; Majeed, M.R.; Israr, M.; Zahid, H.; Ilyas, M.; Nasir, M. Development, quality assessment and nutritive valorization of *Spirulina platensis* in yogurt spread. *Food Sci. Appl. Biotechnol.* **2022**, *5*, 106. [CrossRef]

136. Sanjari, S.; Sarhadi, H.; Shahdadi, F. Investigating the Effect of *Spirulina platensis* Microalgae on Textural and Sensory Properties of Baguette Bread. *J. Nutr. Food Secur.* **2018**, *3*, 218–225. [CrossRef]
137. Fradinho, P.; Niccolai, A.; Soares, R.; Rodolfi, L.; Biondi, N.; Tredici, M.R.; Sousa, I.; Raymundo, A. Effect of *Arthrospira platensis* (spirulina) incorporation on the rheological and bioactive properties of gluten-free fresh pasta. *Algal Res.* **2020**, *45*, 101743. [CrossRef]
138. Şahin, O.I. Functional and sensorial properties of cookies enriched with *SPIRULINA* and *DUNALIELLA* biomass. *J. Food Sci. Technol.* **2020**, *57*, 3639–3646. [CrossRef]
139. Luo, A.; Feng, J.; Hu, B.; Lv, J.; Liu, Q.; Nan, F.; Oliver Chen, C.-Y.; Xie, S. *Arthrospira* (*Spirulina*) platensis extract improves oxidative stability and product quality of Chinese-style pork sausage. *J. Appl. Phycol.* **2018**, *30*, 1667–1677. [CrossRef]
140. Vieira, M.V.; Oliveira, S.M.; Amado, I.R.; Fasolin, L.H.; Vicente, A.A.; Pastrana, L.M.; Fuciños, P. 3D printed functional cookies fortified with *Arthrospira platensis*: Evaluation of its antioxidant potential and physical-chemical characterization. *Food Hydrocoll.* **2020**, *107*, 105893. [CrossRef]
141. Hussein, A.; Ibrahim, G.; Kamil, M.; El-Shamarka, M.; Mostafa, S.; Mohamed, D. Spirulina-Enriched Pasta as Functional Food Rich in Protein and Antioxidant. *Biointerface Res. Appl. Chem.* **2021**, *11*, 14736–14750. [CrossRef]
142. Haghdoost, A.; Golestan, L.; Hasani, M.; Noghabi, M.S.; Shahidi, S.A. Assessment of the potential of algae phycobiliprotein nanoliposome for extending the shelf life of common carp burgers during refrigerated storage. *Fish. Aquat. Sci.* **2022**, *25*, 276–286. [CrossRef]
143. Niccolai, A.; Venturi, M.; Galli, V.; Pini, N.; Rodolfi, L.; Biondi, N.; D'Ottavio, M.; Batista, A.P.; Raymundo, A.; Granchi, L.; et al. Development of new microalgae-based sourdough "crostini": Functional effects of *Arthrospira platensis* (spirulina) addition. *Sci. Rep.* **2019**, *9*, 19433. [CrossRef] [PubMed]
144. Mohammadi-Gouraji, E.; Soleimanian-Zad, S.; Ghiaci, M. Phycocyanin-enriched yogurt and its antibacterial and physicochemical properties during 21 days of storage. *LWT* **2019**, *102*, 230–236. [CrossRef]
145. Ersyah, D.; Jaziri, A.A.; Setijawati, D. Effect of Spirulina (*Arthrospira platensis*) Powder on The Physico-chemical and Sensory Characterization of Dry Noodle. *J. Aquac. Fish Health* **2022**, *11*, 277–288. [CrossRef]
146. De Oliveira, T.T.B.; dos Reis, I.M.; de Souza, M.B.; da Silva Bispo, E.; Fonseca Maciel, L.; Druzian, J.I.; Lordelo Guimarães Tavares, P.P.; de Oliveira Cerqueira, A.; dos Santos Boa Morte, E.; Abreu Glória, M.B.; et al. Microencapsulation of *Spirulina* sp. LEB-18 and its incorporation in chocolate milk: Properties and functional potential. *LWT* **2021**, *148*, 111674. [CrossRef]
147. van der Spiegel, M.; Noordam, M.Y.; van der Fels-Klerx, H.J. Safety of Novel Protein Sources (Insects, Microalgae, Seaweed, Duckweed, and Rapeseed) and Legislative Aspects for Their Application in Food and Feed Production. *Compr. Rev. Food Sci. Food Saf.* **2013**, *12*, 662–678. [CrossRef] [PubMed]
148. Kleinübing, S.J.; Vieira, R.S.; Beppu, M.M.; Guibal, E.; Silva, M.G.C.D. Characterization and evaluation of copper and nickel biosorption on acidic algae *Sargassum Filipendula*. *Mater. Res.* **2010**, *13*, 541–550. [CrossRef]
149. Desideri, D.; Cantaluppi, C.; Ceccotto, F.; Meli, M.A.; Roselli, C.; Feduzi, L. Essential and toxic elements in seaweeds for human consumption. *J. Toxicol. Environ. Health Part A* **2016**, *79*, 112–122. [CrossRef] [PubMed]
150. Rzymski, P.; Niedzielski, P.; Kaczmarek, N.; Jurczak, T.; Klimaszyk, P. The multidisciplinary approach to safety and toxicity assessment of microalgae-based food supplements following clinical cases of poisoning. *Harmful Algae* **2015**, *46*, 34–42. [CrossRef]
151. Gromek, W.; Kołdej, N.; Kurowski, M.; Majsiak, E. Spirulina (*Arthrospira platensis*): Antiallergic Agent or Hidden Allergen? A Literature Review. *Foods* **2024**, *13*, 1052. [CrossRef]
152. Athané, A.; Demol, J.; Brosset-Vincent, S.; Aguenou, C.; Krisa, S.; Courtois, A.; Griffiths, H.; Cagnac, O. The safety evaluation of phycocyanin-enriched *Galdieria sulphuraria* extract using 90-day toxicity study in rats and in vitro genotoxicity studies. *Toxicol. Res. Appl.* **2020**, *4*, 239784732092999. [CrossRef]
153. Jensen, G.S.; Drapeau, C.; Lenninger, M.; Benson, K.F. Clinical Safety of a High Dose of Phycocyanin-Enriched Aqueous Extract from *Arthrospira* (*Spirulina*) platensis: Results from a Randomized, Double-Blind, Placebo-Controlled Study with a Focus on Anticoagulant Activity and Platelet Activation. *J. Med. Food* **2016**, *19*, 645–653. [CrossRef]
154. Fernand, F.; Israel, A.; Skjermo, J.; Wichard, T.; Timmermans, K.R.; Golberg, A. Offshore macroalgae biomass for bioenergy production: Environmental aspects, technological achievements and challenges. *Renew. Sustain. Energy Rev.* **2017**, *75*, 35–45. [CrossRef]
155. Liu, D.; Keesing, J.K.; Xing, Q.; Shi, P. World's largest macroalgal bloom caused by expansion of seaweed aquaculture in China. *Mar. Pollut. Bull.* **2009**, *58*, 888–895. [CrossRef] [PubMed]
156. Radulovich, R.; Umanzor, S.; Cabrera, R.; Mata, R. Tropical seaweeds for human food, their cultivation and its effect on biodiversity enrichment. *Aquaculture* **2015**, *436*, 40–46. [CrossRef]
157. Lotze, H.K.; Milewski, I.; Fast, J.; Kay, L.; Worm, B. Ecosystem-based management of seaweed harvesting. *Bot. Mar.* **2019**, *62*, 395–409. [CrossRef]
158. Kumari, P.; Shukla, S.P.; Rathi Bhuvaneswari, G.; Kumar, S.; Xavier, M.; Kumar, M. High value pigment production and carbon sequestration through wastewater grown *Spirulina* (*Arthrospira*) platensis: A green technology for wastewater utilization. *Waste Manag. Bull.* **2023**, *1*, 1–10. [CrossRef]
159. Yu, P.; Wu, Y.; Wang, G.; Jia, T.; Zhang, Y. Purification and bioactivities of phycocyanin. *Crit. Rev. Food Sci. Nutr.* **2017**, *57*, 3840–3849. [CrossRef] [PubMed]

160. Nikolic, M.R.; Minis, S.; Macvanin, M.; Stanic-Vucinic, D.; Cirkovic Velickovic, T. Analytical Protocols in Phycobiliproteins Analysis. In *Pigments from Microalgae Handbook*; Jacob-Lopes, E., Queiroz, M., Zepka, L., Eds.; Springer International Publishing: Berlin/Heidelberg, Germany, 2020; pp. 179–201.
161. Ruiz-Ruiz, F.; Benavides, J.; Rito-Palomares, M. Scaling-up of a B-phycoerythrin production and purification bioprocess involving aqueous two-phase systems: Practical experiences. *Process Biochem.* **2013**, *48*, 738–745. [CrossRef]
162. Torres-Acosta, M.A.; Ruiz-Ruiz, F.; Aguilar-Yáñez, J.M.; Benavides, J.; Rito-Palomares, M. Economic analysis of pilot-scale production of B-phycoerythrin. *Biotechnol. Prog.* **2016**, *32*, 1472–1479. [CrossRef] [PubMed]
163. Lauceri, R.; Chini Zittelli, G.; Torzillo, G. A simple method for rapid purification of phycobiliproteins from *Arthrospira platensis* and *Porphyridium cruentum* biomass. *Algal Res.* **2019**, *44*, 101685. [CrossRef]
164. Jaouen, P.; Lépine, B.; Rossignol, N.; Royer, R.; Quéméneur, F. Clarification and concentration with membrane technology of a phycocyanin solution extracted from *Spirulina platensis*. *Biotechnol. Tech.* **1999**, *13*, 877–881. [CrossRef]
165. Ramos, A.; Acién, F.G.; Fernández-Sevilla, J.M.; González, C.V.; Bermejo, R. Large-scale isolation and purification of C-phycocyanin from the cyanobacteria *Anabaena marina* using expanded bed adsorption chromatography. *J. Chem. Technol. Biotechnol.* **2010**, *85*, 783–792. [CrossRef]
166. Denis, C.; Massé, A.; Fleurence, J.; Jaouen, P. Concentration and pre-purification with ultrafiltration of a R-phycoerythrin solution extracted from macro-algae *Grateloupia turuturu*: Process definition and up-scaling. *Sep. Purif. Technol.* **2009**, *69*, 37–42. [CrossRef]
167. Martins, M.; Soares, B.P.; Santos, J.H.P.M.; Bharmoria, P.; Torres Acosta, M.A.; Dias, A.C.R.V.; Coutinho, J.A.P.; Ventura, S.P.M. Sustainable Strategy Based on Induced Precipitation for the Purification of Phycobiliproteins. *ACS Sustain. Chem. Eng.* **2021**, *9*, 3942–3954. [CrossRef]
168. Araújo, R.; Vázquez Calderón, F.; Sánchez López, J.; Azevedo, I.C.; Bruhn, A.; Fluch, S.; Garcia Tasende, M.; Ghaderiardakani, F.; Ilmjärv, T.; Laurans, M.; et al. Current Status of the Algae Production Industry in Europe: An Emerging Sector of the Blue Bioeconomy. *Front. Mar. Sci.* **2021**, *7*, 626389. [CrossRef]
169. Hsieh-Lo, M.; Castillo, G.; Ochoa-Becerra, M.A.; Mojica, L. Phycocyanin and phycoerythrin: Strategies to improve production yield and chemical stability. *Algal Res.* **2019**, *42*, 101600. [CrossRef]
170. Ragaza, J.A.; Hossain, M.S.; Meiler, K.A.; Velasquez, S.F.; Kumar, V. A review on Spirulina: Alternative media for cultivation and nutritive value as an aquafeed. *Rev. Aquac.* **2020**, *12*, 2371–2395. [CrossRef]
171. Lucakova, S.; Branyikova, I.; Branyik, T.; Matoulkova, D.; Krausova, G. Wastewater from the demineralization of cheese whey for cost-efficient cultivation of spirulina. *J. Appl. Phycol.* **2022**, *34*, 89–99. [CrossRef]
172. Lim, H.R.; Khoo, K.S.; Chew, K.W.; Chang, C.-K.; Munawaroh, H.S.H.; Kumar, P.S.; Huy, N.D.; Show, P.L. Perspective of Spirulina culture with wastewater into a sustainable circular bioeconomy. *Environ. Pollut.* **2021**, *284*, 117492. [CrossRef] [PubMed]
173. Cui, H.; Yang, Z.; Lu, Z.; Wang, Q.; Liu, J.; Song, L. Combination of utilization of CO_2 from flue gas of biomass power plant and medium recycling to enhance cost-effective *Spirulina* production. *J. Appl. Phycol.* **2019**, *31*, 2175–2185. [CrossRef]
174. Thevarajah, B.; Nishshanka, G.K.S.H.; Premaratne, M.; Nimarshana, P.H.V.; Nagarajan, D.; Chang, J.-S.; Ariyadasa, T.U. Large-scale production of *Spirulina*-based proteins and c-phycocyanin: A biorefinery approach. *Biochem. Eng. J.* **2022**, *185*, 108541. [CrossRef]

International Journal of *Molecular Sciences*

MDPI

Article

Effect of Addition of Tannin Extract from Underutilized Resources on Allergenic Proteins, Color and Textural Properties of Egg White Gels

Yoko Tsurunaga [1,*], Mika Ishigaki [2], Tetsuya Takahashi [1], Shiori Arima [1], Sae Kumagai [3], Yoshimasa Tsujii [4,5] and Shota Koyama [4,5]

1 Faculty of Human Science, Shimane University, Matsue 690-8504, Shimane, Japan
2 Institute of Agricultural and Life Sciences, Academic Assembly, Shimane University, Matsue 690-8504, Shimane, Japan
3 Graduate School of Human and Social Sciences, Shimane University, Matsue 690-8504, Shimane, Japan
4 Kewpie Research Division for Egg Innovation, Tokyo University of Agriculture, Setagaya City 156-8502, Tokyo, Japan
5 Faculty of Applied Biosciences, Tokyo University of Agriculture, Setagaya City 156-8502, Tokyo, Japan
* Correspondence: ytsurunaga@hmn.shimane-u.ac.jp; Tel.: +81-852-32-9806

Abstract: Tannins, present in numerous plants, exhibit a binding affinity for proteins. In this study, we aimed to exploit this property to reduce the concentration of allergenic egg white proteins. Tannins were extracted, using hot water, from the lyophilized powder of underutilized resources, such as chestnut inner skin (CIS), young persimmon fruit (YPF), and bayberry leaves (BBLs). These extracts were then incorporated into an egg white solution (EWS) to generate an egg white gel (EWG). Allergen reduction efficacy was assessed using electrophoresis and ELISA. Our findings revealed a substantial reduction in allergenic proteins across all EWGs containing a 50% tannin extract. Notably, CIS and BBL exhibited exceptional efficacy in reducing low allergen levels. The addition of tannin extract resulted in an increase in the total polyphenol content of the EWG, with the order of effectiveness being CIS > YPF > BBL. Minimal color alteration was observed in the BBL-infused EWG compared to the other sources. Additionally, the introduction of tannin extract heightened the hardness stress, with BBL demonstrating the most significant effect, followed by CIS and YPF. In conclusion, incorporating tannin extract during EWG preparation was found to decrease the concentration of allergenic proteins while enhancing antioxidant properties and hardness stress, with BBL being particularly effective in preventing color changes in EWG.

Keywords: allergenic proteins; color; textural properties; egg white; polyphenol; tannin; chestnut inner skin; young persimmon fruit; bayberry leaf

Citation: Tsurunaga, Y.; Ishigaki, M.; Takahashi, T.; Arima, S.; Kumagai, S.; Tsujii, Y.; Koyama, S. Effect of Addition of Tannin Extract from Underutilized Resources on Allergenic Proteins, Color and Textural Properties of Egg White Gels. *Int. J. Mol. Sci.* **2024**, *25*, 4124. https://doi.org/10.3390/ijms25074124

Academic Editors: Francesco Caruso, Vassilis Athanasiadis and Theodoros G. Chatzimitakos

Received: 5 February 2024
Revised: 18 March 2024
Accepted: 4 April 2024
Published: 8 April 2024

1. Introduction

Chicken eggs are a prevalent source of food allergies. The "National Survey on Immediate Type Food Allergy" by Japan's Consumer Affairs Agency in 2011 identified chicken eggs as the leading source, constituting 39.0% of all reported food allergies. Notably, this allergy affects a significant portion of the population, with 57.6% of 0-year-olds and 39.1% of 1-year-olds reporting allergies to chicken eggs [1]. Eggs are widely utilized in cooking due to their nutritional value and versatility. They are commonly used in various food products such as bread, confectionery, processed meat (ham), and seasonings (mayonnaise). The conventional approach to egg allergy management involves eliminating or significantly reducing egg consumption from the diet. However, this restriction poses a substantial challenge to daily life, and the risk of inadvertent exposure remains, potentially leading to severe anaphylactic reactions. Several studies have explored methods to decrease the allergenicity of chicken eggs. The primary allergens in eggs are ovomucoid (OVM) and ovalbumin (OVA), constituting approximately 11% and 54% of the egg white (EW) proteins, respectively [2]. OVA is less allergenic and more degradable when the proteins are denatured by heating [3]. However, OVM is thermostable and does not lose its allergenic activity [4]. While heating can reduce the allergenicity of OVA, OVM remains thermostable and allergenic even after cooking. Research has shown that removing OVM from EW significantly reduces allergenic activity, highlighting its crucial role in egg allergies [5]. Various methods, such as pulsed electrolysis and electrolysis treatment, have been explored to reduce the IgE-binding capacity of egg allergens [6,7]. However, these methods require cost, labor, and effort, and are not systems that can be implemented immediately at processing sites. As described above, egg allergy impacts many people, and hence, reducing the allergenicity of chicken eggs is crucial. Although research is ongoing on this topic, the main method for avoiding the effects of chicken egg allergy is the elimination diet [8]. However, this places a heavy burden on the patient and their family. This study developed a new hypoallergenic method to treat chicken egg allergy. Tsurunaga et al. conducted research on chestnut peels discarded during food processing [9], persimmon fruits discarded during picking and cultivation [10], and bayberry leaves (BBLs) discarded during pruning [11]. The results showed that they contained large amounts of tannins, which are known for having high protein-binding [12] and antioxidant activities, and polyphenols [13]. Tannins can be divided into two fundamentally different chemical structures: hydrolyzable and condensed tannins. Hydrolyzable tannins are polymers of gallic acid and ellagic acid covalently linked to esters. Condensed tannins are ether-covalently bonded polymers of flavan-3-ol. Out of these tannins, condensed tannins bind more strongly to proteins [14], and it has been reported that condensed tannins are abundant in CIS, YPF, and BBLs [15–17]. In a previous study, we developed a technique to reduce the allergenicity of processed wheat products by adding tannins to wheat, taking advantage of their protein-binding properties [18].

In this study, we aimed to create a new and simple hypoallergenic method for treating chicken egg allergy and used tannins found in chestnut peels, persimmon fruits, and BBLs (pruned material) due to their known protein-binding and antioxidant properties. This was performed by preparing an EWG by adding tannin extracts to an EWS and evaluating allergen reduction, antioxidant properties, textural properties, and the impact of tannin addition on the inherent coagulability of egg white.

2. Results and Discussion

2.1. Characteristics of EWS and Tannin Materials Solution

Table 1 presents the total polyphenol content (TPC), soluble tannin content (STC), image, color, pH, and Brix results for raw EWS, CIS, YPF, and BBL solutions. TPC exhibited the order YPF > CIS > BBL > EWS, with EWS surprisingly displaying a high value of 411.5 mg catechin (CTN) eq/100 mL, potentially attributed to the Folin–Ciocalteu reagent reacting with tyrosine, tryptophan, and cysteine residues, resulting in a deep blue color. The STC values were 391.3 ± 13.3, 331.5 ± 15.5, and 322.5 ± 27.9 CTN eq/100 mL for CIS, YPF, and BBL, respectively, with smaller differences between materials than observed for TPC. YPF had a higher L* value, and CIS had higher a* value than the other materials. BBL showcased a higher b* value, indicating a yellowish color, and was more transparent than the CIS and YPF solutions. The pH ranged from 4.0 to 4.8 for CIS, YPF, and BBL, and 7.2 for EWS, indicating neutrality. The Brix values followed the order YPF > CIS > EWS > BBL, with YPF registering as high as 10.1, likely due to its known higher content of carbohydrates, such as fructose, glucose, and sucrose [19].

Table 1. TPC, STC, image, color, pH, and Brix data of CIS, YPF, BBL, and EWS.

	TPC (mg CTN eq/100 mL)	STC (mg CTN eq/100 mL)	Image	L*	a*	b*	pH	Brix
CIS solution	1290.8 ± 19.2 [c]	391.3 ± 13.3 [b]	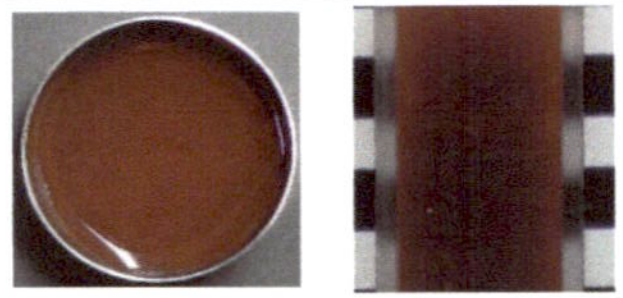	39.3 ± 0.0 [b]	14.7 ± 0.0 [d]	11.6 ± 0.1 [c]	4.1 ± 0.0 [a]	8.6 ± 0.0 [c]
YPF solution	1868.5 ± 40.2 [d]	331.5 ± 15.5 [b]	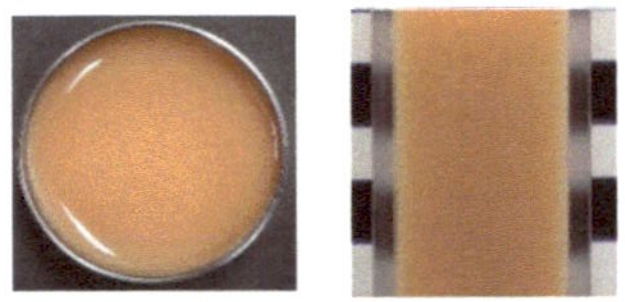	49.5 ± 0.1 [d]	12.2 ± 0.1 [b]	8.9 ± 0.1 [b]	4.8 ± 0.0 [b]	10.1 ± 0.0 [d]
BBL solution	964.8 ± 20.7 [b]	322.5 ± 27.9 [b]	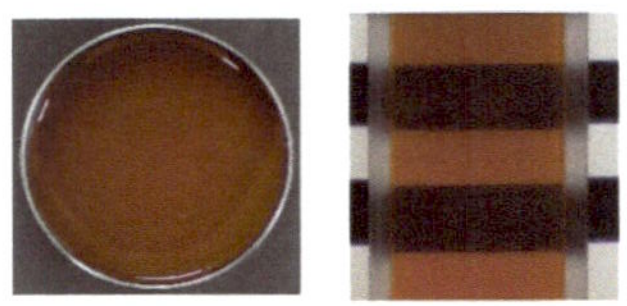	36.6 ± 0.1 [a]	12.9 ± 0.2 [c]	14.5 ± 0.2 [d]	4.0 ± 0.0 [a]	4.7 ± 0.0 [a]
EW solution	411.5 ± 20.0 [a]	0 [a]	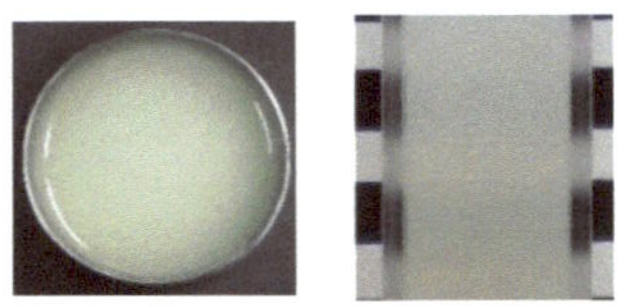	41.6 ± 0.1 [c]	0.1 ± 0.1 [a]	8.4 ± 0.1 [a]	7.2 ± 0.0 [c]	7.1 ± 0.0 [b]

All results were obtained by Tukey's test for multiple comparisons. Different letters indicate significant differences at $p < 0.05$. Data are expressed as mean ± SE ($n = 6$). TPC: total polyphenol content, CTN: catechin, STC: soluble tannin content, CIS: chestnut inner skin, YPF: young persimmon fruit, BBL: bayberry leaf, EW: egg white, EWS: egg white solution.

2.2. Evaluation of EW Allergen

2.2.1. Sodium Dodecyl Sulfate-Polyacrylamide Gel Electrophoresis (SDS-PAGE) Analysis

SDS-PAGE was conducted to assess the interactions between EW proteins and tannins, and bands of writing proteins were referenced from previous studies [20]. In the electrophoretic pattern of EWS, distinct bands for ovotransferrin (OVT), OVA, OVM, lysozyme (LYZ), and aggregates within the molecular weight range from 80 to 250 kDa were evident (Figure 1A). No apparent differences were observed in the band patterns of EW proteins with or without the presence of tannin extract. However, a faint band corresponding to aggregates larger than 250 kDa was noticeable in the EWS containing CIS and YPF.

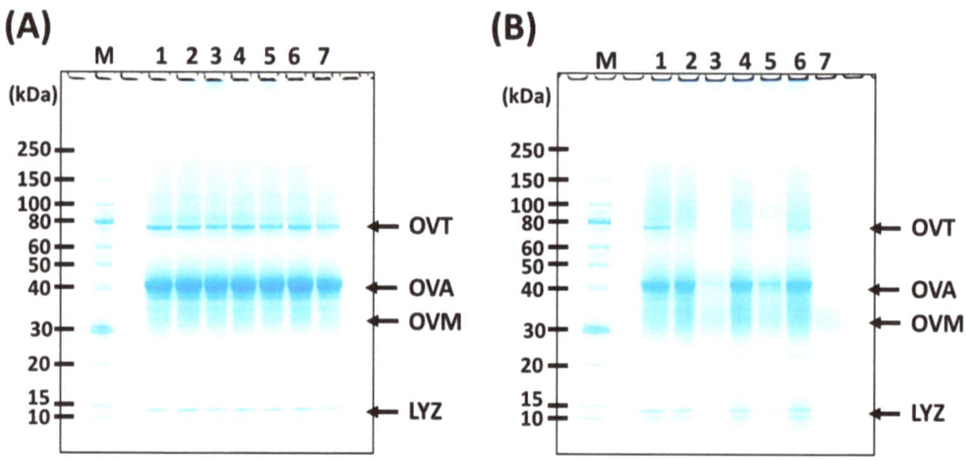

Figure 1. SDS-PAGE patterns of tannin-treated EW proteins. (**A**) EWS; (**B**) EWG. M, molecular mass standards. 1, control; 2, 10% CIS; 3, 50% CIS; 4, 10% YPF; 5, 50% YPF; 6, 10% BBL; 7, 50% BBL. EW, egg white; EWS, egg white solution; EWG, egg white gel; CIS, chestnut inner skin; YPF, young persimmon fruit; BBL, bayberry leaf. OVT, ovotransferrin; OVA, ovalbumin; OVM, ovomucoid; LYZ, lysozyme.

The band patterns of the EWG varied depending on the concentration and type of tannin extract (Figure 1B). Notably, the band intensity of OVT significantly decreased in the EWGs with 10% tannin extracts compared to those without tannin, and no OVT band was observed in the EWG with 50% tannin extracts. Furthermore, the band intensities of OVA, OVM, and LYZ were lower in the EWGs containing 50% CIS and BBL than in those with no tannins or 50% YPF. Specifically, the EWG with 50% BBL displayed no monomeric bands, except for trace amounts of OVM.

Under these experimental conditions, the majority of non-covalent and disulfide bonds were cleaved by the reagents present in the sample buffer. Hence, the observed band attenuation was likely attributed to aggregation via covalent bonds rather than disulfide bonds [20,21]. Some studies have explored the formation of covalent bonds between proteins and tannins [22,23]. Typically, this reaction involves the conversion of the phenolic hydroxyl group of tannins to quinone through alkaline or polyphenol oxidase treatment, followed by addition to the amino acid residues of proteins [22]. According to Pizzi [23], tannins and protein hydrolysates form covalent bonds when heated at 80 °C for 60 min in a neutral solution [23], a condition nearly identical to that in this study. Other potential bonds include lanthionine and lysinoalanine bonds formed between EW proteins during gelation [20]. However, the beta elimination of cysteine, a key reaction for lanthionine or lysinoalanine bonding, was unlikely to be affected by tannins. Thus, during heat-induced gelation, tannins and EW proteins formed aggregates with covalent bonds, indicating a process beyond mere mixing with tannin extracts. Notably, BBL served as a crosslinker that effectively aggregated most EW proteins, including OVA.

2.2.2. Enzyme-Linked Immunosorbent Assay (ELISA)

The protein content of the EWG samples, measured using the FASPEK ELISA II® series for albumin (referred to as Faspek value) (Morinaga Institute of Biological Science, Yokohama, Japan), is illustrated in Figure 2. In comparison to the controls, the Faspek values were significantly reduced in all tannin-supplemented groups (except for the 10% YPF) with more pronounced reductions observed with higher amounts of tannin extract addition ($p < 0.05$). A comparative analysis of the three tannin extracts employed in this study (CIS, YPF, and BBL) revealed that CIS and BBL were more effective than YPF. Hence, the results of the ELISA were found to be consistent with the electrophoresis results. Specifically, the 50% CIS sample demonstrated a value of 47.6 mg/g (62.8% reduction), and the 50% BBL sample exhibited a value of 51.8 mg/g (59.5% reduction), in contrast to the control Faspek value of 127.8 mg/g. Conversely, in the case of the 10% YPF (131.1 mg/g) sample, there was no significant difference from the control value (127.8 mg/g). The 50% YPF (80.5 mg/g), 10% CIS (92.8 mg/g), and 10% BBL (80.0 mg/g) samples displayed comparable Faspek values with no significant differences. OVA constitutes the majority (54% *w*/*w*) of EW [24], and is composed of 385 amino acids with a molecular weight of 45 kDa and a pI of 4.5 [25]. OVA undergoes three-dimensional structural changes through various treatments, including heating [26]. Such changes involve the formation of new disulfide bonds or the addition of side chains, resulting in irreversible modifications, such as aggregation, which are commonly observed during the heat treatment of globular proteins [27]. Conformational changes induced by heat treatment decrease the allergenicity of food proteins by either destroying or shielding specific epitopes or altering protein digestibility [28,29]. For instance, Jiménez-Saiz et al. [30] reported a reduction in the amount of rabbit IgG and human IgE binding to OVA after heating at 90 °C for 15 min. Claude et al. [31] also demonstrated that larger aggregates, when analyzed using sera from egg-allergic patients or sera from OVA-sensitized mice, exhibited lower IgE- and IgG-binding capacities compared to natural OVA. As mentioned, allergy reductions of OVA in EWG through heat treatment have been widely reported. In addition, various other hypoallergenic techniques for eggs have been reported [24]. It has been reported that when egg albumin was treated under pressurized thermal conditions, SDS-PAGE staining spread at 130 °C (0.3 MPa) and no specific bands were observed, and, above 150 °C (0.5 MPa), the staining bands were fuzzy and shifted to the low-molecular-weight side [32], which indicates that heating under normal pressure conditions (100 °C) does not change the OVA content, making hypoallergenization difficult. Another report examined the reduction of egg allergens in cakes containing gamma-irradiated egg whites [33]. It was found that the OVA concentration without gamma radiation was 432.88 mg/g, whereas it was 14.27 and 8.78 mg/g in the 10 and 20 kGy-irradiated samples, respectively, indicating a very high efficacy [33]. However, we believe that its practical application is challenging because special irradiation equipment is required, and the use of radiation for food products may be restricted. Furthermore, a study examined the immunogenic and structural properties of OVA under pulsed-electric-field treatment; when the OVA samples were treated with a high electric-field strength (>25 kV/cm, 180 μs) or for long durations (>60 μs, 35 kV/cm), the results revealed that OVA aggregation significantly reduced the IgG-binding ability, and, in particular, it decreased by approximately 30% after treatment at 35 kV/cm and 180 μs [6]. This technique also requires special equipment. However, we revealed the potential of a new and simple allergen reduction technology for EW by adding tannin extracts. Moreover, this technique facilitates the application of tannin-rich but underutilized resources. Based on these points, the technique of adding tannin extracts is a promising approach; however, further research is required to completely reduce allergens.

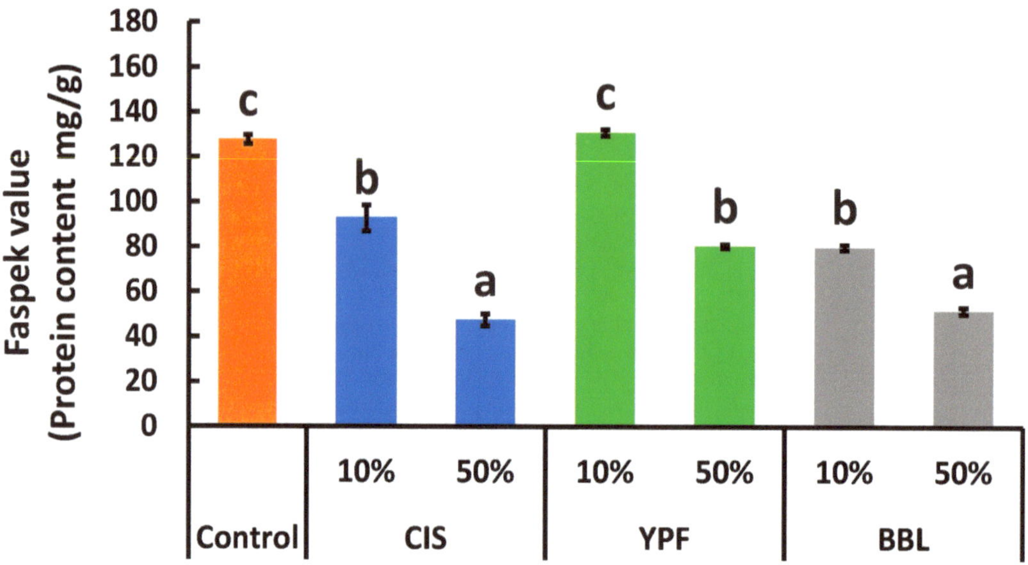

Figure 2. Faspek values of tannin-treated egg white gels. All results were obtained using Tukey's test for multiple comparisons. Faspek, FASPEK ELISA II® series for egg white albumin. Different letters indicate significant differences at $p < 0.05$. Data are expressed as the mean ± SE ($n = 4$). CIS, chestnut inner skin; YPF, young persimmon fruit; BBL, bayberry leaf.

2.3. Fourier Transform Infrared Spectroscopy (FT-IR)

To investigate the structural changes of EWG caused by adding tannins, FT-IR spectroscopic analysis was carried out. Figure 3 depicts the IR absorbance spectra in the 2000–500 cm^{-1} region for three types of tannins. The band assignment for these tannins is summarized in Table 2 [34–36]. The spectra show characteristic patterns owing to the molecules with aromatic rings and phenol groups. The spectral variations of the EWG caused by adding three types of tannins are shown in the upper column of Figure 4. The absorbance spectrum of the pure EWG is defined as a control and was compared to the spectra of the mixed samples of tannins at 10% and 50% concentrations. Spectral variations caused by changes in the ratio of the two compositions were clearly observed. To uncover the mechanisms of the intermolecular interactions between the EWG and tannins that induce aggregations, the spectra were analyzed in detail. Namely, we investigated whether the spectral variations could be explained simply by the spectral changes caused by mixing two compositions, or by the further variations caused by the molecular structural changes between the two compositions.

Table 2. Band assignment of FT-IR bands.

Wavenumber (cm^{-1})	Functional Group
1750–1700	C=O str.
1650–1480	C=C str. of aromatic rings
1400–1300	C-OH def.
1300–1160	C-OH str.
1285	C-O str. of pyran ring
1200–950	C-H def.

FT-IR, fourier transform infrared spectroscopy; str., stretching; def., deformation.

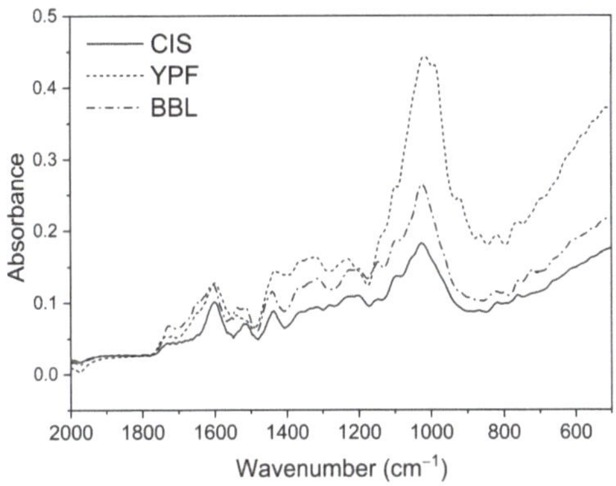

Figure 3. IR absorbance spectra in the 2000–500 cm^{-1} region of three types of tannin. IR, infrared spectroscopy; CIS, chestnut inner skin; YPF, young persimmon fruit; BBL, bayberry leaf.

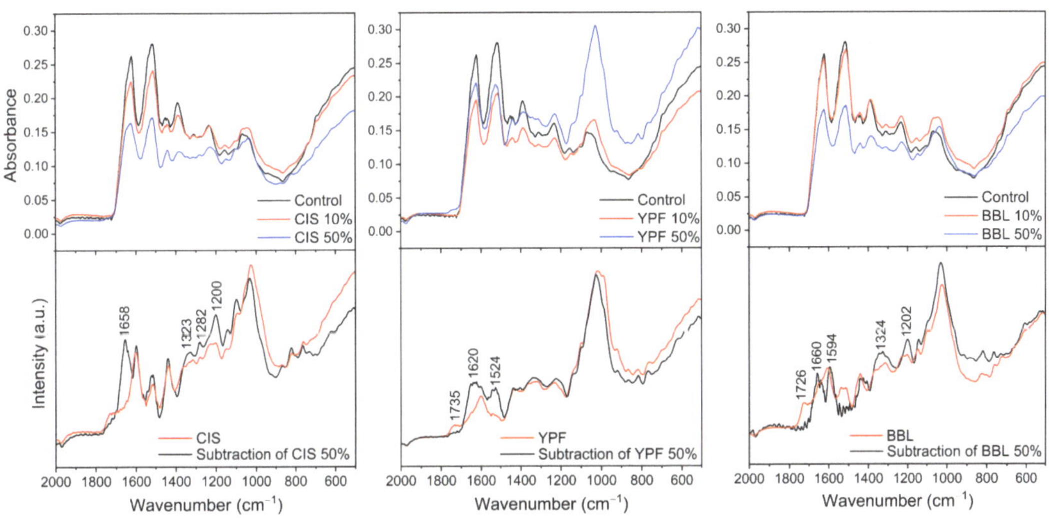

Figure 4. Upper column: comparison of IR absorbance spectra of pure EWG (control) and EWG with CIS, YPF, and BBL at 10 and 50% concentrations. Bottom column: subtraction spectra resulting from the subtraction of the pure EWG spectra from the spectra of the mixed EWG and tannin samples at 50% concentration, and the pure tannins. IR, infrared spectroscopy; EWG, egg white gel; CIS, chestnut inner skin; YPF, young persimmon fruit; BBL, bayberry leaf.

First, the spectrum of the pure EWG was subtracted from that of each mixed sample. The subtracted spectra were expected to show similar patterns to those of the pure tannins added if the molecular structure not changes due to the interaction between the EWG and the tannin molecule. Conversely, if the molecular structure changes due to the interaction between the EWG and the tannin molecule, then the subtracted spectrum is likely to have a different pattern to that of the pure tannin. The bottom column of Figure 4 compares the subtracted spectra for each tannin at 50% concentration with those of each pure tannin. Overall, the subtracted spectra show similar spectral patterns to those of each tannin. However, double bands appeared in the 1600–1500 cm^{-1} region, and the bands at approximately 1320 and 1220 cm^{-1} had remarkably intensified. These variations in the spectra of the mixed samples were highly similar to those observed for the transition of phenol into the quinone–phenolate structure [37]. That is, the results of the spectroscopic analysis were found to be consistent with the electrophoresis results.

2.4. Total Polyphenol Content

Figure 5 depicts the TPC values of the EWG, which were significantly higher ($p < 0.05$) in all tannin extract-supplemented treatments compared to the control. Specifically, the TPC value of the 10% CIS sample (was 4.7 times (437.3 mg/g) higher than the control (93.1 mg/g), while for the 50% CIS sample, an 8.2 times (760.8 mg/g) increase over the control was recorded. The addition of tannin extract resulted in an EWG with elevated TPC levels (Figure 5) and allergen reduction (Figures 1 and 2). CIS polyphenols are known for their antioxidant and antibacterial effects [9], YPF polyphenols exhibit bile acid-binding activity and hypoglycemic effects in mice [38], and BBL polyphenols have been reported to demonstrate anti-obesity activity in a rat model of high-fat diet-induced obesity [39]. The incorporation of tannin material extract imparts these health functionalities to the EWG, enhancing its health benefits.

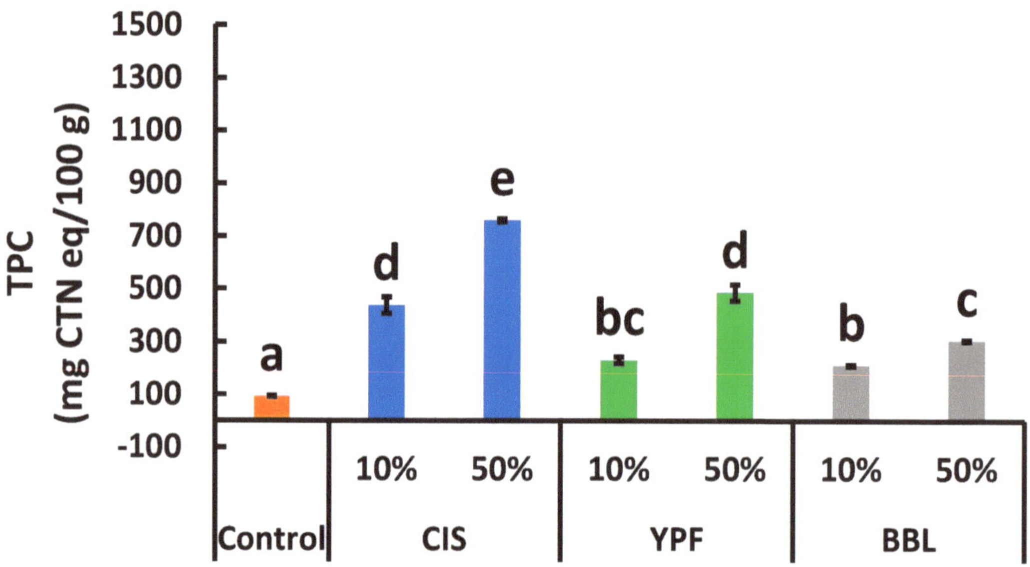

Figure 5. TPC of tannin-treated EWG. All results were obtained using Tukey's test for multiple comparisons. Different letters indicate significant differences at $p < 0.05$. Data are expressed as the mean ± SE ($n = 6$). TPC; total polyphenol content; CIS, chestnut inner skin; YPF, young persimmon fruit; BBL, bayberry leaf; EWG, egg white gel.

2.5. Appearance and Color

As depicted in the digital camera image in Figure 6, the addition of YPF and CIS resulted in both the EWS and EWG adopting a darker reddish hue than the control, and this effect became more pronounced with an increased tannin extract addition ratio. This trend was particularly prominent in the case of the CIS sample. In contrast, the BBL sample exhibited minimal change in appearance with the addition of the extract compared to the YPF and CIS samples (Figure 6). The BBL solution, being more transparent than the CIS and YPF solutions (Table 1), likely contributed to the subdued change in the appearance of the EWG. This subtle impact on the EWG's appearance is a crucial factor to consider when assessing quality.

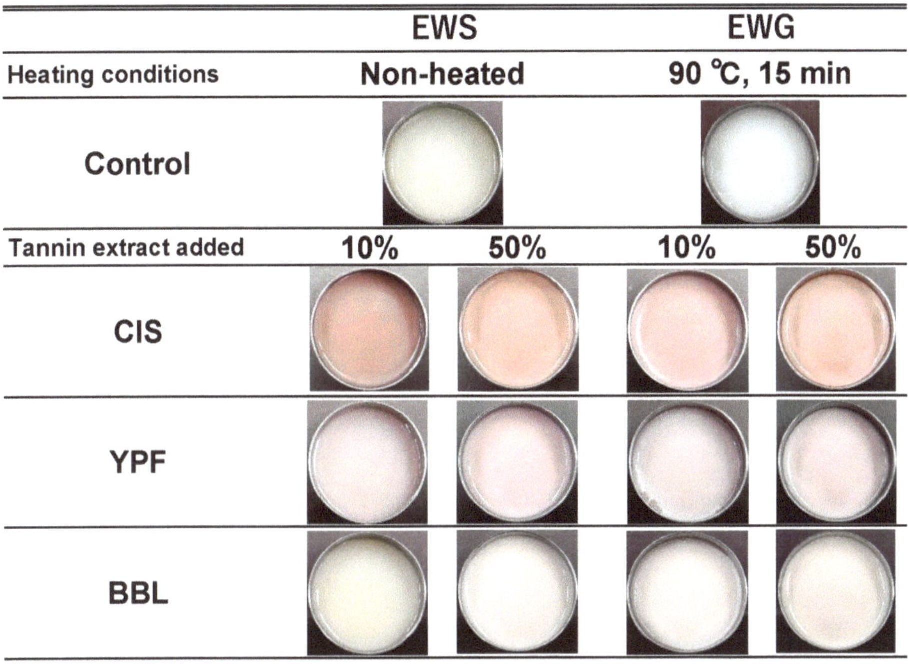

Figure 6. Images of tannin-treated EWS and EWG. EWS, egg white solution; EWG, egg white gel; CIS, chestnut inner skin; YPF, young persimmon fruit; BBL, bayberry leaf.

The L* value signifies lightness, a* indicates the color from green to red, and b* indicates the intensity of the color from blue to yellow. The L*, a*, and b* values for the CIS extract were 39.3 ± 0.0, 14.7 ± 0.0, and 11.6 ± 0.1, respectively; for the YPF extract, they were 49.5 ± 0.1, 12.2 ± 0.1, and 8.9 ± 0.1; and, for the BBL extract, they were 36.6 ± 0.1, 12.9 ± 0.2, and 14.5 ± 0.2 (Table 1). Comparing the effects of adding tannin extract on the L*, a*, and b* values, it was evident that the a* and b* values were conspicuously influenced by the addition of the tannin extract. The broad trends of the L*, a*, and b* values for each treatment were similar for the pre-heated EWS and post-heated EWG (Figure 7). The a* value of the EWS was 4.47 and 13.2 in the 10% and 50% CIS samples, respectively, compared to the control (0.12) (Figure 7B). In the EWG, the a* value was 12.15 and 16.33 in the 10% and 50% CIS samples, respectively, compared to the control (−6.79) (Figure 7E). While the control group showed a decrease in the a* value due to the transition from EWS to EWG caused by heating, the CIS replacement group exhibited an increase in a*, resulting in enhanced redness due to gelation (Figure 7B,E). A similar trend was observed with YPF, although not as prominently as with CIS. The a* value of the BBL sample was least affected by the addition of the tannin extract (Figure 7B,E). Given that the a* value of the CIS extract was higher (Table 1), the reddish coloration of the EWS and EWG with CIS addition could be attributed to the CIS solution as a raw material.

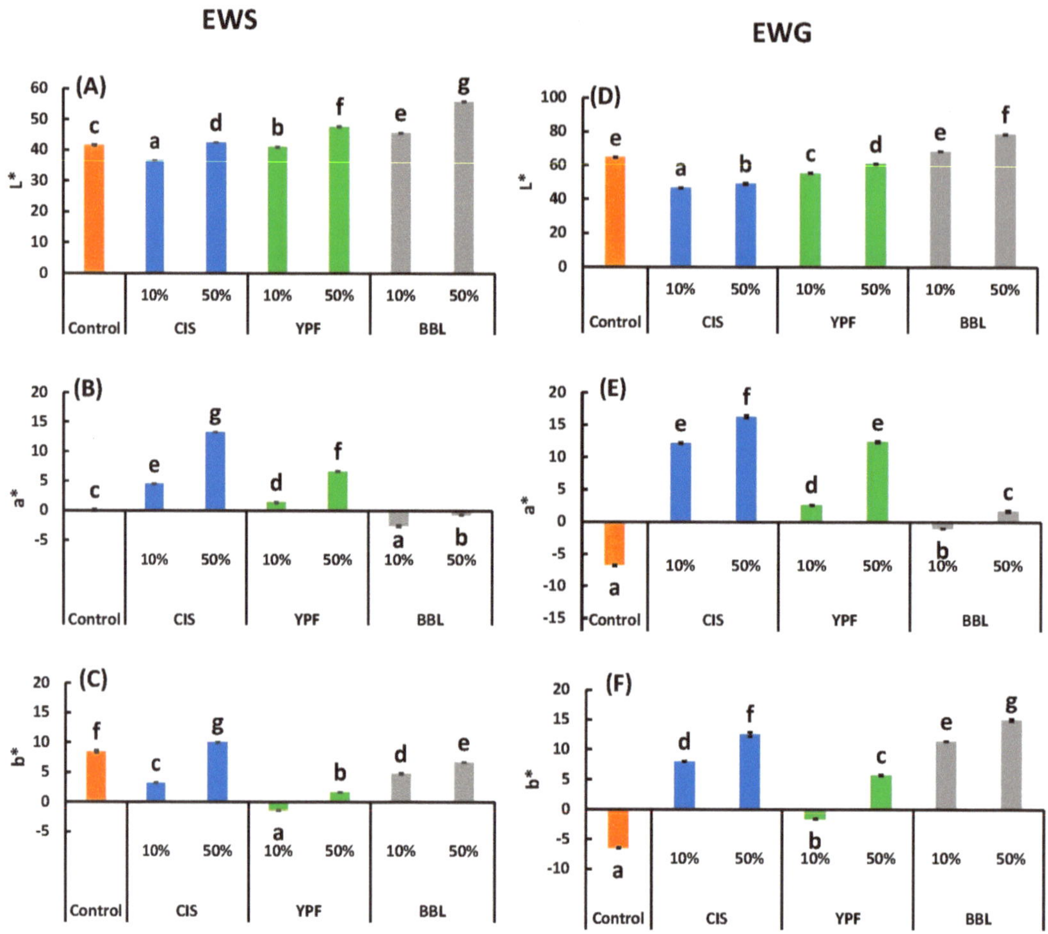

Figure 7. (**A**) L*, (**B**) a*, and (**C**) b* of tannin-treated EWS, and (**D**) L*, (**E**) a*, and (**F**) b* of EWG. Different letters indicate significant differences at $p < 0.05$. Data are expressed as the mean ± SE ($n = 10$). All results were obtained using Tukey's test for multiple comparisons. EWS, egg white solution; EWG, egg white gel; CIS, chestnut inner skin; YPF, young persimmon fruit; BBL, bayberry leaf.

The b* value of the EWG was 8.01 and 12.54 for the 10% and 50% CIS samples, −1.56 and 5.78 for the 10% and 50% YPF samples, and 11.39 and 14.95 for the 10% and 50% BBL samples, compared to the control (−6.72), indicating that the CIS and BBL samples produced more yellow gels (Figure 7F). The transition from EWS to EWG by heating decreased the b* value of the control from 8.43 to −6.52, whereas the BBL-added sample exhibited a remarkable increase in b* value, resulting in a more pronounced yellow color due to gelatinization (Figure 7C,F). A similar trend was noted in the CIS sample, albeit to a lesser extent than in the BBL sample. The higher b* values of the CIS and BBL solutions (Table 1) suggest that the yellowish tint of the EWG in the CIS and BBL treatments was attributed to the raw material tannin extracts.

To summarize, the BBL solution, while enhancing the yellowish color (Figure 7F), had a noticeably smaller effect on the change in the appearance of the EWG (Figure 6). Considering its effect on the change in color of the EWG, BBL stands out as a superior tannin extract material compared to CIS and YPF, due to its minimal effect on appearance (Figure 6).

2.6. Textural Properties and Scanning Electron Microscopy (SEM)

Figure 8 presents the results of the physical property measurements of the EWG. The hardness stress, an indicator of food hardness, was significantly higher ($p < 0.05$) in all tannin extract addition treatments than in the control, except for in the 10% CIS sample. The control registered at 4807 ± 135 Pa, while the 50% CIS sample exhibited 21,161 ± 372 Pa (4.4 times that of the control), the 50% YPF sample recorded 10,059 ± 90 Pa (2.1 times that of the control), and the 50% BBL substitution reached 25,077 ± 278 Pa (5.2 times that of the control) (Figure 8A). When tannin extracts of the same material were added, the hardness stress was significantly higher at the 50% addition amount than at the 10% amount for all CIS, YPF, and BBL samples ($p < 0.05$). These results indicate that the higher the amount of tannin extract added, the harder the EWG, with the effect being more pronounced in the order of BBL > CIS > YPF (Figure 8A).

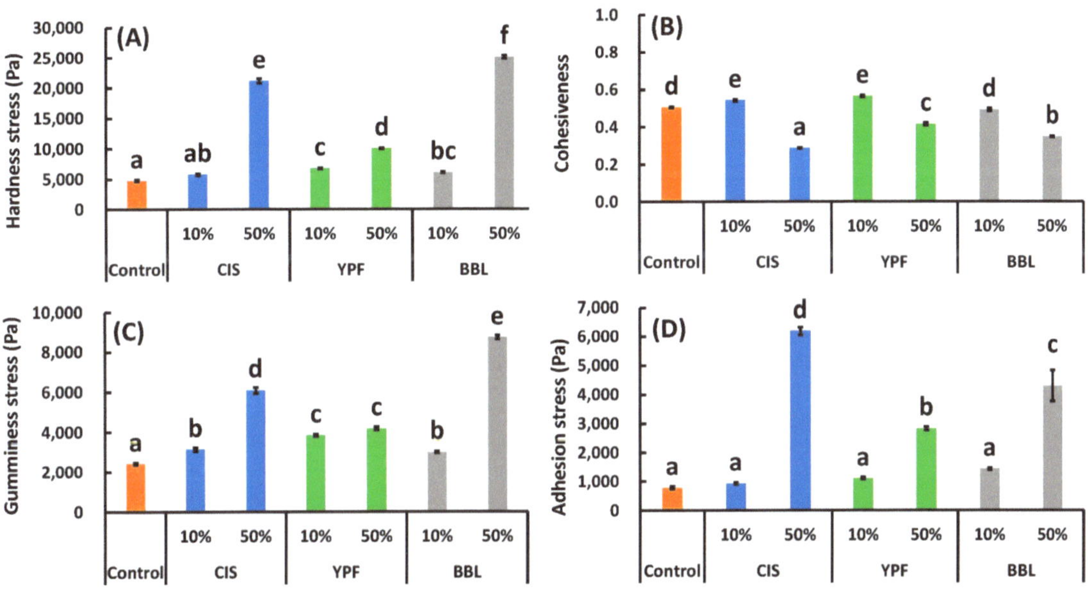

Figure 8. (**A**) Hardness stress, (**B**) cohesiveness, (**C**) gumminess stress, and (**D**) adhesion stress of tannin-treated egg white gels. All results were obtained using Tukey's test for multiple comparisons. Different letters indicate significant differences at $p < 0.05$. Data are expressed as the mean ± SE ($n = 5$). CIS, chestnut inner skin; YPF, young persimmon fruit; BBL, bayberry leaf.

Cohesiveness is a numerical value that indicates the percentage of elasticity remaining after chewing. When compared with the control, CIS, YPF, and BBL were equivalent to or higher than the control with a 10% addition, and the cohesiveness values were significantly lower than the control with a 50% addition (Figure 8B) ($p < 0.05$).

Gumminess stress is the product of hardness stress and cohesiveness, indicating the chewiness of the second bite. Compared to the control, all tannin extract addition treatments resulted in significantly higher gumminess stress values ($p < 0.05$). Gumminess stress was particularly high at 50% CIS and 50% BBL. The control registered at 2428 ± 63 Pa, whereas CIS 50% was 6091 ± 149 Pa, and BBL 50% was 8741 ± 102 Pa, respectively, 2.5 and 3.6 times higher than the control (Figure 8C).

The results of adhesion, indicating stickiness in the mouth, are shown in Figure 8D. The 10% tannin extract addition showed no significant difference ($p > 0.05$) compared to the control in CIS, YPF, and BBL, whereas the 50% addition showed significantly higher values ($p < 0.05$) (Figure 8D). The adhesion value of the control was 783 ± 47 Pa, whereas that of the 50% CIS sample was 6188 ± 134 Pa (7.9 times that of the control), the 50% YPF sample was 2827 ± 63 Pa (3.6 times that of the control), and the 50% BBL sample was 4297 ± 533 Pa (5.5 times that of the control). These results suggest that the addition of 50% tannin extracts makes the EWG firmer, more sticky in the mouth, and chewier the second time around.

Next, SEM observations were made of the microstructure of the EWG, which has a crucial effect on the texture properties. In the control, a dense and regular network structure of about 20–40 μm was observed, separated by a thin film (Figure 9). When the 10% tannin extracts group was compared with the control, the network structure was larger and more varied in size in the CIS sample than in the control, while a network structure similar to the control was observed in the YPF and BBL samples. On the other hand, the addition of 50% tannin extracts showed a remarkable difference compared to the control and 10% addition samples. The membranous part forming the network was thicker with the 50% addition of both CIS, YPF, and BBL than with the control and 10% additions. Differences in tannin material were also observed; the 50% CIS addition EWG had a random network structure, as did the 10% CIS EWG addition. The EWGs with 50% YPF and 50% BBL formed denser networks compared to the respective 10% additions. The tannin material with the greatest effect on textural properties was the 50% addition of CIS and BBL, but the microstructure of the tannins was found to be different in this study. In other words, with 50% CIS addition, the films forming the network were thicker and formed a random network, whereas, with 50% BBL addition, a regular and dense network was formed. Generally, gels with a dense network become hard; however, it was not clear why the gel became strong at 50% CIS in this case.

Tsurunaga [40] prepared jelly-like confections by adding astringent persimmon or non-astringent persimmon to soy milk and found that astringent persimmon jelly had a higher fracture stress than non-astringent persimmon jelly. This difference was attributed to the complex formation between the tannins and soymilk proteins [40]. There have also been reports on the binding properties of EW proteins to tannins; Chen et al. [41] observed the characteristics of the complex of OVA and tannic acid using dynamic light scattering and interfacial tension measurements. Shen et al. [42] reported that EW with tea polyphenols decreased in α-helix content and increased in β-sheet content; Zhou et al. [43] stated that the elastic modulus of fish surimi increased as the concentration of tea polyphenols bound to EW increased. Xue et al. [44] also found that, in EW treated with tea polyphenols followed by heat treatment, the stability of the gel structure improved with increasing amounts of tea polyphenols. These results are consistent with the increase in hardness stress and gumminess stress with the addition of the tannin extract in this experiment (Figure 8A,C). Xue et al. [44] reported that the binding of tea polyphenols to EW proteins is mainly maintained by ionic and disulfide bonds. However, we considered the binding of tannins to the EW protein to be covalent based on the SDS-PAGE and FT-IR results in this study. Since the molecular weights and structures of tea polyphenols and tannins are very different, the binding mode of tannins to proteins is considered to be different.

From the above, it is suggested that the addition of tannin extracts (CIS, YPF, and BBL) can reduce allergens in EWG and make EWG harder in textural properties. In particular, the addition of the BBL solution caused little change in the appearance of the EWG, indicating that BBL was the highest-performing material in terms of EW allergen reduction effect and quality. Because tannins have an astringent taste, there is concern that their addition would alter the taste. In a sensory evaluation, we have confirmed that the astringency disappears in a small number of people. However, because the threshold for astringency varies greatly from person to person, sensory testing must be conducted for a larger number of people. Future investigation should include clinical trials on hypoallergenicity in humans and taste evaluation on a larger number of people. In the future, we hope to expand the application area of these gels by achieving hypoallergenicity while maintaining the quality of egg whites.

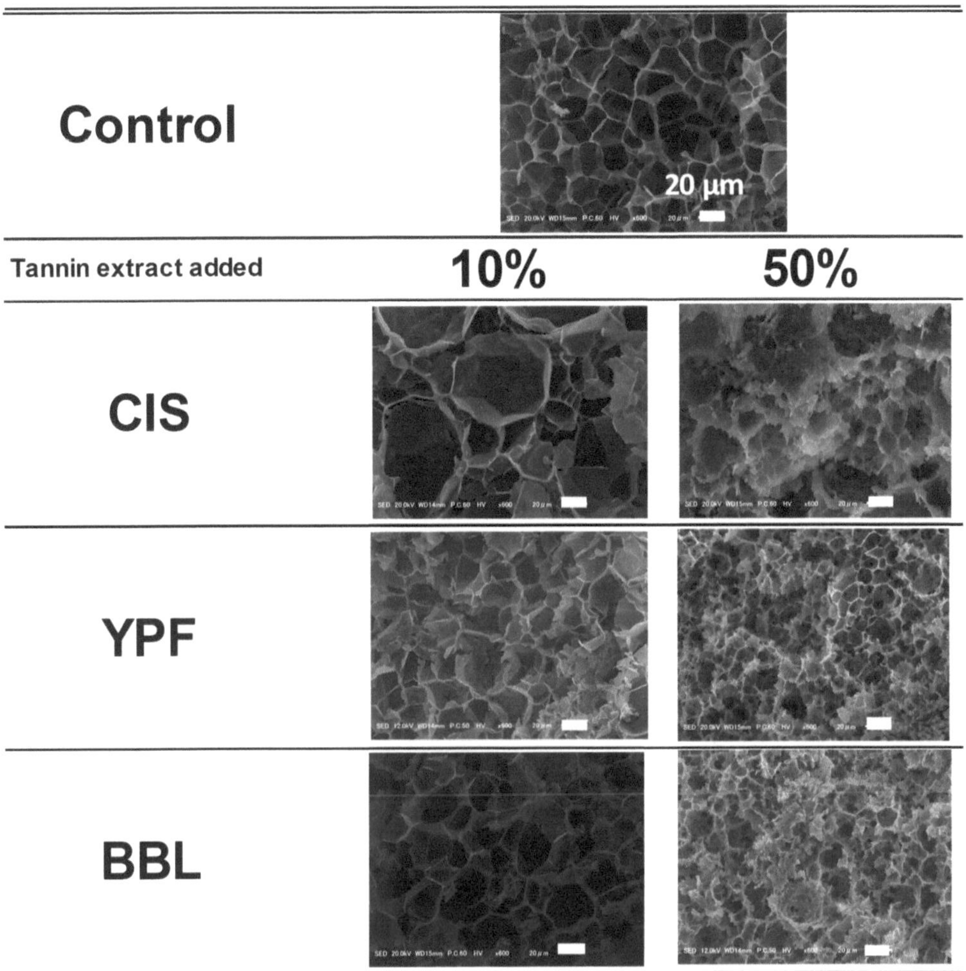

Figure 9. SEM image of tannin-treated EWG. SEM, scanning electron microscopy; EWG, egg white gel; CIS, chestnut inner skin; YPF, young persimmon fruit; BBL, bayberry leaf.

3. Materials and Methods

3.1. Ingredients of EW and Tannin Extract, and Preparation

The EW sample was EW-type W (52870, Kewpie Tamago Co., Tokyo, Japan) consisting of dried egg whites produced by a hygienic spray-drying method and subsequently stored in a freezer until further analysis. Egg white powder produced by the spray-drying method is widely used in industry and is of the same quality as that produced from raw eggs. Chestnut (*Castanea crenata*) inner skin (CIS), young persimmon (*Diospyros kaki*) fruit (YPF), and BBLs (*Morella rubra*) were selected for tannin extract production. To ensure the exclusive collection of the CIS, the 'Porotan' cultivar (Tsukuba City, Ibaraki, Japan), known for its superior peeling ability, was employed. The CIS underwent drying at 60 °C for 12 h in a constant air temperature oven (DN-61, Yamato Scientific Co., Tokyo, Japan). The YPF, sourced from the 'Saijo' cultivar of persimmon at the Shimane Agricultural Technology Center (Izumo City, Shimane, Japan), was freeze-dried after the removal of calyces and seeds. The BBLs, acquired from Shimane University (Matsue City, Shimane, Japan), underwent freeze-drying before use. Each sample was crushed with an Oster blender (Osaka Chemical Co., Ltd., Osaka, Japan) and passed through a 1.0 mm sieve. The processed samples were sealed in aluminum packs and stored at −25 °C.

3.2. Preparation of Tannin Extract

Subsequently, 45 g of the dried powder was placed in a 500 mL heat-resistant bottle, and 300 mL of distilled water was added. Hot water extraction was carried out at 120 °C for 20 min using an autoclave (LSX-300, Tommy Seiko, Tokyo, Japan). The centrifugation process at 1000× *g* for 3 min (LCX-100; Tomy Seiko Co., Tokyo, Japan) removed the powder, and the resulting supernatant served as the tannin extract.

3.3. EWG Production Method

The dried EW powder, weighing 5.5 g, was introduced into 94.5 g of distilled water, with constant stirring to prevent lump formation. Following the dissolution of the EW powder, the EWS was allowed to stand for 15 min. Subsequently, 15 g of the EW solution was carefully pipetted into a stainless-steel dish (40 mm × 15 mm) (ST-40, Yamaden Co., Ltd., Tokyo, Japan) to avoid bubble formation. The dish was then placed in a steam oven range (Panasonic Corporation, NE-BS1600-K, Tokyo, Japan) and heated at 90 °C for 15 min, followed by an additional 3 min of residual heat to gelatinize the product. The resultant samples were cooled in a refrigerator for 1 h before analysis. The EWGs obtained through this procedure were considered controls, lacking any tannin extract. In the tannin extract-supplemented group, either 10 (equivalent to 10% of the total weight) or 50 g (equivalent to 50% of the total weight) of distilled water was substituted with the tannin extract.

3.4. Evaluation of Immunoreactivity of EW Proteins

3.4.1. SDS-PAGE Analysis

Sodium dodecyl sulfate-polyacrylamide gel electrophoresis (SDS) was conducted following the method outlined by Laemmli [45] with slight modifications. The sample buffer, consisting of 2-mercaptoethanol added to commercially available 2x Laemmli sample buffer (BIO RAD# 161-0737), was mixed in equal volumes with the protein solution and heated at 95 °C for 5 min. Electrophoresis utilized 10–20% e PAGEL (ATTO Co., Tokyo, Japan) and was carried out at 300 V. A total of 5.0 µL of molecular weight markers (XL-lader broad; AproScience Co., Tokushima, Japan) and the loading samples were poured into 8.5 µL wells. One-step coomassie brilliant blue (CBB) (AproScience, Tokushima, Japan) was employed for gel staining.

3.4.2. ELISA

ELISA kits from the FASPEK ELISA II® series for albumin (Morinaga Institute of Biological Science, Kanagawa, Japan) were employed in this study to quantify EW albumin content. ELISA was officially introduced as an analytical method in Japan in 2002 [46]. The FASPEK KIT II® utilizes polyclonal antibodies to detect specific purified proteins or individual proteins of specific components, with a focus on albumin in the context of eggs [46]. Analytical procedures were conducted in accordance with the instructions provided in the kit [47], which required 19 mL of the extract to be added per gram of sample; however, in our case, the protein content derived from the antigen in the sample was substantially higher, which could lead to insufficient extraction. Therefore, we added 19.9 mL of extract to 0.1 g of the sample. The samples were appropriately diluted using specified reagents to fall within the assay range. The sample solution was added to the antibody-coated plate and allowed to stand at room temperature for 1 h. After washing, the antibody solution was added and allowed to stand at room temperature for 30 min. After the second washing operation, the enzyme–substrate solution was added and allowed to stand for 20 min at room temperature under shielded light. The absorbances were measured at 450 nm (650 nm was the reference wavelength). Four measurements were taken for each treatment group.

3.5. TPC and STC

The TPC was determined using the method described by Chung et al. [48], which is a modification of the Folin–Ciocalteu method [49]. The CIS, YPF, and BBL extract samples were appropriately diluted using distilled water for the TPC and STC analyses. The STC was assessed using polyvinyl polypyrrolidone (PVPP), a tannin-complexing agent [50,51]. PVPP was introduced into the extract for TPC measurements to insolubilize the tannins. The amount of PVPP (10, 20, 30, 60, 90, 120 mg) required to insolubilize the tannin was preliminarily tested, and it was confirmed that the addition of over 30 mg of PVPP to 1000 μL of the extract diluted with water showed no further insolubilization. Therefore, the amount of PVPP to be added was set at 30 mg per 1000 μL of diluted extract. The STC was then calculated by deducting the TPC value of the extract post-PVPP addition from the initial TPC value.

For the extraction of the EWG, 60% ethanol (v/v) was added to a 5.0 g sample, and homogenization was performed using a homogenizer (AHG-160D, AS ONE Co., Osaka, Japan) at 1000 rpm for 2 min. The supernatant was then centrifuged for 3 min (1000× g) and diluted accordingly. The TPC and STC values were expressed using catechin (CTN) as a standard, with six measurements taken for each treatment.

3.6. External Appearance and Color

The external appearance was captured with a digital camera (WG-40W; Ricoh, Tokyo, Japan). To determine the color of the samples, a spectrophotometer (CR-13; Konica Minolta, Tokyo, Japan) was employed. A disposable cell (optical path length 10 mm) was used to measure the color of the EWS, CIS, YPF, and BBL extracts; the EWG was measured directly on the gel surface. Each treatment underwent ten measurements.

3.7. SEM

The surface and cross-sectional structure of the EWG were examined through SEM. The EWG samples were cut into cubes ($5 \times 5 \times 5$ mm^3) and fixed in a 2.5% (v/v) glutaraldehyde solution (prepared in 0.1 M, pH 7.2 phosphate buffer) for 24 h [52]. Following fixation, the samples underwent three washes with a phosphate buffer and were then lyophilized. After lyophilization, the samples were affixed to a sample stand for SEM (Nissin EM Corporation, Type-HM, Tokyo, Japan) using double-sided carbon tape (Nissin EM Corporation, 8 mm × 20 m, Tokyo, Japan). Gold deposition was carried out, and SEM observations (JSM-IT800SHL, JEOL Ltd., Tokyo, Japan) were conducted at an acceleration voltage of 20 kV with a magnification of 600×.

3.8. pH

The pH values of the EWS, CIS, YPF, and BBL extracts were gauged using a pH meter (LAQUA F-72, Horiba, Ltd., Kyoto, Japan). Each treatment group underwent five measurements.

3.9. Textural Properties

To assess the textural properties of the EWG sample in a stainless-steel dish (40 × 15 mm) (ST-40, Yamaden Co., Ltd., Tokyo, Japan), an RE2-33005B Creep Meter (Yamaden Co., Ltd., Tokyo, Japan) was employed. The measurement conditions included a 20 N load cell, SPEED 10 mm/s, and a strain rate of 66.67% with a No. 56 (φ20 mm) acrylic resin plunger attached to an L40 acrylic resin extension plunger. Measurements were conducted in a stainless-steel petri dish, and the Texture Analysis software Ver. 2.2 (Yamaden) was used to determine the hardness stress (Pa), cohesiveness, and adhesion (Pa) of each sample. Five EWGs were measured for each treatment, and presented as mean ± SE.

3.10. FT-IR

For the FT-IR measurements, lyophilized EWG powder was utilized to eliminate the impact of moisture [53]. FT-IR spectra were recorded on an A-Cary 630 FT-IR spectrometer

(Agilent Technologies, Los Angeles, CA, USA) with a wavenumber range of 4000–400 cm^{-1} and a resolution accuracy of 2 cm^{-1} [54]. A horizontally attenuated total reflectance device (ATR) was employed to measure the spectra of the samples.

3.11. Statistical Analysis

Statistical analysis of the data was conducted using SPSS software (Version 28.0, SPSS, Chicago, IL, USA), and the results were expressed as mean ± SE. For multiple comparisons, data were assessed using a one-way analysis of variance (ANOVA) followed by Tukey's test, with the significance level set at 5%.

Author Contributions: Conceptualization, Y.T. (Yoko Tsurunaga); data curation, Y.T. (Yoko Tsurunaga), S.K. (Sae Kumagai), S.K. (Shota Koyama), Y.T. (Yoshimasa Tsujii) and M.I.; funding acquisition, Y.T. (Yoko Tsurunaga); investigation, Y.T. (Yoko Tsurunaga), S.A., S.K. (Sae Kumagai) and T.T.; writing—original draft, Y.T. (Yoko Tsurunaga), S.K. (Sae Kumagai), S.K. (Shota Koyama) and M.I.; writing—review and editing, Y.T. (Yoko Tsurunaga), S.K. (Shota Koyama) and M.I. All authors have read and agreed to the published version of the manuscript.

Funding: This study was supported by KAKENHI, a Grant-in-Aid for Challenging Exploratory Research (21K19713) and Grant-in-Aid for Scientific Research (B) (21H00808), and the Kieikai Research Foundation.

Institutional Review Board Statement: Not applicable.

Informed Consent Statement: Not applicable.

Data Availability Statement: Data is contained within the article.

Acknowledgments: We thank Ohata for providing YPF, and Akari Fujii, Yuji Hayakumo, and Michiko Hirose for providing technical assistance.

Conflicts of Interest: The authors declare no conflicts of interest.

Abbreviations

OVM	Ovomucoid
OVA	Ovalbumin
OVT	Ovotransferrin
LYZ	Lysozyme
EW	Egg white
BBL	Bayberry leaf
EWG	EW gel
EWS	EW solution
CIS	Chestnut inner skin
YPF	Young persimmon fruit
CTN	Catechin
SE	Standard error
ANOVA	One-way analysis of variance
ATR	Attenuated total reflectance device
SEM	Scanning electron microscopy
FT-IR	Fourier transform infrared spectroscopy
TPC	Total polyphenol content
STC	Soluble tannin measurement
IR	Infrared spectroscopy
ELISA	Enzyme-Linked Immunosorbent Assay
PVPP	Polyvinyl polypyrrolidone

References

1. Ebisawa, M.; Ito, K.; Fujisawa, T.; The Japanese Society of Pediatric Allergy and Clinical Immunology; The Japanese Society of Allergology. Japanese guidelines for food allergy 2017. *Allergol. Int.* **2017**, *66*, 248–264. [CrossRef] [PubMed]
2. Mine, Y.; Yang, M. Recent advances in the understanding of egg allergens: Basic, industrial, and clinical perspectives. *J. Agric. Food Chem.* **2008**, *56*, 4874–4900. [CrossRef]

3. Golias, J.; Schwarzer, M.; Wallner, M.; Kverka, M.; Kozakova, H.; Srutkova, D.; Klimesova, K.; Sotkovsky, P.; Palova-Jelinkova, L.; Ferreira, F. Heat-induced structural changes affect OVA-antigen processing and reduce allergic response in mouse model of food allergy. *PLoS ONE* **2012**, *7*, e37156. [CrossRef]
4. Gu, J.; Matsuda, T.; Nakamura, R. Antigenicity of ovomucoid remaining in boiled shell eggs. *J. Food Sci.* **1986**, *51*, 1448–1450. [CrossRef]
5. Urisu, A.; Ando, H.; Morita, Y.; Wada, E.; Yasaki, T.; Yamada, K.; Komadac, K.; Torii, S.; Gotoe, M.; Wakamatsu, T. Allergenic activity of heated and ovomucoid-depleted egg white. *J. Allergy Clin. Immunol.* **1997**, *100*, 171–176. [CrossRef] [PubMed]
6. Yang, W.; Tu, Z.; Wang, H.; Zhang, L.; Gao, Y.; Li, X.; Tian, M. Immunogenic and structural properties of ovalbumin treated by pulsed electric fields. *Int. J. Food Prop.* **2017**, *20*, S3164–S3176. [CrossRef]
7. Kido, J.; Matsumoto, T. Attenuated allergenic activity of ovomucoid after electrolysis. *Allergy Asthma Immunol. Res.* **2015**, *7*, 599–604. [CrossRef] [PubMed]
8. Burks, A.W.; Tang, M.; Sicherer, S.; Muraro, A.; Eigenmann, P.A.; Ebisawa, M.; Fiocchi, A.; Chiang, W.; Beyer, K.; Wood, R. ICON: Food allergy. *J. Allergy Clin. Immunol.* **2012**, *129*, 906–920. [CrossRef]
9. Tsurunaga, Y.; Takahashi, T. Evaluation of the antioxidant activity, deodorizing effect, and antibacterial activity of 'Porotan' chestnut by-products and establishment of a compound paper. *Foods* **2021**, *10*, 1141. [CrossRef]
10. Tsurunaga, Y.; Takahashi, T.; Kanou, M.; Onda, M.; Ishigaki, M. Removal of astringency from persimmon paste via polysaccharide treatment. *Heliyon* **2022**, *8*, e10716. [CrossRef]
11. Shimosaki, S.; Tsurunaga, Y.; Itamura, H.; Nakamura, M. Anti-allergic effect of the flavonoid myricitrin from Myrica rubra leaf extracts in vitro and in vivo. *Nat. Prod. Res.* **2011**, *25*, 374–380. [CrossRef] [PubMed]
12. Obreque-Slier, E.; López-Solís, R.; Peña-Neira, Á.; Zamora-Marín, F. Tannin–protein interaction is more closely associated with astringency than tannin–protein precipitation: Experience with two oenological tannins and a gelatin. *Int. J. Food Sci. Technol.* **2010**, *45*, 2629–2636. [CrossRef]
13. Gu, H.-F.; Li, C.-M.; Xu, Y.-J.; Hu, W.-F.; Chen, M.-H.; Wan, Q.-H. Structural features and antioxidant activity of tannin from persimmon pulp. *Food Res. Int.* **2008**, *41*, 208–217. [CrossRef]
14. Yamauchi, K.; Soyano, M.; Kobayashi, M.; Kamatari, Y.O.; Mitsunaga, T. Protein aggregation model to explain the bioactivity of condensed tannins. *Food Chem.* **2023**, *416*, 135870. [CrossRef] [PubMed]
15. Hu, M.; Yang, X.; Chang, X. Bioactive phenolic components and potential health effects of chestnut shell: A review. *J. Food Biochem.* **2021**, *45*, e13696. [CrossRef] [PubMed]
16. Murali, P.; Shams, R.; Dar, A.H. Insights on nutritional profile, nutraceutical components, pharmacological potential, and trending utilization of persimmon cultivars: A review. *Food Chem. Adv.* **2023**, *3*, 100431. [CrossRef]
17. Yang, H.; Ye, X.; Liu, D.; Chen, J.; Zhang, J.; Shen, Y.; Yu, D. Characterization of unusual proanthocyanidins in leaves of bayberry (*Myrica rubra* Sieb. et Zucc.). *J. Agric. Food Chem.* **2011**, *59*, 1622–1629. [CrossRef] [PubMed]
18. Tsurunaga, Y.; Arima, S.; Kumagai, S.; Morita, E. Low Allergenicity in Processed Wheat Flour Products Using Tannins from Agri-Food Wastes. *Foods* **2023**, *12*, 2722. [CrossRef] [PubMed]
19. Senter, S.; Chapman, G.; Forbus, W., Jr.; Payne, J. Sugar and nonvolatile acid composition of persimmons during maturation. *J. Food Sci.* **1991**, *56*, 989–991. [CrossRef]
20. Koyama, S.; Kodama, D.; Tsujii, Y.; Handa, A. Soluble-protein-aggregate-assisted improvements in heat-induced gel properties: Effect of genipin-mediated crosslinks on egg white protein. *LWT* **2023**, *184*, 115079. [CrossRef]
21. Koyama, S.; Oka, D.; Tsujii, Y.; Takano, K.; Handa, A. Effects of lanthionine and lysinoalanine on heat-induced gelation of egg white. *Food Sci. Technol. Res.* **2020**, *26*, 789–795. [CrossRef]
22. Le Bourvellec, C.; Renard, C.M. Interactions between polyphenols and macromolecules: Quantification methods and mechanisms. *Crit. Rev. Food Sci. Nutr.* **2012**, *52*, 213–248. [CrossRef] [PubMed]
23. Pizzi, A. Covalent and Ionic Bonding between Tannin and Collagen in Leather making and shrinking: A MALDI-ToF study. *J. Renew. Mater.* **2021**, *9*, 1345–1364. [CrossRef]
24. Ma, X.; Liang, R.; Xing, Q.; Lozano-Ojalvo, D. Can food processing produce hypoallergenic egg? *J. Food Sci.* **2020**, *85*, 2635–2644. [CrossRef] [PubMed]
25. Anyushin, A.V.; Sap, A.; Quanten, T.; Proost, P.; Parac-Vogt, T.N. Selective hydrolysis of ovalbumin promoted by Hf (IV)-substituted Wells-Dawson-type polyoxometalate. *Front. Chem.* **2018**, *6*, 614. [CrossRef]
26. Davis, P.; Williams, S. Protein modification by thermal processing. *Allergy* **1998**, *53*, 102–105. [CrossRef] [PubMed]
27. Nicolai, T.; Durand, D. Controlled food protein aggregation for new functionality. *Curr. Opin. Colloid Interface Sci.* **2013**, *18*, 249–256. [CrossRef]
28. Berin, M.C.; Sampson, H.A. Food allergy: An enigmatic epidemic. *Trends Immunol.* **2013**, *34*, 390–397. [CrossRef] [PubMed]
29. Kosti, R.; Triga, M.; Tsabouri, S.; Priftis, K. Food allergen selective thermal processing regimens may change oral tolerance in infancy. *Allergol. Immunopathol.* **2013**, *41*, 407–417. [CrossRef]
30. Jiménez-Saiz, R.; Belloque, J.; Molina, E.; López-Fandino, R. Human immunoglobulin E (IgE) binding to heated and glycated ovalbumin and ovomucoid before and after in vitro digestion. *J. Agric. Food Chem.* **2011**, *59*, 10044–10051. [CrossRef]
31. Claude, M.; Lupi, R.; Bouchaud, G.; Bodinier, M.; Brossard, C.; Denery-Papini, S. The thermal aggregation of ovalbumin as large particles decreases its allergenicity for egg allergic patients and in a murine model. *Food Chem.* **2016**, *203*, 136–144. [CrossRef] [PubMed]

32. Okon, K.; Yoshida, T.; Hattori, M.; Matsuda, H.; Osada, M. Preparation of hypoallergenic ovalbumin by high-temperature water treatment. *Biosci. Biotechnol. Biochem.* **2021**, *85*, 2442–2449. [CrossRef] [PubMed]
33. Seo, J.-H.; Lee, J.-W.; Lee, Y.-S.; Lee, S.-Y.; Kim, M.-R.; Yook, H.-S.; Byun, M.-W. Change of an egg allergen in a white layer cake containing gamma-irradiated egg white. *J. Food Prot.* **2004**, *67*, 1725–1730. [CrossRef] [PubMed]
34. Edelmann, A.; Lendl, B. Toward the optical tongue: Flow-through sensing of tannin– protein interactions based on FTIR spectroscopy. *J. Am. Chem. Soc.* **2002**, *124*, 14741–14747. [CrossRef] [PubMed]
35. Jensen, J.S.; Egebo, M.; Meyer, A.S. Identification of spectral regions for the quantification of red wine tannins with Fourier transform mid-infrared spectroscopy. *J. Agric. Food Chem.* **2008**, *56*, 3493–3499. [CrossRef] [PubMed]
36. Kuzniarz, A. Infrared spectrum analysis of some flavonoids. *Acta Pol. Pharm. Drug Res.* **2014**, *58*, 415–420.
37. Machida, K.; Lee, H.; Uno, T. Resonance Raman spectra of sulfophthalein dyes in aqueous solutions. *J. Raman Spectrosc.* **1979**, *8*, 172–176. [CrossRef]
38. Matsumoto, K.; Kadowaki, A.; Ozaki, N.; Takenaka, M.; Ono, H.; Yokoyama, S.I.; Gato, N. Bile acid-binding ability of kaki-tannin from young fruits of persimmon (*Diospyros kaki*) in vitro and in vivo. *Phytother. Res.* **2011**, *25*, 624–628. [CrossRef]
39. Zhou, X.; Chen, S.; Ye, X. The anti-obesity properties of the proanthocyanidin extract from the leaves of Chinese bayberry (*Myrica rubra* Sieb. et Zucc.). *Food Funct.* **2017**, *8*, 3259–3270. [CrossRef]
40. Tsurunaga, Y. Formation of a protein-tannin complex to remove astringency during processing of Western-style persimmon jelly. In Proceedings of the VI International Symposium on Persimmon 1195, Valencia, Spain, 16 October 2016; pp. 177–182.
41. Chen, Y.; Hu, J.; Yi, X.; Ding, B.; Sun, W.; Yan, F.; Wei, S.; Li, Z. Interactions and emulsifying properties of ovalbumin with tannic acid. *LWT* **2018**, *95*, 282–288. [CrossRef]
42. Shen, F.; Niu, F.; Li, J.; Su, Y.; Liu, Y.; Yang, Y. Interactions between tea polyphenol and two kinds of typical egg white proteins—Ovalbumin and lysozyme: Effect on the gastrointestinal digestion of both proteins in vitro. *Food Res. Int.* **2014**, *59*, 100–107. [CrossRef]
43. Zhou, X.; Chen, T.; Lin, H.; Chen, H.; Liu, J.; Lyu, F.; Ding, Y. Physicochemical properties and microstructure of surimi treated with egg white modified by tea polyphenols. *Food Hydrocoll.* **2019**, *90*, 82–89. [CrossRef]
44. Xue, H.; Zhang, G.; Han, T.; Li, R.; Liu, H.; Gao, B.; Tu, Y.; Zhao, Y. Improvement of gel properties and digestibility of the water-soluble polymer of tea polyphenol-egg white under thermal treatment. *Food Chem.* **2022**, *372*, 131319. [CrossRef] [PubMed]
45. Laemmli, U.K. Cleavage of structural proteins during the assembly of the head of bacteriophage T4. *Nature* **1970**, *227*, 680–685. [CrossRef] [PubMed]
46. Akiyama, H.; Adachi, R. Japanese food allergy-labeling system and comparison with the international experience; detection and thresholds. *Food Saf.* **2021**, *9*, 101–116. [CrossRef] [PubMed]
47. Morinaga Institute of Biological Science, Inc. FASPEK (Ovalbumin) ELISA Kit II. 2017. Available online: https://www.miobs.com/product/tokutei/faspek2/dl/manual_tori04.pdf (accessed on 3 April 2024).
48. Chung, H.S.; Kim, H.S.; Lee, Y.G.; Seong, J.H. Effect of deastringency treatment of intact persimmon fruits on the quality of fresh-cut persimmons. *Food Chem.* **2015**, *166*, 192–197. [CrossRef]
49. Swain, T.; Hillis, W. The phenolic constituents of Prunus domestica. I.—The quantitative analysis of phenolic constituents. *J. Sci. Food Agric.* **1959**, *10*, 63–68. [CrossRef]
50. Toth, G.B.; Pavia, H. Removal of dissolved brown algal phlorotannins using insoluble polyvinylpolypyrrolidone (PVPP). *J. Chem. Ecol.* **2001**, *27*, 1899–1910. [CrossRef]
51. Besharati, M.; Taghizadeh, A. Effect of tannin-binding agents (polyethylene glycol and polyvinylpyrrolidone) supplementation on in vitro gas production kinetics of some grape yield byproducts. *Int. Sch. Res. Not.* **2011**, *2011*, 780540. [CrossRef]
52. Deng, C.; Shao, Y.; Xu, M.; Yao, Y.; Wu, N.; Hu, H.; Zhao, Y.; Tu, Y. Effects of metal ions on the physico-chemical, microstructural and digestion characteristics of alkali-induced egg white gel. *Food Hydrocoll.* **2020**, *107*, 105956. [CrossRef]
53. Wu, S.; Yang, R. Effect of high-pressure processing on polyphenol oxidase, melanosis and quality in ready-to-eat crabs during storage. *LWT* **2023**, *178*, 114607. [CrossRef]
54. Higgins, F.; Rein, A. Quantitative Analysis of Copolymers Using the Cary 630 FTIR Spectrometer. *Danbury Agil. Technol.*. 2011. Available online: https://gcms.labrulez.com/labrulez-bucket-strapi-h3hsga3/5990_8676_EN_App_Note_630_ATR_Copolymers_8af79525f5/5990-8676EN_AppNote_630_ATR_Copolymers.pdf (accessed on 3 April 2024).

International Journal of
Molecular Sciences

MDPI

Review

Polyphenol Extraction from Food (by) Products by Pulsed Electric Field: A Review

Vassilis Athanasiadis, Theodoros Chatzimitakos *, Konstantina Kotsou, Dimitrios Kalompatsios, Eleni Bozinou and Stavros I. Lalas

Department of Food Science & Nutrition, University of Thessaly, Terma N. Temponera str., 43100 Karditsa, Greece; vaathanasiadis@uth.gr (V.A.); kkotsou@agr.uth.gr (K.K.); dkalompatsios@uth.gr (D.K.); empozinou@uth.gr (E.B.); slalas@uth.gr (S.I.L.)

* Correspondence: tchatzimitakos@uth.gr; Tel.: +30-24410-64783; Fax: +30-24410-64781

Abstract: Nowadays, more and more researchers engage in studies regarding the extraction of bioactive compounds from natural sources. To this end, plenty of studies have been published on this topic, with the interest in the field growing exponentially. One major aim of such studies is to maximize the extraction yield and, simultaneously, to use procedures that adhere to the principles of green chemistry, as much as possible. It was not until recently that pulsed electric field (PEF) technology has been put to good use to achieve this goal. This new technique exhibits many advantages, compared to other techniques, and they have successfully been reaped for the production of extracts with enhanced concentrations in bioactive compounds. In this advancing field of research, a good understanding of the existing literature is mandatory to develop more advanced concepts in the future. The aim of this review is to provide a thorough discussion of the most important applications of PEF for the enhancement of polyphenols extraction from fresh food products and by-products, as well as to discuss the current limitations and the prospects of the field.

Keywords: polyphenols; flavonoids; antioxidants; PEF; non-thermal technique; green extraction; electroporation; fresh food products; food by-products

Citation: Athanasiadis, V.; Chatzimitakos, T.; Kotsou, K.; Kalompatsios, D.; Bozinou, E.; Lalas, S.I. Polyphenol Extraction from Food (by) Products by Pulsed Electric Field: A Review. *Int. J. Mol. Sci.* **2023**, *24*, 15914. https://doi.org/10.3390/ijms242115914

Academic Editors: Elena Azzini and Antonio González-Sarrías

Received: 30 September 2023
Revised: 30 October 2023
Accepted: 1 November 2023
Published: 2 November 2023

1. Introduction

Polyphenols are naturally present in plant-based foods and show an extensive variety of complicated chemical structures [1–3]. They are composed of a phenolic ring which serves as the fundamental monomer [4]. The primary classes of polyphenols include phenolic acids, flavonoids, stilbenes, and lignans [5]. Flavonoids consist of flavones, flavonols, flavanones, isoflavones, flavanols, and anthocyanins, whereas hydroxybenzoic and hydroxycinnamic acids are types of phenolic acids [6]. These compounds are present in the human diet and are mainly derived from plant sources, including fruits, vegetables, grains, and coffee [7]. Polyphenols are a wide array of bioactive compounds that naturally occur in food sources derived from plants [8]. They are recognized for their potential as preventive agents against chronic illnesses, such as cardiovascular diseases and diabetes [9]. The four major types of polyphenols are phenolic acids, flavonoids, stilbenes, and lignans [10,11]. A visual representation of these compounds is illustrated in Figure 1. Flavonoids are highly prevalent in the context of dietary intake [12,13]. Catechin is found in a variety of fruits and beverages, most notably in tea [14]. Citrus fruits are known for their high content of hesperidin [15], whereas red fruits and berries are primarily characterized by their cyanidin content [16]. Fruits, such as apples, are known to possess proanthocyanidins and quercetin [17]. Proanthocyanidins are also present in grapes and cocoa [18], while quercetin can be found in onions and tea [19]. Finally, it should be noted that the soybean plant is primarily characterized by the presence of daidzein [20]. Polyphenolic compounds, such as lignans, are predominantly found in grains and flaxseed [21]. Due to the significant impact of polyphenols on human health, a multitude of research studies have been conducted to examine the physiological effects displayed by these compounds [22–25]. The consumption of foods rich in polyphenols has been demonstrated to contribute to a reduction in the incidence of several health conditions, such as colon cancer, liver disorders [26], cardiovascular diseases [27], and obesity [28].

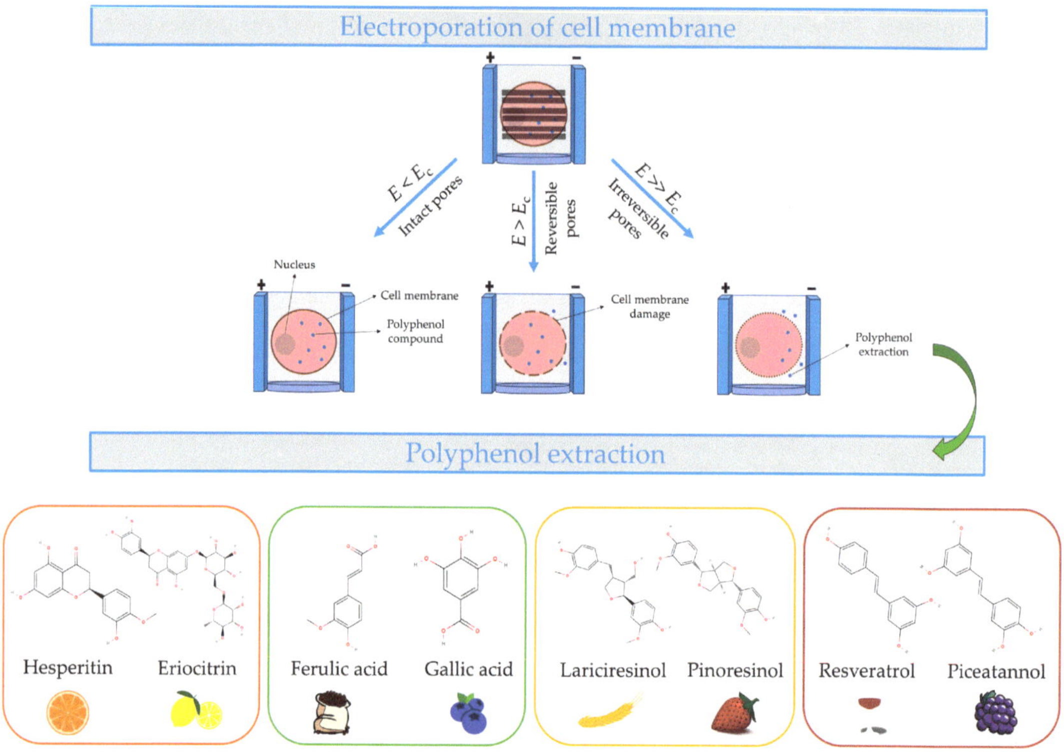

Figure 1. Electroporated cell membrane with bioactive compounds within several electric field strengths. Impact of critical electric field strength in the cell membrane. The four different types of polyphenols found in plant-based foods are transferred from the cell to the extraction solvent.

The investigation of diverse conventional methodologies has been conducted to extract bioactive compounds from specific fresh food products or food waste materials [29–31]. Common extraction techniques include soaking, maceration, infusion, percolation, and Soxhlet extraction [32]. The efficacy of these techniques is influenced by various factors, including the choice of solvent, the solvent's solvation capacity, the degree of agitation, and the temperature [29]. Traditional extraction methods are associated with several challenges, including extended processing durations, reduced extraction efficiencies, excessive solvent usage, possible degradation of thermolabile bioactive compounds, and the utilization of hazardous chemicals [32,33]. A selective extraction is not adequately served by conventional extraction techniques [34,35]. Furthermore, the extracted products, such as proteins and polysaccharides, may not be of high quality if these methods are used [36]. Pulsed electric field (PEF) is a processing method that implies a higher polyphenol extraction [37]. The PEF treatment is a non-thermal technique employed for food preservation, which involves the application of short bursts of electrical power to inactivate microorganisms while minimizing any detrimental impact on the food's quality [38]. This implies that the PEF treatment aspires to enhance the accessibility of consumer-grade, polyphenol-abundant food products of superior quality [39].

The use of PEF to diffuse, osmose, press, and dry food waste and by-products has gained popularity [40]. It reduces the negative effects of regular heating methods [41,42]. Since PEF can electroporate cell membranes, it is also used as a pretreatment to boost recoveries of bioactive compounds, such as polyphenols, carotenoids, and proteins [43,44]. When applied to water, the PEF technique exhibited lower temperatures, lower solvent consumption, and improved constituent extraction rates [45]. The extraction yield could be increased with reduced energy costs, and heat-sensitive substances could be preserved; all of these by also incorporating a "green" extraction method [46–48]. The PEF method exhibits greater environmental sustainability and economic efficiency due to its reduced overall energy consumption and lower energy requirements per unit of processed product [49]. For this reason, various sectors of the food industry have experimented with PEF-based maceration over the past decade [50]. Apart from fresh fruits and vegetables, biomass waste generated in the agricultural and food industries is increasingly seen as a valuable bioresource that can be converted into useful products. Large quantities of wastes, including processing residues, are generated as a result of agro-industrial activity [51].

In the past few years, there has been a growing global interest among researchers in the field of processing through PEF [52,53]. Ongoing research is being conducted to investigate the impact of PEF processing on the nutritional characteristics of various food products, including fruits and vegetables [54–56]. By conducting an extensive literature review, our aim was to shed light on the potential impact of innovative technologies, such as PEF, in several products of the food industry. The main focus of the current study on PEF-assisted extraction was the investigation of polyphenols obtained from fresh foods as well as their by-products. Specifically, the implementation of PEF and how they could effectively enhance sustainable practices by optimizing the extraction of polyphenols was explored. Given the pressing necessity for long-term strategies aimed at enhanced global food production and waste management, this review aims to provide a comprehensive perspective on the functionality of PEF technology. Furthermore, in this review, the existing challenges and potential opportunities within this particular field were also analyzed.

2. Review Methodology

The study utilized three electronic databases, i.e., Google Scholar, Scopus, and Science Direct, to conduct a comprehensive search for research studies associated with the recovery of polyphenols from fresh fruits, vegetables, plants, crops, algae, as well as their byproducts, using PEF technology. The following terms were used to find articles published between 2012 and 2023 that met the review's criteria: ("polyphenols" OR "fruit" OR "vegetable" OR "plants" OR "wheat" OR "cereal" OR "cereal grains" OR "algae" OR "food byproducts" OR "antioxidant activity" AND "Pulsed Electric Field" OR "PEF" AND "polyphenols" OR polyphenolic compounds AND "extraction" OR "isolation").

3. The Impact of PEF Parameters

The most important factor in PEF is electric field strength [57]. In most cases, a stronger electric field will result in a greater amount of electroporation and cell substance transfer [58]. The extraction efficiency is measured by employing the electric field strength, which affects the targeted molecule's surface tension, diffusivity, solubility, and viscosity [59]. In the PEF treatment, electrical impulses with a large voltage amplitude are applied. Items placed inside the chamber are exposed to electrical pulses with peak voltages of up to 80 kV/cm and durations of only some μs [60]. Depending on the effects desired and the characteristics of the food being processed, the process conditions can be altered, such as the electric field strength, pulse frequency, pulse width, shape of the pulse wave, and exposure time (which is affected by the flow rate and volume of fluid in the electrode chamber) [61].

The effectiveness of cell disintegration in releasing intracellular substances is significantly impacted by the pulse width. The membrane's disruption is proportional to the frequency and pulse width of the applied high-voltage pulses [62]. The duration of exposure to electrical field strength for food samples is between 100 μs and 10 ms, which is sufficient to damage plant tissue at an electric field intensity of 0.5–5.0 kV/cm. The flow/speed of the food sample, along with pulse width and pulse frequency, all have an important part in calculating the total treatment time [63,64]. The specific energy of a treatment also depends on the strength of the electrical field strength, the duration of the treatment, and the electrical resistance of the treatment chamber. The treatment zone's dimensions and configuration, as well as the conductivity of the sample food, are used to regulate the resistance [65,66].

The extraction of PEF is significantly impacted by temperature. The PEF extraction method does not require high temperatures and can therefore typically be carried out at room temperature. The extraction process is negatively impacted by temperatures above 90 °C because the viscosity of solvents decreases [67].

4. The Principle of Electroporation

Electroporation is a process where cells become more permeable to outside substances after being subjected to PEF through the application of brief, powerful electric pulses [68]. Electroporation precise mechanism determination remains a challenging task [69]. Attempts to explain the mechanism of reversible electroporation and electrical membrane breakdown have been proposed using experiments performed on hypothetical structures [70,71]. As a result of this imposing process at the membrane interfaces, the membrane potential rises when a biological cell is subjected to an external electric field. In Figure 1, a fundamental situation of a spherical biological cell is illustrated. The emergence of reversible or irreversible pore space in the membrane can be induced by exceeding a critical value (E_c) of the electric field strength [72]. Permeabilization of the cell membrane can be either temporary or permanent, depending on the strength of the external electric fields and the frequency of the pulses [73]. Cell membranes undergo reversible permeabilization when the applied electric field is below the critical value, or when only a small number of pulses are applied [74]. However, irreversible electroporation occurs with stronger PEF treatment, leading to cell membrane disruption and cell death [75]. The process of cell disintegration is commonly acknowledged as a direct method for examining the structural characteristics of samples that have undergone different processing techniques. The cell disintegration index (Z_p) is a metric used to measure the proportion of permeabilized cells in plant tissues by examining the frequency-dependent conductivity of both intact and permeabilized cells. The numerical value of Z_p varies between zero, representing intact tissues, and one, indicating tissues in which all cells have undergone permeabilization [76].

5. Applications of PEF in Fresh Food Products and By-Products

5.1. Fruits

5.1.1. Prunus Fruits

The extraction of polyphenol-rich compounds from defatted apricot (*Prunus armeniaca*) kernel biomass was studied by Makrygiannis et al. [77]. To improve extraction efficiency, the use of deep eutectic solvents (DES) and PEF integration were investigated. The samples were subjected to PEF treatment for 15 min at an electric field strength of 1 kV/cm. The pulses had a frequency of 1000 μs and lasted for 10 μs each. Samples were pretreated with PEF for 15 min, and extracted for 3 h at 60 °C. The defatted apricot kernel biomass was stirred with water or DES (glycerol: choline chloride 2:1 *w/w*) for 15 min. According to the findings, a boost of 88% in total polyphenol content (TPC) was achieved by applying PEF before the extraction process. In a similar way, DES employment led to roughly a 70% improvement in TPC. When the two methods were combined, a 173% increase (12 mg gallic acid equivalent (GAE)/g dw) was observed. The best method for extracting bioactive compounds from defatted apricot kernels was found to be DES with PEF prior to extraction. The levels of total flavonoid content (TFC), ferric-reducing antioxidant power (FRAP), and antioxidant activity (AA) all followed a similar trend. As a result of using the aforementioned values, values for TFC, FRAP, and AA were 10 mg rutin equivalent (RtE)/g dw, 18 mol ascorbic acid equivalent (AAE)/g dw, and 12 mol AAE/g dw, respectively. When compared to water-only control extraction, the respective increases were 150%, 80%, and 71%. This research showed that bioactive compound extraction could be improved by using low-voltage PEF treatment in conjunction with a variety of green solvents, including DES.

Recently, *Prunus spinosa*, commonly known as blackthorn or sloe, possesses numerous health benefits attributed to the antioxidant and antibacterial properties present in the fruit. The research by Kotsou et al. [78] aimed to investigate the impact of different extraction parameters, such as extraction time, temperature, and solvent, on the extractability of polyphenols in *P. spinosa* fruit through US and PEF treatments along with conventional extraction (stirring). The electric field strength was set at 1.0 kV/cm with a pulse period of 1 ms, and 10 μs pulse length. In order to enhance the parameters and evaluate their impact on the antioxidant properties of the extracts, a response surface methodology was employed. The results showed that PEF treatment along with US and stirring was the most efficient way to recover a high quantity of total polyphenols. TPC was predicted by response surface methodology at 23.5 mg GAE/g dw. PEF was observed to have little to no impact on neochlorogenic acid and total anthocyanins recovery, where sole stirring and US with stirring were more efficient, respectively. However, partial least square analysis revealed that the combination of PEF along with US and stirring, with extraction for 30 min utilizing 25% v/v aqueous ethanol as solvent at 80 °C, the polyphenol extraction was enhanced. Indeed, the experimental values showed that TPC was measured at 30.74 mg GAE/g, neochlorogenic acid at 4.13 mg/g, and total anthocyanins at 125 μg cyanidin equivalent (CyE)/g. TPC was increased by 27% from a sole stirring extraction extract and by 57% than US with stirring extract. This research provides important new information about how to optimize the extraction process and how *P. spinosa* fruit could be used in the future of food science and medicine.

Cherry (*Prunus avium* L.) is one of the widespread fruits owing to its sweet taste and intense color associated with anthocyanins [79,80]. In addition to bolstering the immune system and protecting against cancer, heart disease, and other oxidative stress-related diseases, anthocyanins and polyphenols have also been shown to reduce inflammation [81,82]. To that end, PEF treatment was studied in order to examine whether the polyphenol recovery would increase after its use.

Sotelo et al. [83] investigated the impact of low or moderate PEF processing on sweet cherries, focusing on the release of anthocyanins and polyphenols. The electric field strength ranged from 0.3–2.5 kV/cm with a pulse width of 20 μs, pulse frequency of 100 Hz, and pulse number from 385–10,000 with water as solvent. The PEF-treated samples were evaluated immediately and 24 h after treatment. The anthocyanin content, specifically cyanidin glucoside, of the cherry samples was the highest when subjected to high-intensity (2.5 kV/cm) PEF treatment. In addition, it was observed that samples measured 24 h later exhibited a significantly higher content of cyanidin glucoside (~2.3 μg/g wet weight) in comparison to the samples taken immediately after PEF treatment (~1.98 μg/g ww) or untreated samples (~1.95 μg/g ww). This observation indicates that there is a time delay required for the release of anthocyanins following electroporation by PEF. However, an interesting finding was observed in polyphenol extraction, where the increasing PEF intensity resulted in a decrease in recovered polyphenols. For instance, rutin was measured at 7.77 μg/g ww which was much higher than both immediately after PEF measurement (5.04 μg/g ww) and after 24 h of PEF (4.75 μg/g ww).

5.1.2. Grapes

Since 2012, there has been a notable increase in the number of investigations conducted on grape and wine by-products, as well as on grapes and wines themselves. Regarding winemaking, the majority of research studies have focused on investigating the relationship between PEF treatment and the levels of polyphenols present.

El Darra et al. [84] explored the most optimum pretreatment techniques PEF, ultrasound (US), and thermal pretreatment to recover polyphenols from red grapes (Cabernet Franc) during fermentation. The color intensity was increased, the anthocyanins content was raised, and the extraction of phenolic compounds was also increased when thermal pretreatment (50 °C, 125 kJ/kg), US (5–15 min, 121–363 kJ/kg), and PEF (0.8 kV/cm, treatment time of 100 ms, 42 kJ/kg and 5 kV/cm, treatment time of 1 ms, 42–53 kJ/kg) were applied to Cabernet Franc grapes before, during, and after the alcoholic fermentation process. The most effective pretreatments were the moderate (0.8 kV/cm) and high (5 kV/cm) intensity of PEF, which increased anthocyanin extraction yield by 51% and 62%, respectively, from the untreated sample (186 mg/L) while the moderate thermal and US pretreatments increased the yield by 20% and 7%, respectively. Throughout the ethanol fermentation process, wines made from PEF-treated Cabernet Franc grapes maintain their deepest color. It is of high importance that energy expenditure during PEF treatment was the lowest, at 40–50 kJ/kg, while this technique was the most effective one. The phenolic, anthocyanin, and tannin contents of wines could be increased through the US and thermal treatments, and the wine color intensity could be increased in comparison to their untreated counterparts.

A couple of years on, El Darra et al. [85] evaluated the extraction of primary polyphenols and composition (co-pigmentation, non-discolored pigments) of recently fermented model wine from the Cabernet Sauvignon variety using three pretreatments: PEF, enzymes treatment, and thermovinification The application of PEF pretreatment (electric field strength at 5 kV/cm, treatment time of 1 ms, specific energy of 48 kJ/kg) resulted in non-significant variations in the wine polyphenol concentration in the span of 16 days. The untreated sample ranged from initially 130.9 to 305 mg/L, whereas the PEF-treated sample polyphenol content ranged from 364.1 to 359.8 mg/L. Freshly fermented model wines exposed to any of the pretreatments showed increased conversion of anthocyanins to derived pigments. Getting a different profile of newly fermented model wines (color attributes and polyphenol content) without adding additives like enzymes, and especially without heating is the main attraction of PEF pretreatment. When comparing PEF to thermovinification, the former uses less energy (W = 48 kJ/kg) and results in a smaller temperature swing (7 °C compared to 50 °C). However, wines should be tested after months or years of bottling in future studies to further in order for the results to be more valid.

Additional research related to winemaking was carried out by Delsart et al. [86], where Cabernet Sauvignon red grapes were studied. The highest electric field strength 4 kV/cm with a treatment time of 1 ms altered the visual appearance of skin extract to more reddish and led to greater extraction of the anthocyanins from ~480 to ~570 mg/L, while the employment of a lower electric field strength at 0.7 kV/cm and longest treatment duration at 200 ms led to a wine that was richer in tannins by 36% (from 2.5 mg/L) when compared to untreated samples. The parietal tannins and skin cell walls were most affected by the PEF treatment which was both the longest in duration and had the highest energy. Consequently, PEF-treated grapes with low electric field strength for a short time affect primarily anthocyanins, while treatment with a lower intensity but for a much longer time affects mainly tannins.

An interesting research by Delsart et al. [87] examined the evolution of the total polyphenolic index (TPI) at three stages of the vinification cycle (t = 1, 4, 7, and 210 days) in Merlot grapes for both treated and untreated with PEF samples. The results proved that the PEF-treated samples with conditions of an electric field strength of 0.7 kV/cm, and a treatment duration of 40 ms, showed a higher TPC at 210 days. For instance, the PEF-treated sample yielded >950 mg/L, whereas the untreated sample yielded ~870 mg/L of anthocyanins. A similar pattern was observed in tannin concentration, where the PEF-treated samples had a ~18% increase from the untreated samples (~2.7 g/L). However, it was observed that increasing PEF intensity seemed to decrease the tannin concentration, probably due to their diffusion. The pre-fermentative maceration times in winemaking can be shortened as a result of PEF treatment, which accelerates the kinetics of phenolic compound extraction. PEF treatment at 0.7 kV/cm and 40 ms was most preferred based on sensory evaluation. Extraction of intracellular components (total polyphenols, tannins, anthocyanins) was consequently improved with an increase in both electric field strength and pulse duration. However, the enhanced extraction process may compromise the final product's quality (sensory attributes). The findings also showed that PEF treatment affected aroma composition, which could have significant consequences for the final product aroma and style.

A research study conducted by Maza et al. [88] deals with the dependency of PEF on polyphenol extraction in Grenache grapes. TPI extraction rates were modeled against electric field strength and energy input for each PEF treatment to identify optimal processing parameters for maximizing TPI. All treatments used an optimal energy input of 4 kJ/kg to achieve a TPI of 50, and they varied from high intensity and short time (8 kV/cm and 45 μs pulse duration) to low intensity and long time (1 kV/cm and 2800 μs). The best PEF treatment conditions were 8 kV/cm and 6.7 kJ/kg where a TPI value of 73.15 was ensured while in the control sample, the TPI was 61.15 (about a 19% increase). Maceration time was decreased by 25–37% when these PEF treatments were applied compared to untreated grape samples. The results of this study also suggest that wineries that want to shorten maceration time and still produce high-quality red wines after fermentation and 12 months of aging should apply PEF treatments of 4 kV/cm and 4–5 kJ/kg to the grapes before maceration.

The same research team also conducted a study [89] on Grenache grapes. Garnacha grapes were studied, and the results were compared after 3 and 6 days of maceration and then after 6, 12, and 24 months of bottling. The flow rate was set at 2500 kg/h with a residence time of 0.09 s in the processing area, whereas the PEF treatment amounted to 3.7 square pulses of 100 μs width at an electric field intensity of 4 kV/cm and total specific energy of 6.2 kJ/kg. On day 0, grapes exposed to the PEF treatment and in which the extraction period was the longest (6 days of extraction) showed the highest values of total anthocyanin content (~500 mg/L), TPI (~60 AU at 280 nm), tannin content (~1.5 g/L of epicatechin), as long as color intensity (~15 AU). In a 24-month stability study of flavonoids (anthocyanins, hydroxycinnamic acids, flavonols, flavanols) it was shown that maceration for 6 days was more efficient than maceration for 3 days and then the control sample, measuring the highest flavonoid concentration. The results of this study suggest that the application of PEF treatments has a significant impact on the extraction of various polyphenols and individual polyphenols. Consequently, the wines produced from PEF-treated grapes showed a higher concentration of these compounds compared to wines made from untreated grapes, despite undergoing an equal duration of maceration.

Red wine varieties seem to have piqued the interest of Comuzzo et al. [90]. In Italy, where PEF treatment was evaluated on the modifications of polyphenols in red grapes (cv. *Rondinella*). Regarding the results, color intensity, TPI, anthocyanins, and total tannins recorded the highest values using flow at 250 L/h, the electric field strength at 1.5 kV/cm, pulse length 10 μs (total specific energy 20 kJ/kg) after both two and twelve months of storage. Specifically, these values were found to be 4.3 (38.7% increase), 44.8 (41.3% increase), 78 mg/L (50% increase), and 2.4 g/L (50% increase) after twelve months of bottle storage when compared to the untreated samples, respectively. The same pattern was observed in the HPLC analysis of anthocyanins. This study revealed that PEF-pretreated grapes were able to produce wines of the low-color red cv. *Rondinella* with significantly higher color intensity and stability. After a year of storage, anthocyanin and tannin concentrations were highest in treatments with a specific energy of 10–20 kJ/kg; however, when operating at lower energy levels (2 kJ/kg), adverse consequences were observed. The utilization of PEF technology holds promise in assisting winemakers in the production of consistent varietal wines, without the need to incorporate other (colored) varieties for color correction, thus safeguarding the unique flavor profiles of these wines.

A study [91] conducted in our laboratory proved that applying the PEF pretreatment method is effective only under appropriate electric field intensity conditions. The fact that an increase in electric field strength failed to show a correlation with an improved total polyphenol recovery was of high importance. For instance, upon a range of electric field strength of 1.2–2.0 kV/cm, an intensity of 1.4 kV/cm, and short pulses of 10 μs in a period of 1 ms to fresh grapes (*Vitis vinifera*) led to TPC value of ~110 mg GAE/g dw (49.15% increase) and quercetin-3-rutinoside concentration of 0.083 mg/g dw (85% increase) when compared to the untreated sample. Therefore, this research opens up new horizons regarding the benefits of PEF pretreatment when the optimum conditions are chosen. In the results, it was mentioned that in the same conditions, two more secondary metabolites increased in large percentages after treatment with PEF, kaempferol-3-glucoside and gallic acid which reached 0.153 mg/g dw (66% increase), and 0.124 mg/g dw (63% increase), respectively.

Grape pomace is a by-product that has been extensively studied for the influence of pulsed electric field on the concentration of polyphenols. A corresponding study by Brianceau et al. [92] investigated the extraction kinetics and the level of polyphenols during a hydroalcoholic extraction at different temperatures in fermented grape pomace. In the current investigation, alongside the examination of PEF conditions, the variable of densification pressure was also analyzed. For instance, it was found a PEF pretreatment using electric field strength at 1.2 kV/cm, energy input at 18 kJ/kg, and density at 1.0 g/cm^3 had a statistically significant ($p < 0.05$) increase in TPC, irrespective of the extraction temperature. It was observed that this density value showed at least 7.5% higher TPC than other values (0.6, 0.8, and 1.3 g/cm^3). However, a further increase in extraction temperature increased in a higher TPC. For example, gallic acid was increased from 4.53 mg/100 g (20 °C) to 7.40 mg/100 g (50 °C) while, correspondingly, the TPC increase from 60.98 mg/100 g to 113.58 mg/100 g. Total polyphenols extraction from fermented red grape pomace can be improved by densification in conjunction with PEF treatment. Grape pomace that has undergone fermentation can be treated with PEF to enable temperature-selective extraction of total anthocyanins. For these reasons, PEF can be used in place of traditional pre-treatments of raw material (like dehydration and grinding), achieving both goals of lowering production costs and increasing extraction selectivity. This study also sheds light on the possibility of extracting selective phytochemicals from a variety of foods.

The objective of the study by Barba et al. [93] was to assess and compare different solvent-free extraction methods for high-value components in fermented grape pomace. The grape pomace was treated with various physical methods, such as US, PEF, and high voltage electric discharges (HVED), which have the potential to cause cellular damage. These treatments were applied to aqueous suspensions of the pomace. PEF conditions required an electric field strength of 13.3 kV/cm with frequency of 0.5 Hz. The effectiveness of these technologies was evaluated in terms of phenolic compound extraction and, more specifically, anthocyanin recovery while maintaining constant Z_p. According to the results, HVED was found to be the most efficient extraction method. With constant Z_p set at 0.8, HVED reached ~300 mg GAE/L, almost twice the value of PEF and US. However, in the same Z_p value, PEF achieved greater anthocyanin recovery than HVED, 63.47 and 40.64 mg/L, respectively. The HVED method demonstrated the highest interest due to its notable impact on polyphenol compound yield. Nevertheless, the selectivity of HVED in terms of anthocyanin recovery was found to be lower compared to that of PEF and US.

To increase the efficiency of the seeds of red grapes (*Vitis vinifera* L.), rich in valuable phenolic compounds, the study by Atanasov et al. [94] discussed the use of low PEF intensity, with electric field strength at 0.86 kV/cm, frequency at 13 Hz, pulse duration 900 μs, pulse interval 75 ms, and treatment time 810 ms. By applying PEF to the red grape samples, similar polyphenol yields could be achieved with lower concentrations of ethanol. For instance, comparable results (~24 mg GAE/g) were measured when an untreated sample was extracted for 120 min with 75% ethanol and a PEF-treated sample was extracted with 20% ethanol. Therefore, it is possible to enhance the release of thermally unstable bioactive compounds under mild processing conditions by optimizing electric field strength in combination with an appropriate solvent system.

Delso et al. [95] assessed the viability of PEF technology as a potential alternative approach for the processing of red grape juice, so they studied the enhancement of juice derived from grapes. The sample was treated with PEF (electric field strength at 5 kV/cm, specific energy at 63.4 kJ/kg, and pulse width 40 μs) and had a TPC 1.5 times higher than that of untreated grapes, 1434.30 mg GAE/L and 916.10 mg GAE/L, respectively. The same pattern was observed in every antioxidant assay (TPI, CI, antioxidant capacity, and 2,2-diphenyl-1-picrylhydrazyl, DPPH•). The study also contained a decontamination process, in which the PEF conditions included electric field strength at 17.4 kV/cm, specific energy at 173.6 kJ/kg, and pulse width of 10 μs. With PEF treatment, the current microbiota was eliminated to an undetectable level (30 CFU/mL) of yeasts, molds, and vegetative mesophilic bacteria. Furthermore, even at abusive refrigeration storage temperatures (10 °C), PEF-treated juices were microbiologically stable for up to 45 days. The PEF decontamination treatment and storage time/temperature did not affect the juice's quality or sensory characteristics, which were similar to those of fresh juice. This study highlights the promising future of PEF as a sustainable, enzymatic and heat-free alternative for the production of polyphenol-enriched and microbially stabilized red grape juice in the juice industry.

In a study by Ricci et al. [96] the impact of PEF treatment was investigated on the extractability of anthocyanins and polyphenols in Sangiovese red grapes. The grapes underwent a pre-fermentative PEF treatment on a pilot scale, with electric field strengths in the range of 0.9–3 kV/cm, generated by the application of short, high-voltage pulses, and specific energies from 10.4–32.5 kJ/kg. The results showed that a PEF treatment with dynamic maceration for 2 h, and static maceration for 12 h led to the highest polyphenol recovery at 439 mg GAE/L, achieving ~55% increase from the untreated sample.

Ziagova et al. [97] focused on the recovery of polyphenols from grape leaves and marc from cv. *Xinomavro*, which was examined for its antioxidant activity and TPC. PEF was combined with US-assisted extraction. The most effective results were achieved when a solid-to-liquid ratio of 1:20 of plant material and water as solvent were used, 5 min of PEF at an electric field strength of 0.5–2 kV/cm, and 30 min of US. Comparing grape leaves and marc, it was found that grape leaves were richer in polyphenols (97 and 31 mg GAE/g dw, respectively). The same pattern was observed in the antioxidant capacity levels were 88% was observed for grape leaves and 31% for grape marc.

The study conducted by Ntourtoglou et al. [98] examines the impact of PEF treatment on stem grapes, specifically focusing on the increase of polyphenols and volatile compounds. A PEF process with a relatively low electric field strength of 1 kV/cm, for a brief duration of 30 min on the grape stems. When PEF was solely used as an extraction technique, the extracted TPC was measured at ~0.05 AU and showed a statistically non-significant increase (only 4%) regardless of the solvent (50% *v*/*v* aqueous methanol or water). With the implementation of US extraction, the samples with 50% *v*/*v* methanol reached a 17% increase, whereas samples with water as a solvent had a 35% increase, revealing the importance of extraction solvent. Regarding volatile compounds, the control sample showed an average concentration of 0.73 mg/Kg. However, when PEF was applied before US extraction, the observed increase in concentration was as significant as 234%. The fact that two more volatile compounds (Benzene, 1-methoxy-4-methyl, and 1,14-Tetradecanediol) were extracted is of high importance. In conclusion, the utilization of PEF as a preliminary treatment method for extracting different volatile and polyphenolic compounds shows significant potential for improving the efficiency of the extraction process. On top of that, through the examination of additional factors related to the extraction process, such as the selection of extraction solvents, duration, and temperature, it may be possible to further optimize the extraction yield.

The most recent study was conducted by Carpentieri et al. [99]. Total polyphenols, tannins, anthocyanins, and flavonoids were extracted from red grape pomace using PEF, and response surface methodology was used to analyze the efficacy of this process in improving the extraction of these crucial intracellular components. Results showed that grape pomace tissue permeability was significantly increased after PEF was applied under optimal processing conditions (electric field strength of 4.6 kV/cm and energy input of 20 kJ/kg). In comparison to the control (PEF-untreated sample), this increased extraction rates of TPC by 15%, TFC by 60%, total anthocyanin content (TAC) by 23%, and tannins content (TC) by 42%. Additionally, it was observed that PEF treatment did not result in the degradation of these compounds.

5.1.3. Apples

An apple is the fruit of the apple tree of the Rosaceae family. It is by far one of the world's most widespread and widely grown fruits since it represents 50% of deciduous fruit trees, with a global yearly output of about 60 million tons. Apples are rich in polyphenols and mainly flavanols (catechins and proanthocyanidins), which are also considered the main category of polyphenols in apples (71−90%), followed by hydroxycinnamic compounds (4−18%), flavonols (1−11%), dihydrochalcones (2−6%) and anthocyanins (1−3%) [100].

The objective of the study from Wiktor et al. [101] was to investigate the impact of PEF treatment on the concentration of particular bioactive compounds from apple tissue. The range of specific energy input varied between 0–80 kJ/kg. Apple tissues were subjected to PEF treatment at different electric field strengths of 0, 1.85, 3, and 5.0 kV/cm, and at varying pulse numbers of 0, 10, 50, and 100. The results showed that at 3 kV/cm and 100 pulses, a slight increase of ~10% in TPC of apple tissues was feasible compared to the untreated sample (426.69 mg chlorogenic acid/100 g dm). The findings of the study suggest that PEF has the potential to improve the efficiency of extracting bioactive compounds from plant tissue.

In a study by Dziadek et al. [102], a PEF treatment with a number of cycles ranging from 4, 6, and 8 (200, 300, and 400 pulses, correspondingly) was used in apple juice. The concentration of polyphenols and antioxidant activity were all measured immediately after the PEF process after 24, 48, and 72 h of refrigeration. The electric field strength was set at a high value of 30 kV/cm. The untreated sample stored for 0 h yielded 337.51 mg/100 mL and it was observed that the application of PEF treatment, irrespective of the pulse quantity, did not yield a statistically significant impact on the overall polyphenol content found in apple juice. Antioxidant activity was measured at 17.4 μmol Trolox/mL and was found to be reduced both immediately after processing and after 24 h of storage, both of which were affected by PEF treatment and the number of pulses. However, increasing PEF pulses were observed to increase polyphenol concentration after 24, 48, and 72 h of refrigeration, from ~233 to 300.90, 280.88, and 295.66 mg/100 mL.

PEF treatment was also studied in apple by-products. A study by Pollini et al. [103] in 2021 examined the effect of various non-conventional extraction methods, such as PEF, ultrasound-assisted extraction (UAE), supercritical fluid extraction (SFE), microwave-assisted extraction (MAE), and pressurized liquid extraction (PLE) on the polyphenol content in apple pomace. Only two PEF pilot treatment experiments were performed to benchmark a short-term, medium-intensity treatment (3 kV/cm and 9 pulses with 450 J energy/pulse) against a longer, low-intensity treatment (2 kV/cm and 100 pulses with 200 J energy/pulse). Hydroethanolic solutions from 30–70% *v/v* were used as extraction solvents. Regardless of the PEF treatment used, the values of polyphenols ranged from 181.4–223.5 μg GAE/g of fresh apple pomace. However, better values were obtained using 30% *v/v* aqueous EtOH as a solvent in both PEF treatments. Nevertheless, it was observed that the apple pomace samples treated with PEF had the lowest results in terms of polyphenol values compared to other treatments.

Recent research on the apple fruit was carried out by Matys et al. [104] who combined the MAE with PEF treatment to dry apples. The electric field strength was set at 1 kV/cm, pulse frequency at 20 Hz, and pulse width at 7 μs. The total amount of rectangular pulses had supplied energy of 1, 3.5, and 6 kJ/kg, whereas microwave power ranged from 100–300 W. The results showed that in high energy input (6 kJ/kg), the cell disintegration index was measured at 0.36. However, when specific energy was set at 3437 kJ/kg and 300 W of microwave power through response surface methodology, the results showed that the desirability was found to be 0.624, whereas total polyphenols were predicted to be 1257 mg GAE/100 g dm.

5.1.4. Pomegranate

Punica granatum, the pomegranate's scientific name, stems from the Latin for seeded apple, *Pomum granatus* [105]. The pomegranate is a plant of the genus Punica, family Punicaceae, and a source of high content of phenolic components with highly bioactive characteristics [106].

PEF pretreatment was applied to the pomegranate peel in a study by Rajha et al. [107], who used a variety of extraction techniques, such as infrared (IR), US, PEF, and high voltage electric discharge (HVED). For all extractions, the temperature was held at 50 °C. The electric field strength was maintained at 10 kV/cm, with the temperature rise during PEF treatment being kept below 5 °C and the total input energy ranging from 90–100 kJ/kg. The polyphenol extractable fraction by PEF showed a mean value of 39.2 mg GAE/g dm after 7 min of extraction. When compared to the HVED treatment, this value was found to be 15.22% lower. In comparison to other treatments, however, it was observed to be 168.97% more effective than the US, 387.5% more effective than IR, and 680% more effective than a water bath treatment. Comparing this method to Ziagova et al. [97] who also used a combination of PEF and US pretreatments, the latter authors measured 208 mg GAE/g dw of pomegranate peels. This research shows that both the HVED and PEF treatments are highly efficient at removing polyphenols from pomegranate peels. PEF is less efficient at polyphenol recovery than HVED. In contrast to HVED, the treatment process in PEF is more selective and causes less damage.

5.1.5. Citrus Fruits

Citrus fruits such as oranges (*Citrus sinensis*), lemons (*Citrus limon*), and pomelos (*Citrus maxima*) are the most valuable fruit crops worldwide. The vibrant color, delicious flavors, and refreshing aromas of citrus fruits have made them a global favorite. Antioxidant bioactive compounds eliminate free radicals, halt lipid peroxidation reactions, and protect against other forms of oxidative damage, all of which are essential to the proper functioning of a cell. Moreover, their antioxidant activity may protect against a wide range of chronic diseases like cancer, diabetes, and heart disease [108].

The processing of citrus fruits is of industrial significance because they are rich in bioactive compounds (such as vitamins, antioxidants, carotenoids, and polyphenols). Juice extraction from various citrus fruits could benefit from PEF being applied to whole fruits. In addition, PEF-treated citrus peels at high electric field strength could significantly increase the concentration of polyphenols in the extracted juice. To that end, a corresponding study was conducted by El Kantar et al. [109], who intended to improve polyphenol extraction and improve the juice. The experimental procedure involved subjecting unharmed fruits and a pile of peels to PEF treatment, with electric field strengths of 3 kV/cm and 10 kV/cm, respectively. The samples were then subjected to 1 h of extraction with 50% *v/v* aqueous ethanol solution. The implementation of PEF increased extracted polyphenols in citrus fruit juice. Orange, lemon, and pomelo TPC were measured at ~70 mg/100 mL (~49% increase), ~60 mg/100 mL (~50% increase), and ~80 mg/100 mL (~60% increase) when compared to the untreated samples. Furthermore, PEF treatment affected major polyphenols concentration from orange (hesperidin) and pomelo (naringin) flavedo, which were increased from 4.852 to 5.073 mg/g of dried mass and from 7.354 to 10.366 mg/g of dm, respectively. An interesting decrease of eriocitrin in lemon flavedo was observed from 3.064 to 1.440 mg/g dm after PEF treatment.

Since orange peels (albedo and flavedo) contain ascorbic acid, carotenoids, and polyphenols, Athanasiadis et al. [110] set out to investigate and optimize the factors affecting the extraction process. They used US or PEF as a pretreatment step prior to extracting the bioactive compounds using an ethanol and water mixture. In this experiment, the authors used an electric field strength of 1 kV/cm, pulse duration of 10 μs, and treatment period of 1 ms (frequency: 1 kHz). The extraction time (15–180 min) and temperature (20–80 °C) were optimized using a response surface methodology. Hesperidin (16.26 mg/g dw) and TPC (34.71 mg GAE/g dw) were improved by PEF treatment. The study may have flaws, such as its narrow focus on a single orange variety. For real-world, large-scale applications, however, the proposed pretreatment methods involving PEF hold much promise. These applications span numerous sectors, including the food and beverage industry, the cosmetics and pharmaceutical industries, and the movement to replace synthetic pigments with natural ones.

Luengo et al. [111] conducted a study to investigate the PEF treatment on the extraction of polyphenols from orange peel during the pressing process. The highest value of Z_p was obtained when various electric field strengths were tested, with a treatment time of 60 s (consisting of 20 pulses, each lasting 3 s). Treatments with PEF consisted of 5–50 pulses of 3 s each (15–150 s) at electric field strengths of 1–7 kV/cm. The treatments varied in the amount of energy they provided, from 0.06–3.77 kJ/kg. Distilled water was used as the extraction solvent. The frequency of the pulses was set to 1 Hz. Following exposure to PEF with an electric field strength of 5 kV/cm, the concentrations of naringin and hesperidin exhibited a significant increase, rising from 1 to 3.1 mg/100 g fw and 1.3 to 4.6 mg/100 g fw, respectively. Furthermore, the extraction yield of polyphenols was determined to be 34.80 mg/100 g fw. The application of PEF to orange peel resulted in a significant increase in the yield of polyphenol extract. Specifically, at electric field strengths of 1, 3, 5, and 7 kV/cm, the yield of polyphenol extract increased by 20, 129, 153, and 159% respectively. The application of PEF treatments at intensities of 1, 3, 5, and 7 kV/cm resulted in a significant enhancement of the antioxidant activity in the extract. Specifically, the antioxidant activity increased by 51, 94, 148, and 192% compared to the untreated sample, respectively. The utilization of PEF technology presents numerous advantages in comparison to alternative techniques employed for enhancing the extraction of polyphenols from orange peels via pressing. These advantages include the elimination of the requirement to dehydrate the sample and the utilization of water as the solvent. In addition, this method presents an environmentally friendly and cost-effective alternative to conventional extraction techniques, such as initial product dehydration and the utilization of substantial quantities of organic solvents.

The Interest of the researchers was also directed towards the peel of lemons. Total polyphenols were extracted from lemon peel residues by pressing, and Peiró et al. [112] studied the effect of different PEF intensities (3–9 kV/cm and 0–300 s treatment time pulses). Since no statistically significant differences were found between 7 and 9 kV/cm, Z_p concluded that 30 pulses of 3 μs (total 90 μs) and an electric field strength of 7 kV/cm are optimal for increasing permeability. Compared to the untreated samples, PEF did not affect the polyphenol extraction yield regardless of the size of the lemon residue (1, 2, or 3 cm). However, 3-cm lemon peels were chosen as the optimal size, and PEF treatment significantly increased TPC, as an average of 160 mg GAE/100 g of dw was up to a 150% increase from the control sample. Pressure and electric field strength were found to significantly increase the concentrations of hesperidin and eriocitrin, the two most abundant polyphenols in lemon residues. Applying the pulsed electric field with 30 pulses of 30 μs and electric field intensity of 7 kV/cm, the extraction efficiency of polyphenols showed a 300% increase, as well as hesperidin, reached 84 mg/100 g fw and the amount of eriocitrin was 176 mg/100 g fw.

The purpose of the study by Chatzimitakos et al. [113] was to determine the most efficient conditions for extracting bioactive compounds from lemon peel waste. Time, temperature, and solvent composition were among the variables tested alongside a number of different extraction methods like stirring, US, and PEF. The electric field strength was set at 1.0 kV/cm with a pulse period of 1 ms, and 10 μs pulse length. The results revealed a negative impact of PEF in TPC yield, rendering sole stirring as the optimum condition for extraction with 50% v/v aqueous ethanol for 120 min at 50 °C. It was also observed that

PEF with US had a deleterious effect on polyphenol extraction compared to conventional extraction. However, PEF treatment prior to stirring with 100% ethanol for 60 min at 80 °C resulted in the highest TFC yield at 7 mg RtE/g.

5.1.6. Quince

Leaves and peels are quince by-products. Recovery of bioactive compounds from quince peels was investigated by Athanasiadis et al. [114]. Extraction parameters, such as solvent, temperature, and time, as well as extraction techniques such as PEF and US, were initially investigated for their effects. Then, these parameters were optimized using response surface methodology to extract bioactive compounds more effectively. The PEF used an electric field strength of 1 kV/cm, had a pulse duration of 10 μs, a pulse period of 1 ms, and a frequency of 1 kHz. Using a frequency of 3 kHz and a bath temperature of 30 °C, US was applied for 20 min. The best extraction conditions were achieved by stirring at 65 °C for 120 min. Using techniques like principal component analysis and partial least squares analysis, authors were able to establish that quince peels contain a large number of bioactive compounds. TPC (43.99 mg GAE/g dw), TFC (3.86 mg RtE/g dw), and chlorogenic acid (2.12 mg/g dw) were some of these. Using FRAP and DPPH$^{\bullet}$ assays, they found that the quince peels had 627.73 and 699.61 mol AAE/g dw, respectively, of antioxidant activity.

5.1.7. Berry Fruits

Lončarić et al. [115] aimed to determine the specific conditions of PEF that would result in a higher concentration of polyphenols and flavonoids in blueberry pomace. Applying PEF-assisted extraction, where ethanol was used as a solvent, with 100 pulses and an intensity of 20 kV/cm, which equals a specific energy input of 41.03 kJ/kg, a vast amount of TPC was extracted (10.52 mg GAE/g dw). Additionally, with the previous PEF conditions, phenolic acid and flavonol concentrations displayed their highest values (625.47 μg/g dw and 157.54 μg/g dw, correspondingly). On the contrary, when methanol was used as a solvent, anthocyanin and flavanol best amount of field noted (1757.32 μg/g dw and 297.86 μg/g dw, respectively). Upon comparison of other green extraction methods, it was evident that the total phenolic content (TPC) of PEF-treated extracts exhibited higher values in comparison to those obtained through HVED (~5 mg GAE/g dw) and US methods (~6 mg GAE/g dw).

The nutritional properties of red raspberry (*Rubus strigosus* var. *Meeker*) and blueberry (*Vaccinium corymbosum* var. *Bluejay*) purees after US and PEF exposure were investigated by Medina-Meza et al. [116]. When PEF-assisted extraction was applied with a high electric field strength of 25 kV/cm, and flow rate at 300 mL/min along with US treatment, no deleterious effects were observed. In relation to red raspberry, solely PEF treatment resulted in non-significant differences ($p < 0.05$) in TFC, whereas when combined with US led to an increase of ~20% from an initial ~150 μg/mL of quercetin. In PEF-increased anthocyanin concentration from ~125 to ~145 mg/L, however, a deleterious effect and decrease of at least 66% was observed with US treatment, which requires further investigation. Regarding TPC content, PEF increased the recovery from ~430 to ~470 μg/mL. In blueberry samples, PEF treatment had roughly any impact on the extraction of flavonoids (~310 μg/mL) but increased the anthocyanins recovery from ~650 to ~750 mg/L. However, another deleterious effect was observed in TPC, where a decrease of ~6% was measured from an initial ~520 μg/mL.

Ozkan et al. [117] investigated non-thermal processing methods of high-pressure processing (HPP) and PEF to preserve polyphenols in cranberrybush (*Viburnum opulus*) puree. This investigation required PEF conditions, including 20 μs pulse width, an electric field strength of 3 kV/cm, pulse frequency of 2 Hz, and specific energy input of 15 kJ/kg. The application of PEF or HPP resulted in enhanced preservation of bioactive compounds, as evidenced by a significant increase in TPC, ranging from 4.1–14% (from ~400 mg GAE/100 g fw). Additionally, both treatments demonstrated improved antioxidant activity in CUPRAC (~7% increase from 1500 mg TE/100 g fw). The results of this study suggested that the PEF process was successful in non-thermal treatments for the extraction of polyphenols from cranberry bushes.

The objective of the study from Gagneten et al. [118] was to enhance the extraction of polyphenols from blackcurrants through the utilization of PEF treatment. The study aimed to analyze the effects of electroporation on biological samples at two different initial temperatures, specifically 10 and 22 °C. A small quantity of the natural fruit juice is used to hydrate the sample. The optimal PEF conditions for achieving the greatest benefits were found to be an electric field strength of 1318 V/cm and a total of 315 pulses. The study examined the influence of electric field strength and treatment duration on the TPC and AA. Under optimum conditions, notable improvements were observed, including a 19% increase in TPC (from 3.18 mg GAE/g extract), a 45% increase in antioxidant activity (AA) (from 1.12 mg GAE/g extract), and a 6% increase in total monomeric anthocyanins (from 1.30 mg cyanidin-3-glucoside/g extract). The application of PEF treatment to blackcurrant juices resulted in a higher efficiency in extracting bioactive compounds, making them a promising choice for incorporation as functional food ingredients.

In light of the demonstrated beneficial health effects of polyphenols, Stübler et al. [119] examined the impact of various processing methods (thermal, PEF, HPP) as well as the incorporation of a vegetable juice (kale) with a relatively high protein content on the stability and bioaccessibility of polyphenols in strawberry puree. PEF conditions required an electric field strength of 11.9 kV/cm with 20 µs pulse width and energy input of 120 kJ/kg. As for the results, it was found that anthocyanins showed an increase both in the strawberry–kale mix and in the puree. Specifically, anthocyanins content increased from almost 32 (control) to 35 mg pelargonidin-3-glucoside/L (PEF-treated) in the mix and 40 (control) to 45 mg pelargonidin-3-glucoside/L (PEF-treated) in the strawberry puree. The industrial significance of the current trend towards healthier eating habits and fast-paced lifestyles is evident in the increased demand for convenient and nutritious food options centered around fruits and vegetables. Insufficient data exists regarding the effects of alternative processing methods on multi-component juice systems, wherein interactions between various components may transpire. Nevertheless, these methods are regarded as a promising means of producing nutritionally rich products. Strawberries, being the berry with the highest consumption rate, offer a substantial quantity of polyphenols and serve as an exemplary system abundant in polyphenolic compounds.

The effects of PEF on tomato juice and fruit were studied. Firstly, the purpose of the study by Vallverdú-Queralt et al. [120] was to examine the impact of using moderate (MIPEF) and high-intensity PEF (HIPEF) treatments in succession on the polyphenol profile of tomato juices. A variety of intensities in the electric field were used to test the effects of different nutrients on tomato fruits. MIPEF conditions required 1 kV/cm of electric field strength and 16 pulses of 4 µs at a frequency of 0.1 Hz, whereas HIPEF included 35 kV/cm and 4 µs pulses with a frequency of 100 Hz. Tomato juices made from untreated tomatoes had an initial TPC of ~148 µg/g fw, while juices from MIPEF- and HIPEF-treated tomatoes had an initial TPC of ~180 and ~155 µg/g fw. The polyphenol amount that was gained was enhanced by 44% with the addition of 30 pulses at 1.2 kV/cm. Therefore, it may be possible to propose the use of MIPEFs and HIPEFs together as a method for increasing the phenolic content of tomato juices. The same research team [121] studied the impact of PEF on TPC and the antioxidant capacity of tomato fruit. Electric field strength from 0.4–2.0 kV/cm and the number of pulses from 5–30 were investigated. At 1.2 kV/cm and 30 pulses, TPC achieved 144.61% relative TPC (44% increase). With the same electric field strength (1.2 kV/cm) but with fewer pulses (18), a higher % relative hydrophilic antioxidant capacity was achieved (131.75%).

5.1.8. Red Prickly Pear

Surano et al. [122] studied the use of PEF in a multiple-needle chamber to enhance the extraction of bioactive compounds from raw *Opuntia ficus-indica* (red prickly pear) fruits. The high concentration of natural pigments and bioactive compounds in prickly pear fruit contributes to its antioxidant capacity and health-promoting properties. In order to accomplish polyphenol recovery with a non-thermal technique, a novel electroporation chamber was designed and implemented, to identify the optimal pulse parameters that would maximize both juice yield and the extraction of bioactive compounds. To achieve optimum PEF conditions, it was necessary to apply pulses with an electric field strength of 1200 V/cm and a frequency of 10 Hz. A mean value of specific energy input at 11.44 kJ/kg was utilized, resulting in a temperature rise of less than 10 °C. PEF-treated samples resulted in a significant increase in juice yield by a factor of 3.3 (from 16.69%) and betalain extraction by a factor of 1.48 (from 19.5 mg/100 g), as compared to the samples that did not undergo PEF treatment. The extracted juice exhibited a notable enhancement in its antioxidant capacity, with an increase of approximately 1.4–1.5 times, measured with three different methods (DPPH•, ABTS, FRAP). Similarly, the TPC of the juice experienced a proportional increase of 1.4 times (from 9.49 mg GAE/100 g). The chamber could accommodate sizable samples without necessitating the inclusion of a conductive medium or the removal of fruit peels. This simplifies the procedure and leads to cost reduction, thereby positioning it as a viable option for industrial applications. The application of PEF treatment on prickly pear fruits results in the production of juice that exhibits enhanced nutritional content and a greater abundance of naturally occurring pigments compared to juice derived from untreated fruits. Table 1 provides a concise summary of the above research on fresh fruits and by-products.

Table 1. Application of PEF on fresh fruit and by-products with the treatment effects.

Sample	PEF Conditions	Treatment Effect	Ref.
Apricot	1 kV/cm, pulse frequency 1000 µs, 10 µs pulse duration	Increases by 88% in TPC (from ~3.5 to ~6.5 mg GAE/g dw) and 100% in TFC (from 3.78 to ~7.5 mg RtE/g dw)	[77]
Blackthorn	1.0 kV/cm, 1 ms pulse period, 10 µs pulse length	Increased TPC value by 27% (from 24.20 to 30.74 mg GAE/g) when compared to stirring,	[78]
Cherry	2.5 kV/cm, 20 µs, 100 Hz, pulse number 385–10,000	Rutin concentration increased by 54% (from 5.04 to 7.77 µg/g ww)	[83]
Grape	5 kV/cm, 1 ms pulse duration, 42–53 kJ/kg	Increase in anthocyanin content by 62% (from 186 to ~301 mg/L)	[84]
	0.7 kV/cm, 200 ms treatment duration	~19% increase in anthocyanins (from ~480 to ~570 mg/L), 36% increase in tannins (from 2.5 to 3.4 mg/L)	[86]
	0.7 kV/cm, 40 ms treatment duration	~10% increase in TPC (from~870 to ~970 mg/L), ~18% increase in tannins (from ~2.7 to ~3.2 g/L)	[87]
	8 kV/cm, 6.7 kJ/kg, 45 µs pulse duration	TPI increased by ~19% from 61.15 to 73.15	[88]
	4 kV/cm, 3.7 pulses of 100 µs width, 6.2 kJ/kg	High values of TPI (~60 AU compared to ~45 AU from untreated samples), anthocyanins (from ~480 to ~500 mg/L), and tannins (from ~1 to ~1.5 g/L) were achieved	[89]
	1.5 kV/cm, 10 µs pulse length, 20 kJ/kg, 250 L/h	TPI: 41.3% increase (from 26.3 to 44.8), anthocyanins: 50% increase (from 39 to 78 mg/L), tannins 50% increase (from 1.2 to 2.4 g/L)	[90]
	1.4 kV/cm, 10 µs pulse duration, 1 ms treatment time	Increases in TPC from to ~56 to ~110 mg GAE/g dw (49.15%), Quercetin-3-rutinoside from 0.012 to 0.083 mg/g dw (85%), Kaempferol-3-glucoside from 0.052 to 0.153 mg/g dw (66%), Gallic acid from 0.045 to 0.0124 mg/g dw (63%)	[91]
	0.9–3 kV/cm, 10.4–32.5 kJ/kg	TPC increased by ~55% (from 197 to 439 mg GAE/L)	[96]
Grape juice	5 kV/cm, 63.4 kJ/kg, 40 µs pulse width	TPC increased by ~56% (from 916 to 1434 mg GAE/L)	[95]
Wine	5 kV/cm, 1 ms treatment duration, 48 kJ/kg	TPC increased by 17–178% (from 130.9 and 305 to 364.1 and 359.8 mg/L)	[85]
Grape stem	1 kV/cm, treatment duration 30 min	PEF only: 4% increased TPC (from 0.048 to 0.05 AU)	[98]
Grape leaf	0.5–2 kV/cm	High TPC value (97 mg GAE/g dw)	[97]

Table 1. *Cont.*

Sample	PEF Conditions	Treatment Effect	Ref.
Grape pomace/seed	1.2 kV/cm, 18 kJ/kg	Increases in gallic acid from 4.53 to 7.40 mg/100 g (63%) and TPC from 60.98 to 113.58 mg/100 g (86%) when increasing temperature from 20 to 50 °C	[92]
	13.3 kV/cm, 0.5 Hz	At Z_p 0.8, PEF (63.47 mg/L) achieved greater anthocyanin recovery than HVED (40.64 mg/L)	[93]
	0.86 kV/cm, 13 Hz, pulse duration 900 μs, 75 ms pulse interval, 810 ms treatment time	Comparable TPC (~24 mg GAE/g) with the control sample, which was extracted with 75% ethanol, whereas the PEF-treated sample was extracted with 20% ethanol	[94]
	4.6 kV/cm, 20 kJ/kg	Increases in TPC from 8.30 to 9.51 mg GAE/g dm (15%), TFC from 36.68 to 58.53 mg QE/g dm (60%), TAC from 0.84 to 1.03 mg C3G/g dm (23%), and in TC from 3.84 to 5.45 mg TC/g dm	[99]
	0.5–2 kV/cm	High TPC value (31 mg GAE/g dw)	[97]
Apple tissue	3 kV/cm, 100 pulses	TPC increased by ~10% from 426.69 to 472.05 mg chlorogenic acid/100 g dm	[101]
Apple juice	30 kV/cm	Non-significant differences in TPC (from 337.51 to 340.70 mg/L), reduction in AA from 17.40 to 16.74 μmol Trolox/mL)	[102]
Apple pomace	30 kV/cm, 17 kJ/kg or20 kV/cm, 100 kJ/kg	TPC: the lowest concentration (220 μg GAE/g) when PEF with 30% *v*/*v* EtOH was used as extraction solvent compared to UAE (800 μg GAE/g), and ASE (~420 μg GAE/g)	[103]
Apple	1 kV/cm, 20 Hz pulse frequency, and 7 μs pulse width	Dry the sample efficiently, TPC measured 1257 mg GAE/100 dm	[104]
Pomegranate peel	10 kV/cm, 90–100 kJ/kg	TPC through PEF measured at 39.2 mg GAE/g dm, ~15% lower by HVED, ~169% higher than the US, ~388% higher than IR, ~680% higher than water bath treatment	[107]
	0.5–2 kV/cm	High TPC value (208 mg GAE/g dw)	[97]
Citrus juice	3 kV/cm	TPC increased by ~49%(orange) from ~36 to ~70 mg/100 mL, ~50% (lemon) from ~30 to ~60 mg/100 mL, ~60% (pomelo) from ~32 to ~80 mg/100 mL	[109]
Citrus peel	10 kV/cm	Increase in major polyphenols in orange (hesperidin, ~5%) from 4.85 to 5.07 mg/g dm, pomelo (naringin, ~41%) from 7.35 to 10.36 mg/g dm, a decrease in major polyphenol of lemon (eriocitrin, ~112%) from 3.06 to 1.44 mg/g dm	
Orange peel	1 kV/cm, 10 μs pulse duration, 1 ms treatment period	TPC increase by 25% (from 27.70 to 34.71 mg GAE/g dw) and hesperidin content by 19% (from 13.67 to 16.26 mg/g dw)	[110]
	1–7 kV/cm, 5–50 pulses of 3 s each	Increased concentrations of naringin from 1 to 3.1 mg/100 g fw (210%), hesperidin from 1.3 to 4.6 mg/100 g fw (253%)	[111]
Lemon peel	7 kV/cm, 90 μs pulse duration	TPC increased by 150% from ~64 to 160 mg GAE/100 g dw, eriocitrin concentration from 30.39 to 176.35 mg/100 g fw, and hesperidin concentration from 15.90 to 84.44 mg/100 g dw both increased by above 400%	[112]
	1.0 kV/cm, 1 ms pulse period, 10 μs pulse length	Negative impact in TPC (277% decrease) compared to conventional extraction from 51.24 to 13.56 mg GAE/g	[113]
Quince peel	1 kV/cm, 1000 Hz, 10 μs pulse duration, 1 ms pulse period	Initial increase through RSM in TPC by 8% (from 32.78 to 35.43 mg GAE/g dw), and a further increase by 34% through the PLS model as TPC reached 43.99 mg GAE/g dw	[114]
Blueberry pomace	20 kV/cm, 41.03 kJ/kg, 100 pulses	Higher values of TPC (10.52 mg GAE/g dw) than HVED (~5 mg GAE/g dw) and US methods (~6 mg GAE/g dw)	[115]
Red raspberry puree	25 kV/cm, 300 mL/min	Non-significant impact on TFC (~150 μg/mL), but increased ~16% total anthocyanin content (from ~125 to ~145 mg/L) and ~9% TPC (from ~430 to ~470 mg/L)	[116]
Blueberry puree		Non-significant impact on TFC (~310 μg/mL), increased ~15% total anthocyanin content (from ~650 to ~750 mg/L) but decreased ~6% TPC (from ~520 μg/mL)	
Cranberrybush puree	3 kV/cm, 2 Hz, 20 μs pulse width	TPC increased by ~4–14% (from initially ~400 mg GAE/100 g fw), CUPRAC antioxidant activity by ~7% (from 1500 mg TE/100 g fw)	[117]

Table 1. *Cont.*

Sample	PEF Conditions	Treatment Effect	Ref.
Blackcurrants	1318 V/cm, 315 pulses	19% increase in TPC (from 3.18 mg GAE/g extract), 45% increase in AA (from 1.12 mg GAE/g extract), and 6% increase in monomeric anthocyanins content (from 1.30 mg cyanidin-3-glucoside/g extract)	[118]
Strawberry puree and juice (kale)	11.9 kV/cm, 120 kJ/kg, 20 µs pulse width	Increase in anthocyanins content from almost ~32 to 35 mg pelargonidin-3-glucoside/L in kale mix by 9%, and from 40 to 45 mg pelargonidin-3-glucoside/L (PEF-treated) in the strawberry puree by 12.5%	[119]
Tomato juice	MIPEF: 1 kV/cm, 0.1 Hz, 16 pulses of 4 µs HIPEF: 35 kV/cm, 100 Hz, 4 µs pulses	MIPEF: TPC increased by 25% from ~148 to ~180 µg/g fw, HIPEF: TPC increased by 5% from ~148 to ~155 µg/g fw	[120]
Tomato fruit	1.2 kV/cm, 30 pulses	TPC increased by 44%, as it had 144.61% relative TPC	[121]
Red prickly pear fruit	1200 V/cm, 11.44 kJ/kg, 10 Hz	PEF-treated samples increased in juice yield by 3.3 (from 16.69%) and betalain extraction by 1.48 (from 19.5 mg/100 g) compared to untreated samples	[122]

5.2. Vegetables

5.2.1. Potato

The fourth most important food grown and consumed in the world, potatoes are a staple that can be found in almost every region of the country [123]. They are rich in carbohydrates and have a small amount of fat. They are also known to contain various nutrients such as vitamins, proteins, and fiber. Bioactive compounds such as anthocyanins and polyphenols are commonly found in the skin and flesh of potatoes [124].

To improve the extraction of polyphenols with significant antioxidant properties from potato peels, Frontuto et al. [125] studied the optimal PEF-assisted extraction conditions. To verify this, they applied electric field strength from 1–5 kV/cm and total specific energy inputs from 1–10 kJ/kg. Total polyphenol yield from PEF pretreated sample extracts was 1295 mg GAE/kg fw (+10% from the control) when using 5 kV/cm fields strength and 10 kJ/kg specific energy output, 52% ethanol as a solvent, 230 min of extraction time, and 50 °C for the subsequent solid-liquid extraction. The HPLC-DAD analysis revealed that the predominant polyphenolic compounds detected were chlorogenic, caffeic, syringic, protocatechuic, and *p*-coumaric acids. There was no indication of substantial degradation of these individual polyphenols as a result of the application of PEF.

5.2.2. Asparagus

Beneficial bioactive compounds have been found in *Asparagus officinalis* root. The primary goal of the research conducted by Symes et al. [126] was to improve the efficiency with which polyphenol and flavonoid extraction was accomplished from the roots of green asparagus by combining PEF and ionic liquids. Antioxidant activity (oxygen radical absorbance capacity, ORAC), FRAP and DPPH$^\bullet$), TPC, and TFC were all measured in this study. When compared to the standard solvent extraction method, the extraction yield was higher when PEF was used under the optimum conditions of electric field strength at 1.6 kV/cm, pulse width of 20 µs, and frequency of 200 Hz. All assays, except for ORAC, showed improvements in PEF-treated samples compared to untreated samples. Extraction yield was increased by 23%, TPC by 5%, TFC by 6%, and FRAP by 4%. Ionic liquids, on the other hand, were discovered to be more efficient than PEF treatment. For instance, ionic liquids were found to have a TFC of 122 mg RE/g. This value was ~80 times greater than the TFC achieved by PEF treatment. While ionic liquids performed better than PEF in asparagus root samples, more research is needed to determine their safety for use in the food industry.

5.2.3. Mushroom

Mushrooms are frequently consumed as a staple in vegetarian diets. They are recognized for their wide variety of health advantages, including their anti-carcinogenic and anti-infectious attributes. Furthermore, the abundant presence of polyphenols in these substances renders them highly versatile for utilization as pharmaceutical agents or dietary supplements [127,128]. The only type of mushroom that was applied to the PEF system with the purpose of greater extraction of polyphenols was *Agaricus bisporus*.

The utilization of chemical or thermal methods for the extraction of valuable compounds is widely employed across various disciplines. The utilization of PEF during the extraction process reduces the likelihood of nutrient damage in the extracted product. To that end, the study by Xue et al. [129] aimed to investigate the impact of continuous PEF treatment on the extraction process of a white button mushroom suspension with a concentration of 9% *w*/*w*. PEF with intensities ranging from 12.4–38.4 kV/cm were applied using bipolar square pulses lasting 2 μs. The mushroom suspension was exposed to electric pulses with a field intensity of 38.4 kV/cm for a duration of 272 μs at 85 °C. Based on estimations, it was determined that the optimal extraction yields would be 98% (7.9 mg/g mushroom) of polysaccharide, 51% (1.6 mg GAE/g mushroom) of total polyphenols, and 49% (2.7 mg/g mushroom) of proteins. However, traditional mushroom extraction methods yielded only 56% (6.4 mg/g of mushroom) polysaccharide, 25% of total polyphenols (1.3 mg GAE/g of mushroom), and 45% (2.6 mg/g of mushroom) proteins after processing the 9% *w*/*w* mushroom suspension at 95 °C for 1 h. For all of these substances, the yield from conventional extraction carried out at the same temperature and for a similar amount of time was negligible. This indicates that a synergistic effect of electric pulses and temperature on the extraction performance is responsible for the improvement in extraction performance observed with PEF and that this improvement cannot be attributed solely to ohmic heat generated during PEF treatment.

5.2.4. Olives

Olives of several cultivars are key in Mediterranean dishes and an essential agricultural crop for European countries such as Greece, Spain, and Italy. Olives are the fruit of the olive tree with the scientific name *Olea europaea*, which means "European olive". They are cultivated throughout the Mediterranean basin as well as in South America, South Africa, India, China, Australia, New Zealand, Mexico, and the United States. Antioxidant compounds, such as polyphenols and flavonoids, are characteristic in olives, especially oleuropein. The content of extra virgin olive oil in polyphenols is 500 mg/L [130]. It has been claimed that the health benefits of olives and olive oil can protect the human organism from a variety of illnesses [131]. For this reason, the effect of PEF pretreatment on the increase of their polyphenol content was evaluated for all parts of the olive plant as well as for extra virgin olive oil. For instance, in the previously mentioned study by Ziagova et al. [97], PEF-treated leaves and unripe fruit yielded TPC of 105 and 12 mg GAE/g dw, respectively.

Andreou et al. [132] improved the recovery of high value-added compounds from olive pomace as a result of the combined use of PEF. Pretreatments with PEF (1.0–6.5 kV/cm, 0.9–51.1 kJ/kg, and 15 μs pulse width) were applied to olive pomace (cv *Tsounati*). Solid-liquid extraction of intracellular compounds with 50% *v*/*v* aqueous ethanol solution for 1 h at 25 °C was then used for analysis. At electric field strengths over 3 kV/cm, the polyphenol concentration increased significantly, reaching as high as 91.6% from the untreated sample (~1500 mg GAE/L). This method may be useful in conjunction with traditional solvent extraction. With PEF pretreatment, olive pomace can be valorized by increasing the yields of intracellular compounds with high antioxidant properties.

The same research team [133] investigated the application of non-thermal processing techniques to maximize the production and quality of virgin olive oil. Different PEF conditions (electric field strength of 0.5–2.0 kV/cm, 0–2500 pulses, energy input of 0.5–57.5 kJ/kg) were applied to olive paste. Quality, bioactive compounds, oxidative stability, and sensory evaluation of olive oils generated under optimum conditions (electric field strength of 1.5 kV/cm, 100 pulses) were assessed using response surface methodology. The olive oil yield from the PEF-pretreated sample was ~3% higher (25.4%), had ~57% more α-tocopherol (66.9 mg/kg oil), and had ~7% more polyphenols (822.3 mg GAE/kg oil) than the yielded oil from the untreated control sample. Consequently, PEF seeks to increase the production of a sustainable and cost-effective product in the olive oil industry.

Extracting polyphenols from olive leaves using the PEF method was evaluated by Pappas et al. [134]. Water, ethanol, and various mixtures thereof were used across a gradient of 25% in this study. The optimal conditions for PEF extraction took 30 min and required an electric field strength of 1 kV/cm. In addition, they explored a range of pulse durations, from 10–100 ns. Results obtained from PEF-treated extracts were compared to those obtained from untreated extracts. With a pulse duration of 10 μs and a 25% *v*/*v* ethanol aqueous solution, PEF was found to have the greatest effect. Significant increases of 31.85% in total polyphenols and 265.67% in specific metabolites were observed when comparing pre- and post-harvest samples. Differential scanning calorimetry revealed that 569 °C was the highest temperature at which oxidation occurred. The higher the oxidation temperature, the more resistant to oxidation the sample. This remarkable temperature was achieved by subjecting the samples to pulses with a duration of 100 μs and a period of 1000 μs. When the above PEF parameters were applied, the main metabolite luteolin-7-*O*-glucoside showed a significant increase of 71.87%, amounting to a total of 0.82 mg/g dw. Under a 100 μs pulse, however, oleuropein alone showed the highest extraction yield.

The same research team conducted an investigation [135] into the efficacy of PEF in concentrating polyphenol extracts, thereby evaluating the potential worth of olive leaves. The researchers thoroughly investigated the optimal methods for enhancing the PEF process in order to improve the extraction of olive leaf compounds. A comprehensive investigation was conducted to examine various parameters of the extraction chamber, including its geometric configuration, electric field intensity, pulse duration, pulse period (and frequency), and extraction duration. The authors employed experimental methods to ensure the optimal duration of PEF-assisted solid-liquid extraction of olive leaves. The implementation of PEF resulted in a significant increase of 38% in the extractability of total polyphenols compared to the untreated control sample. The TPC reached a value of 25.35 mg GAE/g dw. Furthermore, it is noteworthy to mention the remarkable 117% increase observed in the concentration levels of certain specific metabolites. The optimum conditions required a 15-min extraction with 25% *v*/*v* ethanol conducted in a rectangular extraction chamber, an electric field strength of 0.85 kV/cm, a pulse period of 100 μs, and a pulse duration of 2 μs. Regarding oxidative stability, the samples subjected to a pulse duration of 10 μs, pulse period of 1000 μs, electric field of 0.85 kV/cm, and extraction time of 30 min exhibited the most pronounced oxidation peak at 488 °C. This value was found to be 16% higher than the control sample when assessed using differential scanning calorimetry. The findings of this study indicate that the use of PEF treatment has a notable impact on enhancing extraction efficiency and improving the physicochemical properties. Table 2 provides a concise summary of the above research conducted on fresh vegetables and their by-products.

Table 2. Application of PEF on fresh vegetables and by-products with the treatment effects.

Sample	PEF Conditions	Treatment Effect	Ref.
Potato peel	5 kV/cm, 10 kJ/kg	Increased TPC by ~10% (from ~1160 to 1295 mg GAE/kg fw)	[125]
Asparagus root	1.6 kV/cm, 200 Hz, 20 μs pulse width	Increased values of extraction yield from 47.7 to 58.8% (23%), TPC from 32.6 to 34.4 mg GAE/g extract (5%), TFC from 0.16 to 0.17 mg RE/g extract (6%), and FRAP from 1363 to 1418 mM $FeSO_4$ E/g extract (4%)	[126]
Mushrooms	38.4 kV/cm, 272 μs duration	Estimated ~26% or 1.6 mg GAE/g higher polyphenol extraction yield	[129]

Table 2. *Cont.*

Sample	PEF Conditions	Treatment Effect	Ref.
Olive	0.5–2 kV/cm	High TPC value (12 mg GAE/g dw)	[97]
Olive pomace	3 kV/cm, 15 μs pulse width	Notable increase in TPC (91.6%) from ~1500 to ~2900 mg/L	[132]
Olive paste	1.5 kV/cm, 100 pulses	Increased recovery yield to 25.4% (by ~3%), TPC (by ~7%) from ~760 mg GAE/Kg oil	[133]
Olive leaf	1 kV/cm, 10 ns pulse duration	Increased TPC (by 31.85%) from 15.74 to 20.75 mg GAE/g dw	[134]
	0.85 kV/cm, 100 μs pulse period, 2 μs pulse duration	TPC increase by 38.5% (from 18.30 to 25.35 mg GAE/g dw)	[135]
	0.5–2 kV/cm	High TPC value (105 mg GAE/g dw)	[97]

5.3. Various Plants, Herbs, Nuts and Seaweeds

A variety of aromatic plants, herbs, nuts, plant parts, and seaweeds were utilized in the process of PEF-assisted extraction. The analysis of all the samples revealed a notable increase in their polyphenol levels, with certain samples exhibiting a more pronounced increase compared to others.

5.3.1. Borage

The objective of the study by Segovia et al. [136] was to evaluate the efficacy of PEF in enhancing the aqueous extraction of polyphenols and antioxidant compounds from borage (*Borago officinalis*) leaves. The range of the applied electric field strength ranged from 0–5 kV/cm. Extractions were exposed to different temperatures (10, 25, and 30 °C) for varying durations (10–60 min), whereas water was used as a solvent. The application of PEF treatments resulted in a significant increase in the TPC and ORAC values of the extracts. Specifically, TPC values were enhanced by a factor ranging from 1.3–6.6 (from 0.3 mg GAE/g fw), while the ORAC values were enhanced by a factor ranging from 2.0–13.7 (from ~10 mg TE/g fw). This procedure enhances the efficiency of the extraction process while concurrently increasing the antioxidant potency of the extracts. Furthermore, it reduces the duration of the extraction process. The correlation between the rise in pulse intensity and the quantity of extracted polyphenols indicates that this technology holds promise for effectively managing this byproduct within the food industry.

5.3.2. Flaxseed

Boussetta et al. [137] investigated the polyphenol extraction from flaxseed hulls using PEF as part of an effort to identify potential uses for oil-seed byproducts. The study examined the influence of various factors, including PEF treatment time, electric field strength, composition of solvent, and duration of rehydration, on the extraction of polyphenols. The extraction efficiency of polyphenols (~314 mg GAE/100 g dm) was achieved by 50% *v*/*v* ethanol, reaching an increase of 42% compared to extraction with 20% *v*/*v* ethanol. However, this ethanol level (20% *v*/*v*) was used as optimal because the authors stated that industrial PEF applications should not exceed this level. After 40 min of rehydration at 150 rpm, the study found that the optimal conditions for a PEF treatment were 20 kV/cm for 10 ms with an energy output of 300 kJ/kg. While 270 mg GAE/100 g of dm was recovered using 0.3 M citric acid, 1000 mg GAE/100 g of dm was recovered using 0.3 M sodium hydroxide. The sodium hydroxide-free sample yielded about 200 mg GAE/100 g dm. Consequently, the use of PEF in conjunction with alkaline hydrolysis gave encouraging results in polyphenol recovery.

5.3.3. Rapeseed

Yu et al. [138] examined the utilization of PEF for the extraction of polyphenols from the stems and leaves of rapeseed (*Brassica napus* L.). The leaves were exposed to a PEF ranging from 0.2–20 kV/cm. The study involved the analysis of rapeseed tissue to assess the extent of Z_p resulting from the application of PEF ranging from 5–20 kV/cm. Different temperatures (20–70 °C), solvent concentrations (5–100% *v*/*v* ethanol), and pH values (2–12) were examined. It was found that the most recovered polyphenols were acquired through exposure to an electric field strength of 5 kV/cm. In both stems and leaves, the optimum extraction conditions required 20 min extraction with 100% ethanol, at pH 7 and 70 °C, achieving 0.17 and 1.25 g/100 g dm, respectively. This study provides evidence to support the potential efficacy of PEF treatment as a novel approach for the valorization of rapeseed stems and leaves. This treatment selectively extracted polyphenols from the plant material, while preserving the proteins in the remaining residues.

A year later, the same research team with Yu et al. [139] investigated the impact of applying PEF before pressing rapeseed green biomass (stems) on the efficiency of polyphenol extraction. The impact of pressure, electric field strength, and pulse number on the overall polyphenol content and juice expression yield were examined. The optimal experimental conditions required electric field strength at 8 kV/cm, a total PEF time of 2 ms, with water as the solvent resulting in a significant increase in the juice expressed yield, from 34% to 81%. The press cake underwent successful dehydration, resulting in an increase in its dry matter content from 8.8% to 53.0% upon recovery. Similarly, TPC demonstrated a significant increase, surging from 0.10 to 0.48 g GAE/100 g dm after PEF pretreatment.

5.3.4. Canola

Teh et al. [140] investigated four parameters, namely ethanol concentration, time, frequency, and voltage, for PEF treatment through Box–Behnken response surface methodology. These parameters were utilized to identify the most effective method for extracting polyphenols from defatted canola seed cake. PEF-assisted extraction yielded maximum results when performed at an electric field strength of 1.1 kV/cm, frequency of 30 Hz, with 10% *v*/*v* ethanol concentration for 10 µs. After PEF and MAE, the defatted canola seed cake was subjected to US. The measured responses included total phenolics, flavonoids, DPPH$^{\bullet}$ scavenging, and FRAP. The highest polyphenol yields (2624.18 mg GAE/100 g fw) were obtained in the optimum PEF conditions. Consequently, PEF is an economically feasible way to extract polyphenols by using low electroporation voltage which resulted in reduced solvent usage and shorter extraction time.

5.3.5. Coffee and Cocoa

In order to maximize polyphenol recovery from cocoa bean shell (CBS) and coffee silver skin (CSS), the study by Barbosa-Pereira et al. [141] explored the use of PEF as a novel treatment technique. The PEF conditions were examined through response surface methodology (electric field strength 1.93–3 kV/cm, 9–16 µs pulse duration). The optimized methodology was used to analyze multiple CBS and CSS samples, after which they were separated according to their country of origin, crop type, and industrial processing. Compared to conventional extraction methods, PEF-assisted extraction resulted in roughly ~12 to 22% greater recovery yields for polyphenols (from ~26–54 mg GAE/g dw) in CBS and in ~25 to 13% recovery yields (from ~8–12 mg GAE/g dw) in CSS. Finally, the results demonstrated that the composition of bioactive compounds from various extracts of CBS and CSS, as well as their antioxidant properties, were affected by the origin, variety, and industrial processing of the raw material. These by-products of natural compounds have potential applications in agriculture, medicine, and the health sector.

5.3.6. Saffron

The study from Neri et al. [142] investigated the feasibility of PEF at an electric field strength of 2 kV/cm and 1.5 kJ/kg as a pretreatment alternative to the hot air drying process to improve the quality and functional properties of saffron (*Crocus sativus* L.). TPC and antioxidant activity were measured after processing and during room-temperature aging. The application of PEF did not result in any significant alteration in TPC of fresh stigmas, averaging ~4 mg GAE/g dm. However, the highest TPC value was observed after the drying and aging process for 10 months, specifically with the application of PEF treatment and drying process (~10 mg GAE/g dm). Although the impact of PEF on the antioxidant activity of fresh stigma was found to be negative, resulting in a decrease of 24% (from an initial ~90 mmol/g dm), it was observed that the application of PEF had a positive influence on the antioxidant activity as the stigma underwent the drying process (~100 µmol/g dm). Regardless of the method of data processing, there was a significant decrease in antioxidant activity of up to 86% associated with aging after 10 months (~18 µmol/g dm). Based on the results of this investigation, it can be suggested that PEF-treated saffron exhibits promising characteristics that make it a suitable candidate for incorporation as a premium component in food applications. This is primarily attributed to the increased extraction yield, which enables the utilization of smaller quantities of the ingredients in food formulations and products. Consequently, this leads to notable cost and time savings. These concepts apply to various domains and commodities, excluding food production.

5.3.7. Wheat Plants

In a study conducted by Ahmed et al. [143], an investigation was conducted to examine the impact of US and PEF on the juice derived from wheat plantlets. The treatments that involved the combination of US and PEF demonstrated elevated levels of TPC, TFC, ORAC assay, and DPPH• activities in comparison to the treatments conducted separately. PEF conditions required 1 kHz of pulse frequency, 80 µs pulse width, 9 kV/cm electric field strength, 335 µs treatment time, 30 °C temperature, and 50 mL/min flow rate. The corresponding increases were 5.35% (from 305.23 µg GAE/g) for TPC, 5.51% (from 178.34 µg CE/g) for TFC, 4.91% (from 1.63 mmol TE/L) for DPPH• assay, and 1.36% (from 5.12 mmol TE/L) for ORAC assay. When both treatments were employed, the increases were 8.59, 14.06, 6.74, and 2.34%, respectively.

5.3.8. Sage

Sage (*Salvia officinalis* L.) leaf phytochemicals were the primary focus of the research conducted by Athanasiadis et al. [144], which aimed to evaluate the efficiency of PEF-assisted extraction. Among the variables tested was the pulse duration of PEF, which varied from 10–100 µs over the course of 30 min. They also investigated the efficacy of several "green" extraction solvents, including ethanol, water, and their respective 25–75% *v*/*v* mixtures. The obtained extracts were evaluated against those obtained without PEF as a standard of comparison. Total polyphenols, isolated polyphenols, volatile compounds, and oxidation resistance were measured to assess extraction efficacy. The highest PEF contribution to total and individual polyphenols, as well as rosmarinic acid extractability, was achieved under conditions of a 25% *v*/*v* aqueous ethanol solvent, a pulse duration of 100 µs, and an electric field strength of 1 kV/cm. The result was an increase of 73.2% in TPC (from ~24 mg GAE/g dm) and 403.1% in rosmarinic acid (from 0.37 mg/g) over the control extract. Differential scanning calorimetry was also able to confirm the results. The oxidation temperature of the PEF-treated extracts was on average 61.5% higher than that of the reference extracts (182 °C). The primary compounds, which accounted for roughly the same percentage of the total composition (65.51 and 67.58%, respectively), were ultimately detected in both the PEF-treated and the reference extracts. These findings indicated that the application of low energy intensities through PEF may result in subtle changes to the odor of the tested extracts.

5.3.9. Drumstick Tree

Freeze-dried leaves of the drumstick tree (*Moringa oleifera*) were the subject of an investigation by Bozinou et al. [46]. The effectiveness of PEF extraction was compared to that of several other methods, such as MAE and UAE, as well as boiling water extraction and plain maceration (the control). The control sample was made using the same volume of freeze-dried leaves that were submerged in the same volume of double-distilled water at room temperature for 40 min. A 7 kV/cm electric field strength was required with pulse duration (PD) between 10 and 100 ms, and pulse interval (PI) between 25 and 100 μs. The PEF-treated sample with 20 ms PD and 100 μs PI achieved the highest TPC (40.24 mg GAE/g dw), at a rate that was 45% higher than that of the control sample. Other assays measuring antioxidant capacity showed the expected pattern, including % scavenging activity and ferric reducing antioxidant power (FRAP). TPC decreased in other PEF-treated samples with long PI (100 μs) and growing PD (50–100 ms). Therefore, it is important to emphasize that the optimal condition for extracting total polyphenols from freeze-dried *M. oleifera* leaves is a combination of low PD and high PI.

5.3.10. Almond

Due to their high nutrient content, almonds are widely regarded as one of the world's most valuable fruits. Additionally, in recent years, there has been a shift in focus toward almond by-products such as skins, shells, and hulls, which are abundant but underutilized. In particular, this technology was employed to create a workable valorization strategy by providing a more environmentally friendly alternative to conventional methods of polyphenols extraction. Considering this, Salgado-Ramos et al. [145] studied the innovative PEF method to valorize almond hull biomass. An electric field strength of 3 kV/cm, frequency of 2 Hz, and pulse duration of 100 ms were used to apply a specific energy of 100 kJ/kg. When compared to the traditional maceration method, PEF resulted in an extraction of Trolox equivalent antioxidant capacity (TEAC) values. In relation to the TEAC results, it was observed that the PEF-treated sample exhibited higher values in comparison to the control sample. Specifically, the PEF-treated sample displayed a value of 13.71 μM TE, while the control sample had a value of 7.78 μM TE. However, the results revealed statistically non-significant differences ($p < 0.05$). Specifically, TPC values in PEF-treated samples recorded 2.72 mg GAE/mL and were marginally higher than those of the control group 2.27 mg GAE/mL. In addition, results in ORAC revealed that PEF had a negative effect on samples, as the control sample exhibited a higher ORAC result (47.74 mM TE) compared to the PEF-treated sample (33.28 mM TE).

5.3.11. Hemp

The research by Teh et al. [146] aimed to assess whether the impact of PEF treatment could improve polyphenol extraction yields from defatted hemp seed (*Cannabis sativa*) cake. A Box–Behnken design of response surface methodology was used to optimize the extraction parameters. Four independent variables, including ethanol concentration (0, 5, and 10% v/v), time (10, 20, and 30 s), frequency (30, 40, and 50 Hz), and voltage (30, 40, and 50 V), were used in a Box–Behnken design to create a model for the observed response. The 900 burst pulses, 20 μs pulse width, and 10 kJ energy were all predetermined values for the PEF process. The results showed that the optimum PEF conditions were from design point 16 which required 5% v/v ethanol concentration, 20 s treatment time, 30 Hz frequency, and 30 V. TPC ranged from 467.5–1013.0 mg GAE/100 g fw, total flavonoids from 6.36–15.13 mg luteolin equivalents (LUE)/100 g fw, DPPH$^\bullet$ % inhibition from 12.65–22.06, and FRAP from 5.54–12.22 μmol Fe^{+2}/g fw. However, by incorporating the composite desirability, optimum PEF conditions required 10% v/v ethanol concentration, 10 s treatment time, 30 Hz frequency, and 30 V. The corresponding values were 1025.57 mg GAE/100 g fw, 15.76 mg LUE/100 g fw, 22.84%, and 12.75 μmol Fe^{+2}/g fw.

5.3.12. Sesame

There is a growing demand in the industry for extraction processes that employ reduced or zero quantities of organic solvents and operate at lower temperatures. The extraction process facilitated by PEF induces cellular damage, which subsequently enhances the diffusion of the product into the solvent. As such, a study by Sarkis et al. [147] examined the impact of PEF on the extraction of sesame cake compounds. The electric field for PEF was 13.3 kV/cm. The pulse lasted 10 μs and occurred at a rate of 0.5 Hz. By employing 10% *v*/*v* ethanol as extraction solvent, PEF recorded a TPC value of ~400 mg GAE/100 g dm, whereas the untreated sample recorded ~320 mg GAE/100 g dm. The results of this study demonstrated that polyphenol extraction using ethanol as a solvent can be reduced when the PEF method is used, as can the need for increased temperatures to enhance diffusion.

5.3.13. Rice

The brown rice bran extraction with the PEF process has been studied for the first time by Quagliariello et al. [148]. PEF conditions had electric field strength of 2 kV/cm, 1000 pulses, and specific energy of 64 kJ/kg increased the brown rice's antioxidant activity by 50% (from ~260 to ~390 μg AAE/g). In addition, several phenolic acids, such as chlorogenic acid and ferulic acid, were increased from 53.5 to 65.7 μg/g and from 16.4 to 20.9 μg/g, respectively. Therefore, it appears that including PEF pretreatment in the solvent extraction process of polyphenols from brown rice is a promising practice that will significantly increase their biological activity.

5.3.14. Spruce

Bouras et al. [149] studied the effect of PEF treatment in polyphenol extraction from Nordic spruce bark (*Picea abies* L.). Norway spruce was used to isolate several polyphenols, including phenylpropanoids, tannins, flavonoids, lignans, and stilbenes. Two PEF treatment protocols at an electric field strength of 20 kV/cm and 1–400 number of pulses and pulse duration of 10 μs were tested to determine the feasibility of PEF treatment. Sodium hydroxide solution (0.01 M) was used as an extraction solvent. The results showed that samples had TPC boosted by PEF treatment. For instance, TPC has increased by over a factor of 8 (from 0.96 to 8.52 g GAE/100 g dm). The PEF treatment did not result in any visible degradation of bark tissue, suggesting it could be a viable alternative to milling that saves energy. The obtained results are promising and open up new avenues for the valorization of wood bark.

5.3.15. Barberry

Barberry is a useful plant in treating various diseases, containing valuable compounds in its pruned waste. The study conducted by Sarraf et al. [150] investigated the quantity of berberine, polyphenols, and antioxidant activity present in barberry fruits, leaves, and stems of varying species, including *Berberis integerrima* and *Berberis thunbergia*. This study used a central composite design of RSM to examine the effects of variables on the extraction of berberine from the stem of *B. integerrima* (time: 2–24 h, temperature: 24–70 °C, and ethanol concentration: 50–90% *v*/*v*). Berberine concentration, DPPH$^{\bullet}$ scavenging capacity, and TPC were used as evaluation criteria. Additionally, pretreatment with PEF-assisted was used prior to extraction with electric field strengths of 250, 1000, and 1250 V/cm, pulse numbers of 50 and 100, and frequency of 1 Hz. The berberine concentration with PEF treatment rose dramatically. The stem of *B. integrrima* was chosen for further study because it had the highest levels of antioxidant activity and berberine content. The ideal maceration conditions were 90% ethanol, 70 °C, and 3.36 h (141.6 min). Among the various techniques, maceration with 100 pulses and a field strength of 1 kV/cm was the most effective. Berberine concentration was 1.86 mg/g, TPC was measured at 11.11 mg GAE/g, and antioxidant activity was 71.84% in the optimal maceration condition. The PEF-assisted method increased berberine content to 2.78 mg/g, TPC to 14.57 mg GAE/g, and antioxidant activity to 78.6%, respectively. The results showed that the stem extract from *B. integrrima* is rich in berberine and antioxidants and could be used in a number of different sectors.

5.3.16. Other Plants

Extracts from medicinal and aromatic plants are widely used for their health benefits to humans. However, it is difficult because the success of the process of extracting polyphenols from plants is highly dependent on the technique used and the operating conditions that are imposed. Ziagova et al. [97] investigated the impact of PEF on several plants and herbs. By implementing an electric field strength of 0.5–2 kV/cm, they measured TPC (mg GAE/g dw) in several samples, such as *Melissa officinalis* L. leaves (155), *Cistus incanus* L. *creticus* leaves (148), and *Aronia melanocarpa* L. fruit (67), and *Crocus sativus* L. petals (147). High extraction yields and the biological stability of bioactive compounds are achieved using the combined process of PEF and US proposed in this study. It has advantages compared to the conventional or the most advanced extraction methods due to the application of short extraction time moderate temperatures and the use of water as the solvent.

Extracting polyphenols from the plants *Rosa canina*, *Calendula officinalis*, and *Castanea sativa* using PEF treatment was the objective of the study by Lakka et al. [151]. Traditionally, these plants have been used not only to make medicinal decoctions but also to add flavor to drinks of all kinds. Electric field strength was applied at intensities between 1.2–2.0 kV/cm in pulses of 10 μs duration. The samples were extracted for 20 min, during which time the PEF period was set to 1 ms. In order to track and assess the extracts, their TPC and individual polyphenolic compounds were calculated in comparison to untreated samples. The PEF process appeared to increase polyphenols extraction from all three plant materials tested. TPC in *R. canina* fruits increased to 63.79% (from ~42 mg GAE/g dw) and eriodictyol-7-*O*-rutinoside recovery increased to 84% (from 0.032 mg/g dw) when 1.4 kV/cm was employed, respectively. Regarding *C. officinalis* the corresponding TPC and isorhamnetin-3-*O*-rutinoside increase were 55.02% (from ~35 mg GAE/g dw) and 73% (from 7.868 mg/g dw) through electric field strength of 1.2 kV/cm. Finally, by employing the same electric field strength (1.2 kV/cm) *C. sativa*, the TPC increase was 48.41% (from ~115 mg GAE/g dw), whereas quercetin 3-*O*-glucoside was increased by 82% (from 1.153 mg/g dw). Understanding the potential of PEF will allow the development of more potent extracts that can be used to fortify medicinal herbal teas, traditional beverages, and even alcoholic beverages.

This study by Carpentieri et al. [152] compared the efficiency of hydroethanolic extraction (0–50% *v*/*v* ethanol in water) for up to 4 h following PEF treatment or US on the cell disintegration of two Mediterranean herb tissues (*Origanum vulgare* L., *Thymus serpyllum* L.). The extraction rate of polyphenols decreased over time, as predicted by Peleg's model (R^2 = 0.898–0.989). When applied before solid-liquid extraction, either PEF or US treatment had the potential to shorten the extraction time and lower the ethanol concentration required to recover the same amount of phenolic compounds. Increased values in TPC of *O. vulgare* (36% from ~100 mg GAE/g dw) and *T. serpyllum* (36% from ~40 mg GAE/g dw). The corresponding values in antioxidant activity (FRAP) were also found to increase (29% from 103.9 μmol Fe^{+2}/g dw) and (47% from 31.1 μmol Fe^{+2}/g dw) of extracts obtained from PEF-pretreated herb samples under optimum PEF conditions (3 kV/cm, 10 kJ/kg). No measurable degradation of individual polyphenols from PEF treatment, as determined by GC-MS analysis.

5.3.17. Algae/Microalgae

Microalgae contain polyphenols and coloring compounds that exhibit antioxidant, antibacterial, and anti-inflammatory properties [153,154]. It is worth mentioning that the PEF system was even applied to four species of these algae. To investigate the effects of various treatments on the brown alga *Laminaria digitata*, crude aqueous extracts were prepared using PEF-assisted extraction in a study by Einarsdóttir et al. [155]. Biomass concentration (0.17–3.28% dw), number of pulses during PEF treatment (12–268 pulses), and initial temperature of algae suspension (12–48 °C) were the three factors used for response surface methodology. An electric field of 7.5 kV/cm and a frequency of 1.2 Hz were applied to the samples. The extraction yield was 15%, and the supernatant yield was 70%. TPC was measured at 4 mg GAE/100 g dw. The lowest biomass concentration also had the highest supernatant yield, polyphenol content, and carbohydrate content. A positive relationship between the temperature rises and the total number of PEF pulses was observed. This research demonstrates that valuable compounds in *L. digitata* can be extracted using PEF-assisted extraction rather than extreme temperatures or organic solvents.

Recovery of polyphenols from the microalgae *Tetraselmis chuii* and *Phaeodactylum tricornutum* was tested using PEF-assisted extraction in combination with aqueous or dimethyl sulfoxide (DMSO) solvents. The study by Kokkali et al. [156] investigated several PEF parameters. The specific energy input was 100 kJ/kg, whereas two PEF treatments were administered (1 kV/cm with 400 pulses and 3 kV/cm with 45 pulses). PEF treatment at 3 kV/cm and 4 h of extraction yielded the highest value regardless of the solvent (water yielded 6.42 GAE/g dw, and DMSO 50% in water yielded 6.70 mg GAE/g dw) for *T. chuii*, DMSO was only effective at enhancing polyphenol extraction from *P. tricornutum*. In *P. tricornutum*, PEF pretreatment with DMSO 50% in water as extraction solvent resulted in the highest extraction yield, with values of ~8 mg GAE/g dw. Finally, PEF shows promise as a potential tool for improving the selective extraction of antioxidant bioactive compounds from microalgae.

Research conducted by Castejón et al. [157] presented three Icelandic species of algae, *Ulva lactuca*, *Alaria esculenta*, and *Palmaria palmata*, and the effect of extraction with a heated water bath, PEF, and the combination of these methods. The PEF conditions required an electric field strength of 8 kV/cm at 1.2 Hz for 10 min with 3 pulses. Several advantages of PEF were revealed, including its non-thermal nature and shorter extraction time (10 vs 45 min), and PEF showed results that were comparable to the conventional method. However, the PEF-treated *Alaria esculenta* sample had the highest TPC of 9.37 mg GAE/g dw when compared to the heat water process (+4.8%) and TFC of 12.43 mg 1QE/g dw, along with the greatest antioxidant capacities. PEF had a negative impact on *P. palmata*, as PEF declined TPC (1.8 mg GAE/g dw, −2.43%) but increased TFC (0.94 mg QE/g dw, +16%). In *U. lactuca*, PEF had a mostly deleterious effect in both TPC (1.59 mg GAE/g dw, −22%) and TFC (3.43 mg QE/g dw, −32%). Therefore, this exploratory research indicates that PEF-produced extracts of Icelandic *Alaria esculenta* may be useful as ingredients in natural cosmetic and cosmeceutical formulations. However, the deleterious impact of PEF should be further investigated.

The last microalgae studied was *Spirulina*, which is listed by the European Commission as a new type of food for daily nutrition. Three different extraction methods were studied by Zhou et al. [158] for the *Spirulina* samples, namely PEF, pressurized liquid extraction (PLE), and a combination of PEF and PLE, with water as extraction solvent. The PEF condition required an electric field strength of 3 kV/cm, 44 pulses, specific energy 99 kJ/kg. In comparison to the conventional control technique of Folch extraction, the combination of PEF and PLE resulted in a significant reduction in extraction time (~165 min) and a significant increase ($p < 0.05$) in TPC values of *Spirulina* extracts, from ~2 to ~20 mg/g dw. Table 3 provides a summary of the above studies on various plants, herbs, nuts, and seaweeds.

Table 3. Application of PEF on various plants, herbs, nuts, seaweeds, and by-products with the treatment effects.

Sample	PEF Conditions	Treatment Effect	Ref.
Borage leaf	0–5 kV/cm, 10–60 min treatment duration	TPC: 1.3–6.6 times increase (from 0.3 mg GAE/g fw), ORAC: 2.0–13.7 times increase (from ~10 mg TE/g fw)	[136]
Rapeseed stem	5 kV/cm	High TPC value (0.17 g/100 g dm)	[138]
Rapeseed leaf		High TPC value (0.25 g/100 g dm)	
Rapeseed stem	8 kV/cm, 2 ms treatment duration	TPC increased by 380% (from 0.10 to 0.48 g GAE/100 g dm)	[139]
Canola seed cake	1.1 kV/cm, 30 Hz, 10 s exposure time	High TPC (2624.18 mg GAE/100 g fw) yielded in a short time	[140]
Cocoa bean shell	1.93–3 kV/cm, 9–16 μs pulse duration	Up to 22% increase in TPC (from ~26–54 mg GAE/g dw)	[141]
Coffee silver skin		Up to 13% increase in TPC (from ~8–12 mg GAE/g dw)	
Saffron	2 kV/cm, 1.5 kJ/kg	Non-significant increase in TPC compared to untreated samples (~4 mg GAE/g dm), significant decrease in AA to ~18 μmol/g dm (~86%) when aging after 10 months	[142]
T. chuii	3 kV/cm, 45 pulses, 100 kJ/kg	High TPC yield (~6.7 mg GAE/g dw)	[156]
P. tricornutum	1 kV/cm, 400 pulses, 100 kJ/kg	High TPC yield (~8 mg GAE/g dw)	
Wheat plantlet	9 kV/cm, 1 kHz, 80 μs pulse width, 335 μs treatment time	Increase in TPC from 305.23 μg GAE/g (5.35%), in TFC from 178.34 μg CE/g (5.51%), in DPPH from 1.63 mmol TE/L (4.91%), and in ORAC from 5.12 mmol TE/L (1.36%)	[143]
Sage leaf	1 kV/cm, 100 μs pulse duration	Increase in TPC by 73.2% from ~24 mg GAE/g dm) and in rosmarinic acid concentration by 403.1% from 0.37 mg/g	[144]
Almond hull	3 kV/cm, 2 Hz, 100 kJ/kg, 100 ms pulse duration	Slight increase in TPC (~19%) from 2.27 to 2.72 mg GAE/mL	[145]
Hemp seed	30 V, 30 Hz, 10 s treatment time	High TPC (1025.57 mg GAE/100 g fw) and TFC (15.76 mg LUE/100 g fw)	[146]
Sesame cake	13.3 kV/cm, 0.5 Hz, 10 μs	TPC increased by ~25% from ~320 to ~400 mg GAE/100 g dm	[147]
Rice	2 kV/cm, 64 kJ/kg, 1000 pulses	TPC increased by ~50% from ~260 to ~390 μg AAE/g	[148]
Spruce bark	20 kV/cm, 10 μs pulse duration, 1–400 pulses	TPC increased 8 times (from 0.96 to 8.52 g GAE/100 g dm)	[149]
Barberry	1.0 kV/cm, 100 pulses	Increase in TPC by 30% (from 11.11 to 14.57 mg GAE/g) and berberine content by 49% (from 1.86 to 2.78 mg/g)	[150]
R. canina	1.4 kV/cm, 10 μs pulse duration	Increase in TPC by 63.79% (from ~42 mg GAE/g dw) and in eriodictyol-7-*O*-rutinoside concentration by 84% (from 0.032 mg/g dw)	[151]
C. officinalis	1.2 kV/cm, 10 μs pulse duration	Increase in TPC by55.02% (from ~35 mg GAE/g dw) and in isorhamnetin-3-*O*-rutinoside concentration by 73% (from 7.868 mg/g dw)	
C. sativa		Increase in TPC by 48.41% (from ~115 mg GAE/g dw) and isorhamnetin-3-*O*-rutinoside concentration by 82% (from 1.153 mg/g dw)	
L. digitata	7.5 kV/cm, 1.2 Hz	High extraction yield (15%), supernatant yield (70%), TPC (4 mg GAE/100 g dw)	[155]

Table 3. *Cont.*

Sample	PEF Conditions	Treatment Effect	Ref.
O. vulgare	3 kV/cm, 10 kJ/kg	Increase in TPC by 36% from ~100 mg GAE/g dw and in FRAP by 29% from 103.9 μmol Fe^{+2}/g dw	[152]
T. serpyllum		Increase in TPC by 36% from ~40 mg GAE/g dw and in FRAP by 47% from 31.1 μmol Fe^{+2}/g dw	
M. officinalis L. leaf	0.5–2 kV/cm	High TPC value (155 mg GAE/g dw)	[97]
C. incanus L. spp. *creticus* leaf		High TPC value (148 mg GAE/g dw)	
C. sativus L. petal		High TPC value (147 mg GAE/g dw)	
A. melanocarpa L. fruit		High TPC value (67 mg GAE/g dw)	
Mixture of *C. sativus* L. petal and *V. vinifera* L. cv. *Xinomavro* fruit		High TPC value (54 mg GAE/g dw)	
Flaxseed hull	20 kV/cm, treatment duration 10 ms, 300 kJ/kg.	PEF: High TPC (1000 mg GAE/100 g) with alkaline hydrolysis compared to acidic hydrolysis (270 mg GAE/100 g dm)	[137]
Drumstick tree leaves	7 kV/cm, 20 ms pulse duration, 100 μs pulse interval	Increased TPC by ~45%, achieving 40.24 mg GAE/g dw	[46]
A. esculenta	8 kV/cm, 1.2 Hz, 10 min treatment duration, 3 pulses	Both TPC and TFC were slightly increased (by ~5% and ~1.5%) from 8.94 mg GAE/g dw and 12.23 mg QE/g dw, respectively	[157]
P. palmata		TPC after PEF decreased by 2.43% (1.8 mg GAE/g dw), TFC increased by 16% (0.94 mg QE/g dw)	
U. lactuca		Both TPC (1.59 mg GAE/g dw) and TFC (3.43 mg QE/g dw) dramatically decreased after PEF treatment (by –22% and –32%, respectively)	
Spirulina	3 kV/cm, 99 kJ/kg, 44 pulses	Significant TPC increase (by ~900%) from ~2 to ~20 mg/g dw	[158]

6. Current Challenges and Limitations

PEF technology has many applications in waste valorization, but its widespread adoption has been hampered by a number of challenges [159]. The initial cost of the PEF system is the major issue for PEF-assisted waste valorization [160]. Treatment chamber electrodes undergo electrochemical changes, so durable and low-cost electrodes are required as well. In addition, food waste mainly exists in solid, semi-solid, and liquid states, so the PEF treatment chamber should be redesigned to maximize extraction efficiency [161]. The utilization of PEF in the context of solid waste is currently in its early stages. Due to the variability in electrical resistivity observed in solid waste, and the outcomes of PEF treatment can be less repeatable, leading to instances of untreated regions alongside areas that may be subjected to excessive treatment [162].

Utilizing food waste is crucial for the sustainability of food industry [163]. At the moment, the food industry is mainly concerned with recycling waste and reducing energy and water consumption [164]. A huge challenge would be a large-scale utilization of these several food by-products for increased collaboration between academia and industry. The application of PEF technique for the recovery of compounds from these wastes could create a high value-added product despite its relatively low market value [165,166]. According to the available literature, more research is needed into the effectiveness of PEF pre-treatment for food wastes before it can be used consistently in industrial scale [49].

The majority of scientific investigations have been carried out using laboratory-scale apparatus, employing small sample volumes and batch flow configurations. Consequently, engineering difficulties exist in scaling up PEF processes for fresh fruits, vegetables, and their corresponding byproducts. Due to the need to consider variations in treatment uniformity, and residence times, the use of such data to pilot-scale or full commercial-scale production is frequently unfeasible [167]. Nevertheless, there has been significant progress in the development of PEF equipment specifically designed for industrial applications. This equipment has demonstrated successful implementation in Europe, where it has been utilized to effectively enhance the shelf life of fruit juices at a remarkable rate of up to 8000 L/h [168]. According to Toepfl et al. [169], the estimated cost range for processing PEF is between 0.02–0.03 USD/L.

Finally, extraction yield, extraction time, and specific energy consumption are three parameters that have been compared across studies of modern pretreatment PEF methods [170,171]. However, several cutting-edge techniques could be combined with PEF in order to enhance polyphenol extraction, therefore they need further investigation. Although, it is of high importance to emphasize that the combination of PEF and US led to deleterious effects on the final product, since less polyphenols were measured in a sample where both methods were performed than in a control sample. The quality standards of the final product should be taken into account when deciding which pretreatment method to use [172].

7. Conclusions and Future Perspectives

Numerous studies have been conducted to investigate the impact of applying PEF on the polyphenol extraction of different samples. The application of PEF extends to a variety of agricultural products, including fruits, vegetables, aromatic and medicinal plants, as well as wine and algae. In the majority of instances, the utilization of PEF-assisted extraction resulted in a substantial increase in the extraction of polyphenols in the examined sample. Also, the lack of evidence supporting the view that PEF is a destructive technique for the specific bioactive compounds under investigation is of high importance. This finding suggests that the adoption of PEF is not only environmentally and economically sustainable, but also provides access to a variety of food options that are abundant in antioxidant compounds.

The PEF pretreatment technique still presents several obstacles that necessitate future resolution. Firstly, extraction kinetics models should be developed and tested, and the mechanisms by which PEF is extracted should be confirmed. Future practical applications will necessitate further research into the mechanism and the creation of a scientific model of mass transfer. Also, PEF applications in the food industry offer a chance to implement large-scale energy-efficient processes that would result in minimally processed products with a higher concentration of bioactive compounds. Despite PEF being a non-destructive method for bioactive compounds, a deleterious effect of PEF with other pretreatment or extraction methods was observed, thus, a thorough investigation of the reason behind that incidence is of high importance. Juice production has been significantly enhanced through the use of PEF on fruits. However, it should be noted that the release of enzymes has caused a general decline in the quality of fruit juices as a result of PEF usage. Further investigation is also needed to determine if and how these treatments can be used to produce safe and stable products that retain their fresh-like bioactive potential. In addition, PEF technology can be used to extract bioactive compounds from wastewater discharged from the dairy, meat, and seafood processing industries, allowing these businesses to maximize the value of this water resource with minimal additional expense. Society and the scientific community would benefit from optimizing PEF treatment conditions for extracting bioactive compounds from the above residues. To fully accomplish the objectives above, it is imperative that the PEF technique be implemented on a large-scale within the industrial sector. Finally, it is crucial that authors provide all relevant experimental details to enable replication and comparison of results.

Author Contributions: Conceptualization, V.A., T.C. and S.I.L.; methodology, V.A. and T.C.; validation, V.A., T.C., D.K., K.K., E.B. and S.I.L.; writing—original draft preparation, D.K. and K.K.; writing—review and editing, V.A., T.C., D.K., K.K., E.B. and S.I.L.; visualization, D.K. and K.K.; supervision, V.A., T.C. and S.I.L. All authors have read and agreed to the published version of the manuscript.

Funding: This research received no external funding.

Institutional Review Board Statement: Not applicable.

Informed Consent Statement: Not applicable.

Data Availability Statement: Not applicable.

Conflicts of Interest: The authors declare no conflict of interest.

References

1. Quideau, S.; Deffieux, D.; Douat-Casassus, C.; Pouységu, L. Plant Polyphenols: Chemical Properties, Biological Activities, and Synthesis. *Angew. Chem. Int. Ed.* **2011**, *50*, 586–621. [CrossRef] [PubMed]
2. Lund, M.N. Reactions of Plant Polyphenols in Foods: Impact of Molecular Structure. *Trends Food Sci. Technol.* **2021**, *112*, 241–251. [CrossRef]
3. Gomez-Pinilla, F.; Nguyen, T.T.J. Natural Mood Foods: The Actions of Polyphenols against Psychiatric and Cognitive Disorders. *Nutr. Neurosci.* **2012**, *15*, 127–133. [CrossRef] [PubMed]
4. El Gharras, H. Polyphenols: Food Sources, Properties and Applications—A Review. *Int. J. Food Sci. Technol.* **2009**, *44*, 2512–2518. [CrossRef]
5. Bešlo, D.; Golubić, N.; Rastija, V.; Agić, D.; Karnaš, M.; Šubarić, D.; Lučić, B. Antioxidant Activity, Metabolism, and Bioavailability of Polyphenols in the Diet of Animals. *Antioxidants* **2023**, *12*, 1141. [CrossRef]
6. Mutha, R.E.; Tatiya, A.U.; Surana, S.J. Flavonoids as Natural Phenolic Compounds and Their Role in Therapeutics: An Overview. *Future J. Pharm. Sci.* **2021**, *7*, 25. [CrossRef]
7. Román, G.C.; Jackson, R.E.; Gadhia, R.; Román, A.N.; Reis, J. Mediterranean Diet: The Role of Long-Chain ω-3 Fatty Acids in Fish; Polyphenols in Fruits, Vegetables, Cereals, Coffee, Tea, Cacao and Wine; Probiotics and Vitamins in Prevention of Stroke, Age-Related Cognitive Decline, and Alzheimer Disease. *Rev. Neurol.* **2019**, *175*, 724–741. [CrossRef]
8. Pandey, K.B.; Rizvi, S.I. Plant Polyphenols as Dietary Antioxidants in Human Health and Disease. *Oxid. Med. Cell. Longev.* **2009**, *2*, 270–278. [CrossRef]
9. Abbas, M.; Saeed, F.; Anjum, F.M.; Afzaal, M.; Tufail, T.; Bashir, M.S.; Ishtiaq, A.; Hussain, S.; Suleria, H.A.R. Natural Polyphenols: An Overview. *Int. J. Food Prop.* **2017**, *20*, 1689–1699. [CrossRef]
10. Manach, C.; Scalbert, A.; Morand, C.; Rémésy, C.; Jiménez, L. Polyphenols: Food Sources and Bioavailability. *Am. J. Clin. Nutr.* **2004**, *79*, 727–747. [CrossRef]
11. Hazafa, A.; Iqbal, M.O.; Javaid, U.; Tareen, M.B.K.; Amna, D.; Ramzan, A.; Piracha, S.; Naeem, M. Inhibitory Effect of Polyphenols (Phenolic Acids, Lignans, and Stilbenes) on Cancer by Regulating Signal Transduction Pathways: A Review. *Clin. Transl. Oncol.* **2022**, *24*, 432–445. [CrossRef] [PubMed]
12. Maleki, S.J.; Crespo, J.F.; Cabanillas, B. Anti-Inflammatory Effects of Flavonoids. *Food Chem.* **2019**, *299*, 125124. [CrossRef] [PubMed]
13. Rufino, A.T.; Costa, V.M.; Carvalho, F.; Fernandes, E. Flavonoids as Antiobesity Agents: A Review. *Med. Res. Rev.* **2021**, *41*, 556–585. [CrossRef]
14. Rothwell, J.A.; Madrid-Gambin, F.; Garcia-Aloy, M.; Andres-Lacueva, C.; Logue, C.; Gallagher, A.M.; Mack, C.; Kulling, S.E.; Gao, Q.; Praticò, G. Biomarkers of Intake for Coffee, Tea, and Sweetened Beverages. *Genes Nutr.* **2018**, *13*, 15. [CrossRef] [PubMed]
15. Miles, E.A.; Calder, P.C. Effects of Citrus Fruit Juices and Their Bioactive Components on Inflammation and Immunity: A Narrative Review. *Front. Immunol.* **2021**, *12*, 712608. [CrossRef]
16. Bortolini, D.G.; Maciel, G.M.; Fernandes, I.d.A.A.; Rossetto, R.; Brugnari, T.; Ribeiro, V.R.; Haminiuk, C.W.I. Biological Potential and Technological Applications of Red Fruits: An Overview. *Food Chem. Adv.* **2022**, *1*, 100014. [CrossRef]
17. Thilakarathna, S.H.; Rupasinghe, H.P.V. Flavonoid Bioavailability and Attempts for Bioavailability Enhancement. *Nutrients* **2013**, *5*, 3367–3387. [CrossRef]
18. Kruger, M.J.; Davies, N.; Myburgh, K.H.; Lecour, S. Proanthocyanidins, Anthocyanins and Cardiovascular Diseases. *Food Res. Int.* **2014**, *59*, 41–52. [CrossRef]
19. Dabeek, W.M.; Marra, M.V. Dietary Quercetin and Kaempferol: Bioavailability and Potential Cardiovascular-Related Bioactivity in Humans. *Nutrients* **2019**, *11*, 2288. [CrossRef]
20. Mayo, B.; Vázquez, L.; Flórez, A.B. Equol: A Bacterial Metabolite from The Daidzein Isoflavone and Its Presumed Beneficial Health Effects. *Nutrients* **2019**, *11*, 2231. [CrossRef]
21. De Silva, S.F.; Alcorn, J. Flaxseed Lignans as Important Dietary Polyphenols for Cancer Prevention and Treatment: Chemistry, Pharmacokinetics, and Molecular Targets. *Pharmaceuticals* **2019**, *12*, 68. [CrossRef] [PubMed]

22. Khurana, S.; Venkataraman, K.; Hollingsworth, A.; Piche, M.; Tai, T.C. Polyphenols: Benefits to the Cardiovascular System in Health and in Aging. *Nutrients* **2013**, *5*, 3779–3827. [CrossRef] [PubMed]
23. Cory, H.; Passarelli, S.; Szeto, J.; Tamez, M.; Mattei, J. The Role of Polyphenols in Human Health and Food Systems: A Mini-Review. *Front. Nutr.* **2018**, *5*, 87. [CrossRef] [PubMed]
24. Viuda-Martos, M.; Fernández-López, J.; Pérez-Álvarez, J.A. Pomegranate and Its Many Functional Components as Related to Human Health: A Review. *Compr. Rev. Food Sci. Food Saf.* **2010**, *9*, 635–654. [CrossRef] [PubMed]
25. Pap, N.; Fidelis, M.; Azevedo, L.; do Carmo, M.A.V.; Wang, D.; Mocan, A.; Pereira, E.P.R.; Xavier-Santos, D.; Sant'Ana, A.S.; Yang, B. Berry Polyphenols and Human Health: Evidence of Antioxidant, Anti-Inflammatory, Microbiota Modulation, and Cell-Protecting Effects. *Curr. Opin. Food Sci.* **2021**, *42*, 167–186. [CrossRef]
26. Bose, M.; Lambert, J.D.; Ju, J.; Reuhl, K.R.; Shapses, S.A.; Yang, C.S. The Major Green Tea Polyphenol, (-)-Epigallocatechin-3-Gallate, Inhibits Obesity, Metabolic Syndrome, and Fatty Liver Disease in High-Fat-Fed Mice. *J. Nutr.* **2008**, *138*, 1677–1683. [CrossRef]
27. McSweeney, M.; Seetharaman, K. State of Polyphenols in the Drying Process of Fruits and Vegetables. *Crit. Rev. Food Sci. Nutr.* **2015**, *55*, 660–669. [CrossRef]
28. Lu, C.; Zhu, W.; Shen, C.L.; Gao, W. Green Tea Polyphenols Reduce Body Weight in Rats by Modulating Obesity-Related Genes. *PLoS ONE* **2012**, *7*, e38332. [CrossRef]
29. Rifna, E.J.; Misra, N.N.; Dwivedi, M. Recent Advances in Extraction Technologies for Recovery of Bioactive Compounds Derived from Fruit and Vegetable Waste Peels: A Review. *Crit. Rev. Food Sci. Nutr.* **2023**, *63*, 719–752. [CrossRef]
30. Nayak, A.; Bhushan, B. An Overview of the Recent Trends on the Waste Valorization Techniques for Food Wastes. *J. Environ. Manag.* **2019**, *233*, 352–370. [CrossRef]
31. Kumar, K.; Yadav, A.N.; Kumar, V.; Vyas, P.; Dhaliwal, H.S. Food Waste: A Potential Bioresource for Extraction of Nutraceuticals and Bioactive Compounds. *Bioresour. Bioprocess.* **2017**, *4*, 18. [CrossRef]
32. Yan, L.-G.; He, L.; Xi, J. High Intensity Pulsed Electric Field as an Innovative Technique for Extraction of Bioactive Compounds—A Review. *Crit. Rev. Food Sci. Nutr.* **2017**, *57*, 2877–2888. [CrossRef] [PubMed]
33. Chemat, F.; Vian, M.A.; Fabiano-Tixier, A.-S.; Nutrizio, M.; Jambrak, A.R.; Munekata, P.E.S.; Lorenzo, J.M.; Barba, F.J.; Binello, A.; Cravotto, G. A Review of Sustainable and Intensified Techniques for Extraction of Food and Natural Products. *Green Chem.* **2020**, *22*, 2325–2353. [CrossRef]
34. Barba, F.J.; Zhu, Z.; Koubaa, M.; Sant'Ana, A.S.; Orlien, V. Green Alternative Methods for the Extraction of Antioxidant Bioactive Compounds from Winery Wastes and By-Products: A Review. *Trends Food Sci. Technol.* **2016**, *49*, 96–109. [CrossRef]
35. Pojić, M.; Mišan, A.; Tiwari, B. Eco-Innovative Technologies for Extraction of Proteins for Human Consumption from Renewable Protein Sources of Plant Origin. *Trends Food Sci. Technol.* **2018**, *75*, 93–104. [CrossRef]
36. Li, K.-Y.; Ye, J.-T.; Yang, J.; Shao, J.-Q.; Jin, W.-P.; Zheng, C.; Wan, C.-Y.; Peng, D.-F.; Deng, Q.-C. Co-Extraction of Flaxseed Protein and Polysaccharide with a High Emulsifying and Foaming Property: Enrichment through the Sequence Extraction Approach. *Foods* **2023**, *12*, 1256. [CrossRef]
37. Liu, Z.; Esveld, E.; Vincken, J.-P.; Bruins, M.E. Pulsed Electric Field as an Alternative Pre-Treatment for Drying to Enhance Polyphenol Extraction from Fresh Tea Leaves. *Food Bioprocess Technol.* **2019**, *12*, 183–192. [CrossRef]
38. Mohamed, M.E.A.; Amer Eiss, A.H. Pulsed Electric Fields for Food Processing Technology. In *Structure and Function of Food Engineering*; Amer Eissa, A., Ed.; InTech: London, UK, 2012; ISBN 978-953-51-0695-1.
39. Quass, D.W. *Pulsed Electric Field Processing in the Food Industry*; A Status Report on Pulsed Electric Field; Electric Power Research Institute: Palo Alto, CA, USA, 1997; pp. 23–35.
40. Bansal, V.; Sharma, A.; Ghanshyam, C.; Singla, M.L.; Kim, K.-H. Influence of Pulsed Electric Field and Heat Treatment on Emblica Officinalis Juice Inoculated with *Zygosaccharomyces bailii*. *Food Bioprod. Process.* **2015**, *95*, 146–154. [CrossRef]
41. Gabrić, D.; Barba, F.; Roohinejad, S.; Gharibzahedi, S.M.T.; Radojčin, M.; Putnik, P.; Bursać Kovačević, D. Pulsed Electric Fields as an Alternative to Thermal Processing for Preservation of Nutritive and Physicochemical Properties of Beverages: A Review. *J. Food Process Eng.* **2018**, *41*, e12638. [CrossRef]
42. Mohamad, A.; Shah, N.N.A.K.; Sulaiman, A.; Mohd Adzahan, N.; Aadil, R.M. Impact of the Pulsed Electric Field on Physicochemical Properties, Fatty Acid Profiling, and Metal Migration of Goat Milk. *J. Food Process. Preserv.* **2020**, *44*, e14940. [CrossRef]
43. Tzima, K.; Brunton, N.P.; Lyng, J.G.; Frontuto, D.; Rai, D.K. The Effect of Pulsed Electric Field as a Pre-Treatment Step in Ultrasound Assisted Extraction of Phenolic Compounds from Fresh Rosemary and Thyme by-Products. *Innov. Food Sci. Emerg. Technol.* **2021**, *69*, 102644. [CrossRef]
44. Martí-Quijal, F.J.; Ramon-Mascarell, F.; Pallarés, N.; Ferrer, E.; Berrada, H.; Phimolsiripol, Y.; Barba, F.J. Extraction of Antioxidant Compounds and Pigments from Spirulina (*Arthrospira Platensis*) Assisted by Pulsed Electric Fields and the Binary Mixture of Organic Solvents and Water. *Appl. Sci.* **2021**, *11*, 7629. [CrossRef]
45. Soquetta, M.B.; Terra, L.d.M.; Bastos, C.P. Green Technologies for the Extraction of Bioactive Compounds in Fruits and Vegetables. *CyTA J. Food* **2018**, *16*, 400–412. [CrossRef]
46. Bozinou, E.; Karageorgou, I.; Batra, G.; Dourtoglou, V.G.; Lalas, S.I. Pulsed Electric Field Extraction and Antioxidant Activity Determination of *Moringa Oleifera* Dry Leaves: A Comparative Study with Other Extraction Techniques. *Beverages* **2019**, *5*, 8. [CrossRef]

47. Chemat, F.; Rombaut, N.; Meullemiestre, A.; Turk, M.; Perino, S.; Fabiano-Tixier, A.-S.; Abert-Vian, M. Review of Green Food Processing Techniques. Preservation, Transformation, and Extraction. *Innov. Food Sci. Emerg. Technol.* **2017**, *41*, 357–377. [CrossRef]
48. Jha, A.K.; Sit, N. Extraction of Bioactive Compounds from Plant Materials Using Combination of Various Novel Methods: A Review. *Trends Food Sci. Technol.* **2022**, *119*, 579–591. [CrossRef]
49. Arshad, R.N.; Abdul-Malek, Z.; Roobab, U.; Qureshi, M.I.; Khan, N.; Ahmad, M.H.; Liu, Z.W.; Aadil, R.M. Effective Valorization of Food Wastes and By-Products through Pulsed Electric Field: A Systematic Review. *J. Food Process. Eng.* **2021**, *44*, e13629. [CrossRef]
50. Baiano, A. Recovery of Biomolecules from Food Wastes—A Review. *Molecules* **2014**, *19*, 14821–14842. [CrossRef]
51. Athanasiadis, V.; Grigorakis, S.; Lalas, S.; Makris, D.P. Methyl β-Cyclodextrin as a Booster for the Extraction for Olea Europaea Leaf Polyphenols with a Bio-Based Deep Eutectic Solvent. *Biomass Convers. Biorefin.* **2018**, *8*, 345–355. [CrossRef]
52. Poojary, M.M.; Lund, M.N.; Barba, F.J. 4—Pulsed Electric Field (PEF) as an Efficient Technology for Food Additives and Nutraceuticals Development. In *Pulsed Electric Fields to Obtain Healthier and Sustainable Food for Tomorrow*; Barba, F.J., Parniakov, O., Wiktor, A., Eds.; Academic Press: Cambridge, MA, USA, 2020; pp. 65–99, ISBN 978-0-12-816402-0.
53. Barba, F.J.; Parniakov, O.; Pereira, S.A.; Wiktor, A.; Grimi, N.; Boussetta, N.; Saraiva, J.A.; Raso, J.; Martin-Belloso, O.; Witrowa-Rajchert, D. Current Applications and New Opportunities for the Use of Pulsed Electric Fields in Food Science and Industry. *Food Res. Int.* **2015**, *77*, 773–798. [CrossRef]
54. Khan, M.K.; Ahmad, K.; Hassan, S.; Imran, M.; Ahmad, N.; Xu, C. Effect of Novel Technologies on Polyphenols during Food Processing. *Innov. Food Sci. Emerg. Technol.* **2018**, *45*, 361–381. [CrossRef]
55. Ranjitha Gracy, T.K.; Sharanyakanth, P.S.; Radhakrishnan, M. Non-Thermal Technologies: Solution for Hazardous Pesticides Reduction in Fruits and Vegetables. *Crit. Rev. Food Sci. Nutr.* **2022**, *62*, 1782–1799. [CrossRef]
56. Onwude, D.I.; Hashim, N.; Janius, R.; Abdan, K.; Chen, G.; Oladejo, A.O. Non-Thermal Hybrid Drying of Fruits and Vegetables: A Review of Current Technologies. *Innov. Food Sci. Emerg. Technol.* **2017**, *43*, 223–238. [CrossRef]
57. Knoerzer, K.; Baumann, P.; Buckow, R. An Iterative Modelling Approach for Improving the Performance of a Pulsed Electric Field (PEF) Treatment Chamber. *Comput. Chem. Eng.* **2012**, *37*, 48–63. [CrossRef]
58. Hossain, M.B.; Aguiló-Aguayo, I.; Lyng, J.G.; Brunton, N.P.; Rai, D.K. Effect of Pulsed Electric Field and Pulsed Light Pre-Treatment on the Extraction of Steroidal Alkaloids from Potato Peels. *Innov. Food Sci. Emerg. Technol.* **2015**, *29*, 9–14. [CrossRef]
59. Ranjha, M.M.A.N.; Kanwal, R.; Shafique, B.; Arshad, R.N.; Irfan, S.; Kieliszek, M.; Kowalczewski, P.Ł.; Irfan, M.; Khalid, M.Z.; Roobab, U. A Critical Review on Pulsed Electric Field: A Novel Technology for the Extraction of Phytoconstituents. *Molecules* **2021**, *26*, 4893. [CrossRef]
60. Lasekan, O.; Ng, S.; Azeez, S.; Shittu, R.; Teoh, L.; Gholivand, S. Effect of Pulsed Electric Field Processing on Flavor and Color of Liquid Foods†. *J. Food Process. Preserv.* **2017**, *41*, e12940. [CrossRef]
61. Nowosad, K.; Sujka, M.; Pankiewicz, U.; Kowalski, R. The Application of PEF Technology in Food Processing and Human Nutrition. *J. Food Sci. Technol.* **2021**, *58*, 397–411. [CrossRef]
62. Goettel, M.; Eing, C.; Gusbeth, C.; Straessner, R.; Frey, W. Pulsed Electric Field Assisted Extraction of Intracellular Valuables from Microalgae. *Algal Res.* **2013**, *2*, 401–408. [CrossRef]
63. Wang, M.-S.; Wang, L.-H.; Bekhit, A.E.-D.A.; Yang, J.; Hou, Z.-P.; Wang, Y.-Z.; Dai, Q.-Z.; Zeng, X.-A. A Review of Sublethal Effects of Pulsed Electric Field on Cells in Food Processing. *J. Food Eng.* **2018**, *223*, 32–41. [CrossRef]
64. Arshad, R.N.; Abdul-Malek, Z.; Munir, A.; Buntat, Z.; Ahmad, M.H.; Jusoh, Y.M.M.; Bekhit, A.E.-D.; Roobab, U.; Manzoor, M.F.; Aadil, R.M. Electrical Systems for Pulsed Electric Field Applications in the Food Industry: An Engineering Perspective. *Trends Food Sci. Technol.* **2020**, *104*, 1–13. [CrossRef]
65. Yang, R.J.; Li, S.Q.; Zhang, Q.H. Effects of Pulsed Electric Fields on the Activity of Enzymes in Aqueous Solution. *J. Food Sci.* **2004**, *69*, FCT241–FCT248. [CrossRef]
66. Buckow, R.; Baumann, P.; Schroeder, S.; Knoerzer, K. Effect of Dimensions and Geometry of Co-Field and Co-Linear Pulsed Electric Field Treatment Chambers on Electric Field Strength and Energy Utilisation. *J. Food Eng.* **2011**, *105*, 545–556. [CrossRef]
67. Zhao, Y.; Zheng, Y.; He, H.; Sun, Z.; Li, A. Effective Aluminum Extraction Using Pressure Leaching of Bauxite Reaction Residue from Coagulant Industry and Leaching Kinetics Study. *J. Environ. Chem. Eng.* **2021**, *9*, 104770. [CrossRef]
68. Novickij, V.; Grainys, A.; Lastauskienė, E.; Kananavičiūtė, R.; Pamedytytė, D.; Kalėdienė, L.; Novickij, J.; Miklavčič, D. Pulsed Electromagnetic Field Assisted In Vitro Electroporation: A Pilot Study. *Sci. Rep.* **2016**, *6*, 33537. [CrossRef]
69. Donsì, F.; Ferrari, G.; Pataro, G. Applications of Pulsed Electric Field Treatments for the Enhancement of Mass Transfer from Vegetable Tissue. *Food Eng. Rev.* **2010**, *2*, 109–130. [CrossRef]
70. Chen, C.; Smye, S.W.; Robinson, M.P.; Evans, J.A. Membrane Electroporation Theories: A Review. *Med. Biol. Eng. Comput.* **2006**, *44*, 5–14. [CrossRef]
71. Soliva-Fortuny, R.; Balasa, A.; Knorr, D.; Martín-Belloso, O. Effects of Pulsed Electric Fields on Bioactive Compounds in Foods: A Review. *Trends Food Sci. Technol.* **2009**, *20*, 544–556. [CrossRef]
72. García, D.; Gómez, N.; Mañas, P.; Raso, J.; Pagán, R. Pulsed Electric Fields Cause Bacterial Envelopes Permeabilization Depending on the Treatment Intensity, the Treatment Medium pH and the Microorganism Investigated. *Int. J. Food Microbiol.* **2007**, *113*, 219–227. [CrossRef]

73. Pagán, R.; Mañas, P. Fundamental Aspects of Microbial Membrane Electroporation. In *Pulsed Electric Fields Technology for the Food Industry: Fundamentals and Applications*; Food Engineering Series; Raso, J., Heinz, V., Eds.; Springer: Boston, MA, USA, 2006; pp. 73–94, ISBN 978-0-387-31122-7.
74. Kolosnjaj-Tabi, J.; Gibot, L.; Fourquaux, I.; Golzio, M.; Rols, M.-P. Electric Field-Responsive Nanoparticles and Electric Fields: Physical, Chemical, Biological Mechanisms and Therapeutic Prospects. *Adv. Drug Deliv. Rev.* **2019**, *138*, 56–67. [CrossRef]
75. Demir, E.; Tappi, S.; Dymek, K.; Rocculi, P.; Gómez Galindo, F. Reversible Electroporation Caused by Pulsed Electric Field—Opportunities and Challenges for the Food Sector. *Trends Food Sci. Technol.* **2023**, *139*, 104120. [CrossRef]
76. Fauster, T.; Schlossnikl, D.; Rath, F.; Ostermeier, R.; Teufel, F.; Toepfl, S.; Jaeger, H. Impact of Pulsed Electric Field (PEF) Pretreatment on Process Performance of Industrial French Fries Production. *J. Food Eng.* **2018**, *235*, 16–22. [CrossRef]
77. Makrygiannis, I.; Athanasiadis, V.; Bozinou, E.; Chatzimitakos, T.; Makris, D.P.; Lalas, S.I. Combined Effects of Deep Eutectic Solvents and Pulsed Electric Field Improve Polyphenol-Rich Extracts from Apricot Kernel Biomass. *Biomass* **2023**, *3*, 66–77. [CrossRef]
78. Kotsou, K.; Stoikou, M.; Athanasiadis, V.; Chatzimitakos, T.; Mantiniotou, M.; Sfougaris, A.I.; Lalas, S.I. Enhancing Antioxidant Properties of *Prunus Spinosa* Fruit Extracts via Extraction Optimization. *Horticulturae* **2023**, *9*, 942. [CrossRef]
79. Di Cagno, R.; Surico, R.F.; Minervini, G.; Rizzello, C.G.; Lovino, R.; Servili, M.; Taticchi, A.; Urbani, S.; Gobbetti, M. Exploitation of Sweet Cherry (*Prunus avium* L.) Puree Added of Stem Infusion through Fermentation by Selected Autochthonous Lactic Acid Bacteria. *Food Microbiol.* **2011**, *28*, 900–909. [CrossRef] [PubMed]
80. Crisosto, C.H.; Crisosto, G.M.; Metheney, P. Consumer Acceptance of "Brooks" and "Bing" Cherries Is Mainly Dependent on Fruit SSC and Visual Skin Color. *Postharvest Biol. Technol.* **2003**, *28*, 159–167. [CrossRef]
81. Serra, A.T.; Duarte, R.O.; Bronze, M.R.; Duarte, C.M.M. Identification of Bioactive Response in Traditional Cherries from Portugal. *Food Chem.* **2011**, *125*, 318–325. [CrossRef]
82. Schreiner, M.; Huyskens-Keil, S. Phytochemicals in Fruit and Vegetables: Health Promotion and Postharvest Elicitors. *Crit. Rev. Plant Sci.* **2006**, *25*, 267–278. [CrossRef]
83. Sotelo, K.A.G.; Hamid, N.; Oey, I.; Pook, C.; Gutierrez-Maddox, N.; Ma, Q.; Ying Leong, S.; Lu, J. Red Cherries (Prunus Avium Var. Stella) Processed by Pulsed Electric Field—Physical, Chemical and Microbiological Analyses. *Food Chem.* **2018**, *240*, 926–934. [CrossRef]
84. El Darra, N.; Grimi, N.; Maroun, R.G.; Louka, N.; Vorobiev, E. Pulsed Electric Field, Ultrasound, and Thermal Pretreatments for Better Phenolic Extraction during Red Fermentation. *Eur. Food Res. Technol.* **2013**, *236*, 47–56. [CrossRef]
85. El Darra, N.; Turk, M.F.; Ducasse, M.A.; Grimi, N.; Maroun, R.G.; Louka, N.; Vorobiev, E. Changes in Polyphenol Profiles and Color Composition of Freshly Fermented Model Wine Due to Pulsed Electric Field, Enzymes and Thermovinification Pretreatments. *Food Chem.* **2016**, *194*, 944–950. [CrossRef] [PubMed]
86. Delsart, C.; Cholet, C.; Ghidossi, R.; Grimi, N.; Gontier, E.; Gény, L.; Vorobiev, E.; Mietton-Peuchot, M. Effects of Pulsed Electric Fields on Cabernet Sauvignon Grape Berries and on the Characteristics of Wines. *Food Bioprocess Technol.* **2012**, *7*, 424–436. [CrossRef]
87. Delsart, C.; Ghidossi, R.; Poupot, C.; Cholet, C.; Grimi, N.; Vorobiev, E.; Milisic, V.; Peuchot, M.M. Enhanced Extraction of Phenolic Compounds from Merlot Grapes by Pulsed Electric Field Treatment. *Am. J. Enol. Vitic.* **2012**, *63*, 205–211. [CrossRef]
88. Maza, M.A.; Martínez, J.M.; Delso, C.; Camargo, A.; Raso, J.; Álvarez, I. PEF-Dependency on Polyphenol Extraction during Maceration/Fermentation of Grenache Grapes. *Innov. Food Sci. Emerg. Technol.* **2020**, *60*, 102303. [CrossRef]
89. Maza, M.A.; Martínez, J.M.; Cebrián, G.; Sánchez-Gimeno, A.C.; Camargo, A.; Álvarez, I.; Raso, J. Evolution of Polyphenolic Compounds and Sensory Properties of Wines Obtained from Grenache Grapes Treated by Pulsed Electric Fields during Aging in Bottles and in Oak Barrels. *Foods* **2020**, *9*, 542. [CrossRef]
90. Comuzzo, P.; Voce, S.; Grazioli, C.; Tubaro, F.; Marconi, M.; Zanella, G.; Querzè, M. Pulsed Electric Field Processing of Red Grapes (Cv. Rondinella): Modifications of Phenolic Fraction and Effects on Wine Evolution. *Foods* **2020**, *9*, 414. [CrossRef]
91. Lakka, A.; Bozinou, E.; Makris, D.P.; Lalas, S.I. Evaluation of Pulsed Electric Field Polyphenol Extraction from *Vitis Vinifera*, *Sideritis Scardica* and *Crocus sativus*. *ChemEngineering* **2021**, *5*, 25. [CrossRef]
92. Brianceau, S.; Turk, M.; Vitrac, X.; Vorobiev, E. Combined Densification and Pulsed Electric Field Treatment for Selective Polyphenols Recovery from Fermented Grape Pomace. *Innov. Food Sci. Emerg. Technol.* **2015**, *29*, 2–8. [CrossRef]
93. Barba, F.J.; Brianceau, S.; Turk, M.; Boussetta, N.; Vorobiev, E. Effect of Alternative Physical Treatments (Ultrasounds, Pulsed Electric Fields, and High-Voltage Electrical Discharges) on Selective Recovery of Bio-Compounds from Fermented Grape Pomace. *Food Bioprocess Technol.* **2015**, *8*, 1139–1148. [CrossRef]
94. Atanasov, S.; Stoylov, B.L.; Saykova, I.; Tchaoushev, S.T. Mass Transfer Intensification in Bioactive Compounds Recovery by Alternative Extraction Methods: Effects of Solvent. *Glob. Nest J.* **2019**, *21*, 30–36. [CrossRef]
95. Delso, C.; Berzosa, A.; Sanz, J.; Álvarez, I.; Raso, J. Two-Step PEF Processing for Enhancing the Polyphenol Concentration and Decontaminating a Red Grape Juice. *Foods* **2022**, *11*, 621. [CrossRef] [PubMed]
96. Ricci, A.; Parpinello, G.P.; Banfi, B.A.; Olivi, F.; Versari, A. Preliminary Study of the Effects of Pulsed Electric Field (PEF) Treatments in Wines Obtained from Early-Harvested Sangiovese Grapes. *Beverages* **2020**, *6*, 34. [CrossRef]
97. Ziagova, M.; Mavromatidou, C.; Samiotis, G.; Amanatidou, E. Total Phenolic Content and Antioxidant Capacity of Greek Medicinal and Aromatic Plant Extracts Using Pulsed Electric Field Followed by Ultrasounds Extraction Process. *J. Food Process. Preserv.* **2022**, *46*, e16639. [CrossRef]

98. Ntourtoglou, G.; Drosou, F.; Chatzimitakos, T.; Athanasiadis, V.; Bozinou, E.; Dourtoglou, V.G.; Elhakem, A.; Sami, R.; Ashour, A.A.; Shafie, A. Combination of Pulsed Electric Field and Ultrasound in the Extraction of Polyphenols and Volatile Compounds from Grape Stems. *Appl. Sci.* **2022**, *12*, 6219. [CrossRef]
99. Carpentieri, S.; Ferrari, G.; Pataro, G. Pulsed Electric Fields-Assisted Extraction of Valuable Compounds from Red Grape Pomace: Process Optimization Using Response Surface Methodology. *Front. Nutr.* **2023**, *10*, 1158019. [CrossRef] [PubMed]
100. Vrhovsek, U.; Rigo, A.; Tonon, D.; Mattivi, F. Quantitation of Polyphenols in Different Apple Varieties. *J. Agric. Food Chem.* **2004**, *52*, 6532–6538. [CrossRef] [PubMed]
101. Wiktor, A.; Sledz, M.; Nowacka, M.; Rybak, K.; Chudoba, T.; Lojkowski, W.; Witrowa-Rajchert, D. The Impact of Pulsed Electric Field Treatment on Selected Bioactive Compound Content and Color of Plant Tissue. *Innov. Food Sci. Emerg. Technol.* **2015**, *30*, 69–78. [CrossRef]
102. Dziadek, K.; Kopeć, A.; Dróżdż, T.; Kiełbasa, P.; Ostafin, M.; Bulski, K.; Oziembłowski, M. Effect of Pulsed Electric Field Treatment on Shelf Life and Nutritional Value of Apple Juice. *J. Food Sci. Technol.* **2019**, *56*, 1184–1191. [CrossRef]
103. Pollini, L.; Cossignani, L.; Juan, C.; Mañes, J. Extraction of Phenolic Compounds from Fresh Apple Pomace by Different Non-Conventional Techniques. *Molecules* **2021**, *26*, 4272. [CrossRef]
104. Matys, A.; Dadan, M.; Witrowa-Rajchert, D.; Parniakov, O.; Wiktor, A. Response Surface Methodology as a Tool for Optimization of Pulsed Electric Field Pretreatment and Microwave-Convective Drying of Apple. *Appl. Sci.* **2022**, *12*, 3392. [CrossRef]
105. Teixeira da Silva, J.A.; Rana, T.S.; Narzary, D.; Verma, N.; Meshram, D.T.; Ranade, S.A. Pomegranate Biology and Biotechnology: A Review. *Sci. Hortic.* **2013**, *160*, 85–107. [CrossRef]
106. Cai, Y.; Yu, Y.; Duan, G.; Li, Y. Study on Infrared-Assisted Extraction Coupled with High Performance Liquid Chromatography (HPLC) for Determination of Catechin, Epicatechin, and Procyanidin B2 in Grape Seeds. *Food Chem.* **2011**, *127*, 1872–1877. [CrossRef]
107. Rajha, H.N.; Abi-Khattar, A.M.; El Kantar, S.; Boussetta, N.; Lebovka, N.; Maroun, R.G.; Louka, N.; Vorobiev, E. Comparison of Aqueous Extraction Efficiency and Biological Activities of Polyphenols from Pomegranate Peels Assisted by Infrared, Ultrasound, Pulsed Electric Fields and High-Voltage Electrical Discharges. *Innov. Food Sci. Emerg. Technol.* **2019**, *58*, 102212. [CrossRef]
108. Zou, Z.; Xi, W.; Hu, Y.; Nie, C.; Zhou, Z. Antioxidant Activity of Citrus Fruits. *Food Chem.* **2016**, *196*, 885–896. [CrossRef]
109. El Kantar, S.; Boussetta, N.; Lebovka, N.; Foucart, F.; Rajha, H.N.; Maroun, R.G.; Louka, N.; Vorobiev, E. Pulsed Electric Field Treatment of Citrus Fruits: Improvement of Juice and Polyphenols Extraction. *Innov. Food Sci. Emerg. Technol.* **2018**, *46*, 153–161. [CrossRef]
110. Athanasiadis, V.; Chatzimitakos, T.; Kotsou, K.; Palaiogiannis, D.; Bozinou, E.; Lalas, S.I. Optimization of the Extraction Parameters for the Isolation of Bioactive Compounds from Orange Peel Waste. *Sustainability* **2022**, *14*, 13926. [CrossRef]
111. Luengo, E.; Álvarez, I.; Raso, J. Improving the Pressing Extraction of Polyphenols of Orange Peel by Pulsed Electric Fields. *Innov. Food Sci. Emerg. Technol.* **2013**, *17*, 79–84. [CrossRef]
112. Peiró, S.; Luengo, E.; Segovia, F.; Raso, J.; Almajano, M.P. Improving Polyphenol Extraction from Lemon Residues by Pulsed Electric Fields. *Waste Biomass Valorization* **2019**, *10*, 889–897. [CrossRef]
113. Chatzimitakos, T.; Athanasiadis, V.; Kotsou, K.; Bozinou, E.; Lalas, S.I. Response Surface Optimization for the Enhancement of the Extraction of Bioactive Compounds from *Citrus Limon* Peel. *Antioxidants* **2023**, *12*, 1605. [CrossRef]
114. Athanasiadis, V.; Chatzimitakos, T.; Bozinou, E.; Kotsou, K.; Palaiogiannis, D.; Lalas, S.I. Optimization of Extraction Parameters for Enhanced Recovery of Bioactive Compounds from Quince Peels Using Response Surface Methodology. *Foods* **2023**, *12*, 2099. [CrossRef]
115. Lončarić, A.; Celeiro, M.; Jozinović, A.; Jelinić, J.; Kovač, T.; Jokić, S.; Babić, J.; Moslavac, T.; Zavadlav, S.; Lores, M. Green Extraction Methods for Extraction of Polyphenolic Compounds from Blueberry Pomace. *Foods* **2020**, *9*, 1521. [CrossRef] [PubMed]
116. Medina-Meza, I.G.; Boioli, P.; Barbosa-Cánovas, G.V. Assessment of the Effects of Ultrasonics and Pulsed Electric Fields on Nutritional and Rheological Properties of Raspberry and Blueberry Purees. *Food Bioprocess Technol.* **2016**, *9*, 520–531. [CrossRef]
117. Ozkan, G.; Stübler, A.S.; Aganovic, K.; Dräger, G.; Esatbeyoglu, T.; Capanoglu, E. Retention of Polyphenols and Vitamin C in Cranberrybush Purée (*Viburnum Opulus*) by Means of Non-Thermal Treatments. *Food Chem.* **2021**, *360*, 129918. [CrossRef] [PubMed]
118. Gagneten, M.; Leiva, G.; Salvatori, D.; Schebor, C.; Olaiz, N. Optimization of Pulsed Electric Field Treatment for the Extraction of Bioactive Compounds from Blackcurrant. *Food Bioprocess Technol.* **2019**, *12*, 1102–1109. [CrossRef]
119. Stübler, A.S.; Lesmes, U.; Juadjur, A.; Heinz, V.; Rauh, C.; Shpigelman, A.; Aganovic, K. Impact of Pilot-Scale Processing (Thermal, PEF, HPP) on the Stability and Bioaccessibility of Polyphenols and Proteins in Mixed Protein- and Polyphenol-Rich Juice Systems. *Innov. Food Sci. Emerg. Technol.* **2020**, *64*, 102426. [CrossRef]
120. Vallverdú-Queralt, A.; Odriozola-Serrano, I.; Oms-Oliu, G.; Lamuela-Raventós, R.M.; Elez-Martínez, P.; Martín-Belloso, O. Changes in the Polyphenol Profile of Tomato Juices Processed by Pulsed Electric Fields. *J. Agric. Food Chem.* **2012**, *60*, 9667–9672. [CrossRef]
121. Vallverdú-Queralt, A.; Oms-Oliu, G.; Odriozola-Serrano, I.; Lamuela-Raventos, R.M.; Martín-Belloso, O.; Elez-Martínez, P. Effects of Pulsed Electric Fields on the Bioactive Compound Content and Antioxidant Capacity of Tomato Fruit. *J. Agric. Food Chem.* **2012**, *60*, 3126–3134. [CrossRef]
122. Surano, B.; Leiva, G.; Marshall, G.; Maglietti, F.; Schebor, C. Pulsed Electric Fields Using a Multiple Needle Chamber to Improve Bioactive Compounds Extraction from Unprocessed *Opuntia Ficus-Indica* Fruits. *J. Food Eng.* **2022**, *317*, 110864. [CrossRef]

123. Aleti, G.; Nikolić, B.; Brader, G.; Pandey, R.V.; Antonielli, L.; Pfeiffer, S.; Oswald, A.; Sessitsch, A. Secondary Metabolite Genes Encoded by Potato Rhizosphere Microbiomes in the Andean Highlands Are Diverse and Vary with Sampling Site and Vegetation Stage. *Sci. Rep.* **2017**, *7*, 2330. [CrossRef]
124. Rasheed, H.; Ahmad, D.; Bao, J. Genetic Diversity and Health Properties of Polyphenols in Potato. *Antioxidants* **2022**, *11*, 603. [CrossRef]
125. Frontuto, D.; Carullo, D.; Harrison, S.M.; Brunton, N.P.; Ferrari, G.; Lyng, J.G.; Pataro, G. Optimization of Pulsed Electric Fields-Assisted Extraction of Polyphenols from Potato Peels Using Response Surface Methodology. *Food Bioprocess Technol.* **2019**, *12*, 1708–1720. [CrossRef]
126. Symes, A.; Shavandi, A.; Bekhit, A.E.A. Effects of Ionic Liquids and Pulsed Electric Fields on the Extraction of Antioxidants from Green Asparagus Roots. *Int. J. Food Sci. Technol.* **2023**, *58*, 3935–3945. [CrossRef]
127. Lee, I.H.; Huang, R.L.; Chen, C.T.; Chen, H.C.; Hsu, W.C.; Lu, M.K. *Antrodia camphorata* Polysaccharides Exhibit Anti-Hepatitis B Virus Effects. *FEMS Microbiol. Lett.* **2002**, *209*, 63–67. [CrossRef] [PubMed]
128. Chaturvedi, V.K.; Agarwal, S.; Gupta, K.K.; Ramteke, P.W.; Singh, M.P. Medicinal Mushroom: Boon for Therapeutic Applications. *3 Biotech* **2018**, *8*, 334. [CrossRef] [PubMed]
129. Xue, D.; Farid, M.M. Pulsed Electric Field Extraction of Valuable Compounds from White Button Mushroom (*Agaricus Bisporus*). *Innov. Food Sci. Emerg. Technol.* **2015**, *29*, 178–186. [CrossRef]
130. Gorzynik-Debicka, M.; Przychodzen, P.; Cappello, F.; Kuban-Jankowska, A.; Gammazza, A.M.; Knap, N.; Wozniak, M.; Gorska-Ponikowska, M. Potential Health Benefits of Olive Oil and Plant Polyphenols. *Int. J. Mol. Sci.* **2018**, *19*, 686. [CrossRef]
131. Tsantili, E.; Evangelou, E.; Kiritsakis, A. Botanical Characteristics of Olive Trees: Cultivation and Growth Conditions—Defense Mechanisms to Various Stressors and Effects on Olive Growth and Functional Compounds. In *Olives and Olive Oil as Functional Foods*; John Wiley & Sons, Ltd.: Hoboken, NJ, USA, 2017; pp. 13–33, ISBN 978-1-119-13534-0.
132. Andreou, V.; Psarianos, M.; Dimopoulos, G.; Tsimogiannis, D.; Taoukis, P. Effect of Pulsed Electric Fields and High Pressure on Improved Recovery of High-Added-Value Compounds from Olive Pomace. *J. Food Sci.* **2020**, *85*, 1500–1512. [CrossRef]
133. Andreou, V.; Kourmbeti, E.; Dimopoulos, G.; Psarianos, M.; Katsaros, G.; Taoukis, P. Optimization of Virgin Olive Oil Yield and Quality Applying Nonthermal Processing. *Food Bioprocess Technol.* **2022**, *15*, 891–903. [CrossRef]
134. Pappas, V.M.; Lakka, A.; Palaiogiannis, D.; Bozinou, E.; Ntourtoglou, G.; Batra, G.; Athanasiadis, V.; Makris, D.P.; Dourtoglou, V.G.; Lalas, S.I. Use of Pulsed Electric Field as a Low-Temperature and High-Performance "Green" Extraction Technique for the Recovery of High Added Value Compounds from Olive Leaves. *Beverages* **2021**, *7*, 45. [CrossRef]
135. Pappas, V.M.; Lakka, A.; Palaiogiannis, D.; Athanasiadis, V.; Bozinou, E.; Ntourtoglou, G.; Makris, D.P.; Dourtoglou, V.G.; Lalas, S.I. Optimization of Pulsed Electric Field as Standalone "Green" Extraction Procedure for the Recovery of High Value-Added Compounds from Fresh Olive Leaves. *Antioxidants* **2021**, *10*, 1554. [CrossRef]
136. Segovia, F.J.; Luengo, E.; Corral-Pérez, J.J.; Raso, J.; Almajano, M.P. Improvements in the Aqueous Extraction of Polyphenols from Borage (*Borago Officinalis* L.) Leaves by Pulsed Electric Fields: Pulsed Electric Fields (PEF) Applications. *Ind. Crops Prod.* **2015**, *65*, 390–396. [CrossRef]
137. Boussetta, N.; Soichi, E.; Lanoisellé, J.; Vorobiev, E. Valorization of Oilseed Residues: Extraction of Polyphenols from Flaxseed Hulls by Pulsed Electric Fields. *Ind. Crops Prod.* **2014**, *52*, 347–353. [CrossRef]
138. Yu, X.; Bals, O.; Grimi, N.; Vorobiev, E. A New Way for the Oil Plant Biomass Valorization: Polyphenols and Proteins Extraction from Rapeseed Stems and Leaves Assisted by Pulsed Electric Fields. *Ind. Crops Prod.* **2015**, *74*, 309–318. [CrossRef]
139. Yu, X.; Gouyo, T.; Grimi, N.; Bals, O.; Vorobiev, E. Pulsed Electric Field Pretreatment of Rapeseed Green Biomass (Stems) to Enhance Pressing and Extractives Recovery. *Bioresour. Technol.* **2016**, *199*, 194–201. [CrossRef]
140. Teh, S.S.; Niven, B.E.; Bekhit, A.E.D.A.; Carne, A.; Birch, J. Optimization of Polyphenol Extraction and Antioxidant Activities of Extracts from Defatted Flax Seed Cake (*Linum Usitatissimum* L.) Using Microwave-Assisted and Pulsed Electric Field (PEF) Technologies with Response Surface Methodology. *Food Sci. Biotechnol.* **2015**, *24*, 1649–1659. [CrossRef]
141. Barbosa-Pereira, L.; Guglielmetti, A.; Zeppa, G. Pulsed Electric Field Assisted Extraction of Bioactive Compounds from Cocoa Bean Shell and Coffee Silverskin. *Food Bioprocess Technol.* **2018**, *11*, 818–835. [CrossRef]
142. Neri, L.; Giancaterino, M.; Rocchi, R.; Tylewicz, U.; Valbonetti, L.; Faieta, M.; Pittia, P. Pulsed Electric Fields (PEF) as Hot Air Drying Pre-Treatment: Effect on Quality and Functional Properties of Saffron (*Crocus Sativus* L.). *Innov. Food Sci. Emerg. Technol.* **2021**, *67*, 102592. [CrossRef]
143. Ahmed, Z.; Faisal Manzoor, M.; Hussain, A.; Hanif, M.; Zia-ud-Din; Zeng, X.A. Study the Impact of Ultra-Sonication and Pulsed Electric Field on the Quality of Wheat Plantlet Juice through FTIR and SERS. *Ultrason. Sonochem.* **2021**, *76*, 105648. [CrossRef]
144. Athanasiadis, V.; Lakka, A.; Palaiogiannis, D.; Pappas, V.M.; Bozinou, E.; Ntourtoglou, G.; Makris, D.P.; Dourtoglou, V.G.; Lalas, S.I. Pulsed Electric Field and *Salvia Officinalis* l. Leaves: A Successful Combination for the Extraction of High Value Added Compounds. *Foods* **2021**, *10*, 2014. [CrossRef]
145. Salgado-Ramos, M.; Martí-Quijal, F.J.; Huertas-Alonso, A.J.; Sánchez-Verdú, M.P.; Barba, F.J.; Moreno, A. Almond Hull Biomass: Preliminary Characterization and Development of Two Alternative Valorization Routes by Applying Innovative and Sustainable Technologies. *Ind. Crops Prod.* **2022**, *179*, 114697. [CrossRef]
146. Teh, S.S.; Niven, B.E.; Bekhit, A.E.D.A.; Carne, A.; Birch, E.J. The Use of Microwave and Pulsed Electric Field as a Pretreatment Step in Ultrasonic Extraction of Polyphenols from Defatted Hemp Seed Cake (*Cannabis Sativa*) Using Response Surface Methodology. *Food Bioprocess Technol.* **2014**, *7*, 3064–3076. [CrossRef]

147. Sarkis, J.R.; Boussetta, N.; Blouet, C.; Tessaro, I.C.; Marczak, L.D.F.; Vorobiev, E. Effect of Pulsed Electric Fields and High Voltage Electrical Discharges on Polyphenol and Protein Extraction from Sesame Cake. *Innov. Food Sci. Emerg. Technol.* **2015**, *29*, 170–177. [CrossRef]
148. Quagliariello, V.; Iaffaioli, R.V.; Falcone, M.; Ferrari, G.; Pataro, G.; Donsì, F. Effect of Pulsed Electric Fields—Assisted Extraction on Anti-Inflammatory and Cytotoxic Activity of Brown Rice Bioactive Compounds. *Food Res. Int.* **2016**, *87*, 115–124. [CrossRef] [PubMed]
149. Bouras, M.; Grimi, N.; Bals, O.; Vorobiev, E. Impact of Pulsed Electric Fields on Polyphenols Extraction from Norway Spruce Bark. *Ind. Crops Prod.* **2016**, *80*, 50–58. [CrossRef]
150. Sarraf, M.; Beig-Babaei, A.; Naji-Tabasi, S. Optimizing Extraction of Berberine and Antioxidant Compounds from Barberry by Maceration and Pulsed Electric Field-Assisted Methods. *J. Berry Res.* **2021**, *11*, 133–149. [CrossRef]
151. Lakka, A.; Bozinou, E.; Stavropoulos, G.; Samanidis, I.; Athanasiadis, V.; Dourtoglou, V.G.; Makris, D.P.; Lalas, S.I. Enhancement of Polyphenols Recovery from *Rosa Canina*, *Calendula Officinalis* and *Castanea Sativa* Using Pulsed Electric Field. *Beverages* **2021**, *7*, 63. [CrossRef]
152. Carpentieri, S.; Mazza, L.; Nutrizio, M.; Jambrak, A.R.; Ferrari, G.; Pataro, G. Pulsed Electric Fields- and Ultrasound-Assisted Green Extraction of Valuable Compounds from *Origanum Vulgare* L. and *Thymus Serpyllum* L. *Int. J. Food Sci. Technol.* **2021**, *56*, 4834–4842. [CrossRef]
153. Alberto, J.; Costa, V.; Catarina, B.; Freitas, B.; Moraes, L.; Zaparoli, M.; Greque, M. Bioresource Technology Progress in the Physicochemical Treatment of Microalgae Biomass for Value- Added Product Recovery. *Bioresour. Technol.* **2020**, *301*, 122727. [CrossRef]
154. Morais Junior, W.G.; Gorgich, M.; Corrêa, P.S.; Martins, A.A.; Mata, T.M.; Caetano, N.S. Microalgae for Biotechnological Applications: Cultivation, Harvesting and Biomass Processing. *Aquaculture* **2020**, *528*, 735562. [CrossRef]
155. Einarsdóttir, R.; Þórarinsdóttir, K.A.; Aðalbjörnsson, B.V.; Guðmundsson, M.; Marteinsdóttir, G.; Kristbergsson, K. The Effect of Pulsed Electric Field-Assisted Treatment Parameters on Crude Aqueous Extraction of *Laminaria digitata*. *J. Appl. Phycol.* **2021**, *33*, 3287–3296. [CrossRef]
156. Kokkali, M.; Martí-Quijal, F.J.; Taroncher, M.; Ruiz, M.J.; Kousoulaki, K.; Barba, F.J. Improved Extraction Efficiency of Antioxidant Bioactive Compounds from *Tetraselmis Chuii* and *Phaedoactylum Tricornutum* Using Pulsed Electric Fields. *Molecules* **2020**, *25*, 3921. [CrossRef] [PubMed]
157. Castejón, N.; Thorarinsdottir, K.A.; Einarsdóttir, R.; Kristbergsson, K.; Marteinsdóttir, G. Exploring the Potential of Icelandic Seaweeds Extracts Produced by Aqueous Pulsed Electric Fields-Assisted Extraction for Cosmetic Applications. *Mar. Drugs* **2021**, *19*, 662. [CrossRef] [PubMed]
158. Zhou, J.; Wang, M.; Berrada, H.; Zhu, Z.; Grimi, N.; Barba, F.J. Pulsed Electric Fields (PEF), Pressurized Liquid Extraction (PLE) and Combined PEF + PLE Process Evaluation: Effects on Spirulina Microstructure, Biomolecules Recovery and Triple TOF-LC-MS-MS Polyphenol Composition. *Innov. Food Sci. Emerg. Technol.* **2022**, *77*, 102989. [CrossRef]
159. Jiménez-Moreno, N.; Esparza, I.; Bimbela, F.; Gandía, L.M.; Ancín-Azpilicueta, C. Valorization of Selected Fruit and Vegetable Wastes as Bioactive Compounds: Opportunities and Challenges. *Crit. Rev. Environ. Sci. Technol.* **2020**, *50*, 2061–2108. [CrossRef]
160. Andreou, V.; Dimopoulos, G.; Dermesonlouoglou, E.; Taoukis, P. Application of Pulsed Electric Fields to Improve Product Yield and Waste Valorization in Industrial Tomato Processing. *J. Food Eng.* **2020**, *270*, 109778. [CrossRef]
161. Niu, D.; Zeng, X.-A.; Ren, E.-F.; Xu, F.-Y.; Li, J.; Wang, M.-S.; Wang, R. Review of the Application of Pulsed Electric Fields (PEF) Technology for Food Processing in China. *Food Res. Int.* **2020**, *137*, 109715. [CrossRef] [PubMed]
162. Aşık-Canbaz, E.; Çömlekçi, S.; Can Seydim, A. Effect of Moderate Intensity Pulsed Electric Field on Shelf-Life of Chicken Breast Meat. *Br. Poult. Sci.* **2022**, *63*, 641–649. [CrossRef]
163. Moretto, G.; Russo, I.; Bolzonella, D.; Pavan, P.; Majone, M.; Valentino, F. An Urban Biorefinery for Food Waste and Biological Sludge Conversion into Polyhydroxyalkanoates and Biogas. *Water Res.* **2020**, *170*, 115371. [CrossRef]
164. Ashokkumar, V.; Flora, G.; Venkatkarthick, R.; SenthilKannan, K.; Kuppam, C.; Mary Stephy, G.; Kamyab, H.; Chen, W.-H.; Thomas, J.; Ngamcharussrivichai, C. Advanced Technologies on the Sustainable Approaches for Conversion of Organic Waste to Valuable Bioproducts: Emerging Circular Bioeconomy Perspective. *Fuel* **2022**, *324*, 124313. [CrossRef]
165. Morone, P.; Koutinas, A.; Gathergood, N.; Arshadi, M.; Matharu, A. Food Waste: Challenges and Opportunities for Enhancing the Emerging Bio-Economy. *J. Clean. Prod.* **2019**, *221*, 10–16. [CrossRef]
166. Bottausci, S.; Midence, R.; Serrano-Bernardo, F.; Bonoli, A. Organic Waste Management and Circular Bioeconomy: A Literature Review Comparison between Latin America and the European Union. *Sustainability* **2022**, *14*, 1661. [CrossRef]
167. Mohd Basri, M.S.; Abdul Karim Shah, N.N.; Sulaiman, A.; Mohamed Amin Tawakkal, I.S.; Mohd Nor, M.Z.; Ariffin, S.H.; Abdul Ghani, N.H.; Mohd Salleh, F.S. Progress in the Valorization of Fruit and Vegetable Wastes: Active Packaging, Biocomposites, By-Products, and Innovative Technologies Used for Bioactive Compound Extraction. *Polymers* **2021**, *13*, 3503. [CrossRef] [PubMed]
168. Golberg, A.; Sack, M.; Teissie, J.; Pataro, G.; Pliquett, U.; Saulis, G.; Stefan, T.; Miklavcic, D.; Vorobiev, E.; Frey, W. Energy-Efficient Biomass Processing with Pulsed Electric Fields for Bioeconomy and Sustainable Development. *Biotechnol. Biofuels* **2016**, *9*, 94. [CrossRef]
169. Toepfl, S. Pulsed Electric Field Food Treatment—Scale up from Lab to Industrial Scale. *Procedia Food Sci.* **2011**, *1*, 776–779. [CrossRef]

170. Sosa-Hernández, J.E.; Escobedo-Avellaneda, Z.; Iqbal, H.M.N.; Welti-Chanes, J. State-of-the-Art Extraction Methodologies for Bioactive Compounds from Algal Biome to Meet Bio-Economy Challenges and Opportunities. *Molecules* **2018**, *23*, 2953. [CrossRef] [PubMed]
171. Waseem, M.; Majeed, Y.; Nadeem, T.; Naqvi, L.H.; Khalid, M.A.; Sajjad, M.M.; Sultan, M.; Khan, M.U.; Khayrullin, M.; Shariati, M.A. Conventional and Advanced Extraction Methods of Some Bioactive Compounds with Health Benefits of Food and Plant Waste: A Comprehensive Review. *Food Front.* **2023**, 1–21. [CrossRef]
172. Naliyadhara, N.; Kumar, A.; Girisa, S.; Daimary, U.D.; Hegde, M.; Kunnumakkara, A.B. Pulsed Electric Field (PEF): Avant-Garde Extraction Escalation Technology in Food Industry. *Trends Food Sci. Technol.* **2022**, *122*, 238–255. [CrossRef]

International Journal of *Molecular Sciences*

MDPI

Article

Functionalization of Polyhydroxyalkanoates (PHA)-Based Bioplastic with Phloretin for Active Food Packaging: Characterization of Its Mechanical, Antioxidant, and Antimicrobial Activities

Seyedeh Fatemeh Mirpoor [1,†], Giuseppe Tancredi Patanè [2,†], Iolanda Corrado [1], C. Valeria L. Giosafatto [1], Giovanna Ginestra [2], Antonia Nostro [2], Antonino Foti [3], Pietro G. Gucciardi [3], Giuseppina Mandalari [2], Davide Barreca [2,*], Teresa Gervasi [4,*] and Cinzia Pezzella [1]

1 Department of Chemical Sciences, University of Naples Federico II, 80126 Naples, Italy; fatemeh.mirpour.fm@gmail.com (S.F.M.); iolanda.corrado@unina.it (I.C.); giosafat@unina.it (C.V.L.G.); cpezzella@unina.it (C.P.)

2 Department of Chemical, Biological, Pharmaceutical and Environmental Science, University of Messina, 98166 Messina, Italy; giuseppe.patane@studenti.unime.it (G.T.P.); giovanna.ginestra@unime.it (G.G.); antonia.nostro@unime.it (A.N.); giuseppina.mandalari@unime.it (G.M.)

3 CNR IPCF, Istituto per i Processi Chimico-Fisici, Viale F. Stagno D'Alcontres 37, 98156 Messina, Italy; antonino.foti@cnr.it (A.F.); pietrogiuseppe.gucciardi@cnr.it (P.G.G.)

4 Department of Biomedical and Dental Sciences and Morphofunctional Imaging, University of Messina, 98125 Messina, Italy

* Correspondence: dbarreca@unime.it (D.B.); teresa.gervasi@unime.it (T.G.)

† These authors contributed equally to this work.

Citation: Mirpoor, S.F.; Patanè, G.T.; Corrado, I.; Giosafatto, C.V.L.; Ginestra, G.; Nostro, A.; Foti, A.; Gucciardi, P.G.; Mandalari, G.; Barreca, D.; et al. Functionalization of Polyhydroxyalkanoates (PHA)-Based Bioplastic with Phloretin for Active Food Packaging: Characterization of Its Mechanical, Antioxidant, and Antimicrobial Activities. *Int. J. Mol. Sci.* **2023**, *24*, 11628. https://doi.org/10.3390/ijms241411628

Academic Editors: Vassilis Athanasiadis and Theodoros G. Chatzimitakos

Received: 30 June 2023
Revised: 15 July 2023
Accepted: 16 July 2023
Published: 19 July 2023

Abstract: The formulation of eco-friendly biodegradable packaging has received great attention during the last decades as an alternative to traditional widespread petroleum-based food packaging. With this aim, we designed and tested the properties of polyhydroxyalkanoates (PHA)-based bioplastics functionalized with phloretin as far as antioxidant, antimicrobial, and morpho-mechanic features are concerned. Mechanical and hydrophilicity features investigations revealed a mild influence of phloretin on the novel materials as a function of the concentration utilized (5, 7.5, 10, and 20 mg) with variation in FTIR e RAMAN spectra as well as in mechanical properties. Functionalization of PHA-based polymers resulted in the acquisition of the antioxidant activity (in a dose-dependent manner) tested by DPPH, TEAC, FRAR, and chelating assays, and in a decrease in the growth of food-borne pathogens (*Listeria monocytogenes* ATCC 13932). Finally, apple samples were packed in the functionalized PHA films for 24, 48, and 72 h, observing remarkable effects on the stabilization of apple samples. The results open the possibility to utilize phloretin as a functionalizing agent for bioplastic formulation, especially in relation to food packaging.

Keywords: phloretin; Listeria monocytogenes; active packaging; food contact materials; bioplastics

1. Introduction

The formulation of eco-friendly biodegradable packaging produced from bio-based resources has received great attention during the last decades as an alternative to traditional widespread petroleum-based food packaging. In addition to their passive protective function, they can also play a pivotal role as a delivery system for bioactive compounds able to increase the mechanical characteristics of the materials and supply protection to packed foods against oxidative damage and bacterial contamination [1,2]. Among biodegradable and biobased polymers, polyhydroxyalkanoates (PHA) are one of the most investigated molecules. They are polyesters of 3-hydroxyalkanoic acids, synthesized by numerous classes of microorganisms as intracellular carbon and energy storage granules, usually under nutrient-limiting conditions [3]. Different bacterial species are known as

PHA producers, displaying diversity in the type of polymer produced and the metabolic pathways that channel 3-hydroxyalkanoic acid precursors in the synthesized polymer. As a fact, PHA has been classified according to its monomer chain length into short chain length (scl)-PHA (C4 and C5), medium chain length (mcl)-PHA ($C \geq 6$), and long chain length (lcl)-PHA ($C > 14$). Among them, polyhydroxybutyrate (PHB), a scl-PHA, is by far the most well-known PHA polymer, accumulated to up to 90% of cell dry weight by native (*Cupriavidus necator* as the workhorse) and recombinant microorganisms. On the other hand, most mcl-PHA are produced from *Pseudomonas* sp., whilst scl-mcl copolymers are produced by many bacterial species. The introduction of other monomeric units in the PHB backbone allows to improve polymer properties in favor of a reduced stiffness, higher elongation to break, and lower melting point. Although up to 13 different routes allowing channeling0specific precursors into PHA have been described, three main pathways related to sugar catabolism, fatty acids oxidation, and synthesis are responsible for the synthesis of scl-PHA, scl-mcl copolymers, and mcl-PHA, depending on the supplied C-source (sugars or lipids) [4]. Being totally produced by various bacterial species through microbial fermentation of different C-sources (both carbohydrate and lipid ones), PHA origin is properly renewable. The formulation of a plethora of PHA-based materials, i.e., copolymers, physical and/or melt-reactive blending with tailored polymers, and the addition of organic and inorganic fillers and plasticizers, has led to the improvement of mechanical features, wide processability windows, and enhanced stability and permeability performances [2]. So far, owing to its special polymer features, PHA with diverse structures and properties has been exploited in several application fields, i.e., as bioplastics for food packaging [3], in tissue engineering for biomedical implants and drug delivery carriers [5], and in the agriculture sector for the controlled delivery of herbicides [6].

Phloretin (dihydronaringenin or phloretol) is a dihydrochalcone, belonging to the class of flavonoids and found in many fruits and vegetables, and it is the most abundant compound identified in apples and in apple-derived products as well as in the kumquat in its glycosylated forms, characterized by the presence of the pharmacophore 2,6-dihydroxyacetophenone, which is responsible for its biological potential. Moreover, the two aromatic phenol rings A and B, the hydroxyl groups in positions 2, 4, 4$'$, and 6, and the carbonyl group in position 1$''$ supply its specific functions. Recently, the broad spectrum of beneficial properties of phloretin for human health, such as anti-inflammatory, antimicrobial, anti-hypertensive, antioxidant, anti-cancer, and other biological activities, have been reported by the authors [7–9]. It is also able to induce change in ion transport across lipid bilayer membranes, to regulate glucose uptake at the intestinal level exhibiting potential food–drug interactions by inhibiting human UDP-glucuronosyltransferases in vitro and anti-metabolic syndrome [10,11].

In this scenario, the aim of this work was to manufacture novel active packaging based on PHAs functionalized with phloretin able to act against food-borne pathogen bacteria and to prevent food spoilage. Amongst food-borne bacteria, *Listeria monocytogenes*, responsible for listeriosis, ranks third in the total number of deaths, exceedingly even *Salmonella* spp. and *Clostridium botulinum* ("Campylobacter and Listeria infections still rising in the EU—say EFSA and ECDC—European Food Safety Authority", www.efsa.europa.eu, accessed on 17 December 2015). Some of its virulence factors include the ability to grow at 4 °C and reproduce inside the host's cells. On the other hand, biodegradable packaging materials endowed with antioxidant activity are widely searched by the food industry to prevent food losses while addressing the circularity criteria [12].

Hence, in this paper phloretin-functionalized PHA materials were prepared for their potential application as food contact materials. The films were characterized in terms of functional as well as biological properties. Furthermore, their positive effect on fruit quality was investigated in apple samples. A summary of the experiment was performed and the aims obtained are depicted in Scheme 1.

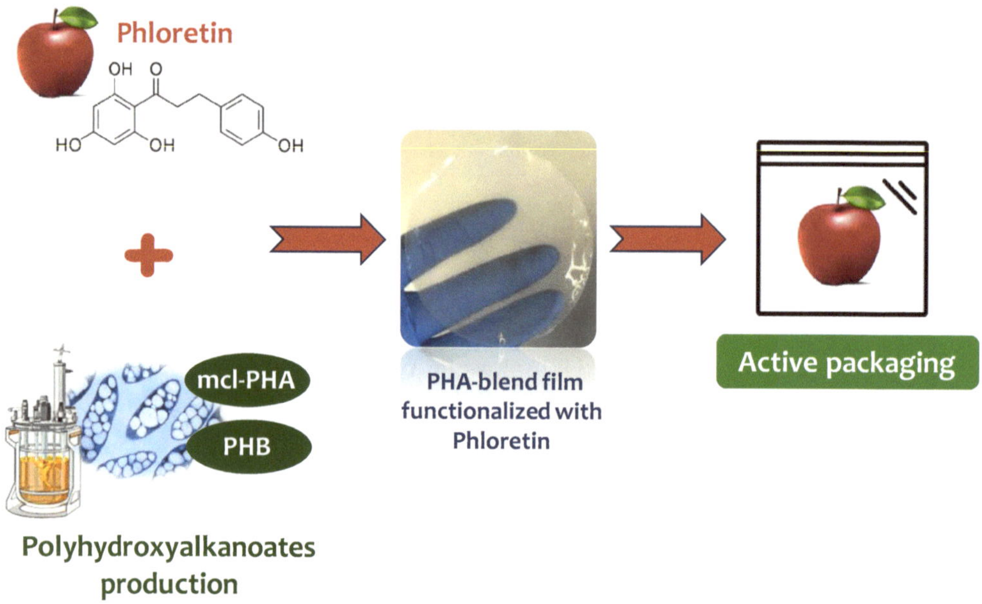

Scheme 1. Graphical representation of the work and of the aims obtained.

2. Results and Discussion

2.1. Evaluation of the Antimicrobial Activity of Phloretin

Determination of the MIC and MBC values of phloretin against several food-borne pathogens was carried out (Table 1). Gram-negative bacteria were overall more resistant compared with the positive ones. The tested dehydrochalcone was particularly effective against *L. monocytogenes* ATCC 13932 and all the *L. monocytogenes* food isolates. The activity reported for *L. monocytogenes* strains was always bacteriostatic rather than bactericidal. These data confirmed our previous results [7] and led us to test phloretin for the subsequent film functionalization.

Table 1. MICs and MBCs of phloretin. Values are expressed as μg mL^{-1} and represent the mean of three determinations.

	MIC	MBC
Enterococcus hirae ATCC 10541	250	>1000
Salmonella enterica serovar Typhimurium ATCC 13311	500	500
S. enterica serovar Typhimurium (clinical isolate)	>1000	>1000
S. enterica (clinical isolate)	500	500
Escherichia coli ATCC 25922	>1000	>1000
E. coli ATCC 10536	>1000	>1000
Listeria monocytogenes ATCC 13932	125–62.5	>1000
L. monocytogenes A241 (1/2a)	125	>1000
L. monocytogenes A216(1/2a)	125	>1000
L. monocytogenes A149 (1/2b)	125–62.5	>1000
L. monocytogenes A240 (1/2b)	250–125	>1000
L. monocytogenes G197 (1/2c)	250–125	>1000
L. monocytogenes G193 (1/2c)	125	>1000
L. monocytogenes G152 (4b)	125–62.5	>1000
L. monocytogenes A222 (4b)	62.5	>1000
L. monocytogenes A256 (1/2a)	250	>1000

Table 1. *Cont.*

	MIC	MBC
L. monocytogenes A84(1/2b)	125	>1000
L. monocytogenes A223 (1/2c)	125	>1000
L. monocytogenes G259 (1/2b)	125–250	>1000
L. monocytogenes G171 (1/2a)	125	>1000
L. monocytogenes G282 (4b)	125	>1000

We have previously demonstrated the efficacy of polyphenols against *L. monocytogenes* strains: A bactericidal effect against a food isolate was detected using an extract of *Hibiscus sabdariffa* L., rich in anthocyanins (cyanidin-3-O-sambubioside and delphinidin-3-O-sambubioside), whereas a bacteriostatic and bactericidal effect was obtained with polyphenols-rich extracts from nuts [13–15]. Due to the dangerous effects of *L. monocytogenes* infections, which can cause listeriosis often with severe outbreaks, and the possibility of contamination during food chain production, packing, and distribution, the obtained results show phloretin as a promising functionalizing agent for the production of bioplastic in food packaging industrial applications, and this may help control contamination in the food chain, from production to end consumers.

2.2. Preparation and Characterization of Phloretin-Functionalized PHA Films

The films were produced by casting 200 mg of PHA polymer blend (90:10 PHB/mcl-PHA) in the absence and in the presence of different concentrations of phloretin (5–7.5–10–20 mg/film). They appeared macroscopically very similar, whitish, and very homogeneous. Their functional and biological characterization is described in the following paragraphs.

2.2.1. Mechanical Properties

The mechanical properties reported in Figure 1 indicate that the tensile strength (TS) of PHA films significantly increased when the films contain 5 and 7.5 mg of phloretin compared to the neat PHA film. However, TS progressively decreased in the presence of a higher amount of this phenolic molecule (10 and 20 mg). The film's Young's modulus (YM) showed a similar behavior, increasing after phloretin grafting in the film with the lower concentration and decreasing at higher amounts of phloretin. However, it should be mentioned that the elongation at break (EAB) of all the films is lower than 2% and the thickness was not affected by different concentrations of phloretin remaining similar for all the films. Similar mechanical behaviors were reported by Figueroa–Lopez et al. [16] and Rubini et al. [17], who demonstrated that the presence of low contents of eugenol and quercetin incorporated in poly(3-hydroxybutyrate-co-3-hydroxyvalerate) and gelatin films were able to improve all the mechanical properties following by worsening this property at higher concentrations. This behavior could be mostly related to the not homogeneous distribution of the additives in the film matrix that should be overcome by means of dispersant agents (e.g., surfactants and plasticizers produced from epoxidized oils) [18].

2.2.2. Water Sensitivity and Opacity

The results reported in Table 2 indicate that moisture content and swelling ratio values of the films decreased as a function of phloretin at a lower concentration, which was significantly lower than the control film and slightly lower than the films containing the higher amount of phloretin. The reduction in moisture content and swelling ratio at lower concentrations of phloretin most likely is due to the hydrophobic properties of this phenol that limited the water retention in the film matrix [19]. However, all the films have low moisture content (less than 6%) and swelling ratio (less than 4%). Although there is not a clear trend as a function of phloretin amount, the addition of molecule to PHA-based films determines an increase in film hydrophobicity as depicted by the increase in contact angle values. This effect can be attributed to the hydrophobic nature of phloretin which positively contributes to the change wettability of the surface. Another important film property is film opacity, as it is one of the key factors that affect the food quality of a packed food and consumers' willingness to choose a certain type of product [20]. From Table 2 it is clear that the films become opaquer as a function of enhancement of the phloretin content.

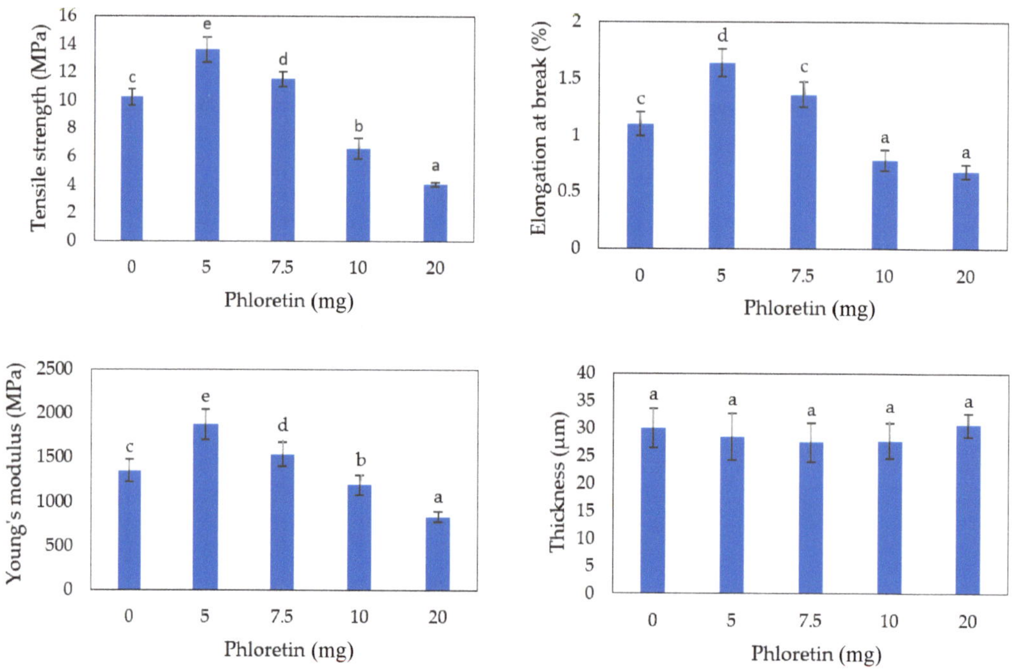

Figure 1. Mechanical properties of PHA-based films as a function of phloretin amount/film matrix (mg/film). Different small letters (a–e) indicate significant differences among the values reported in each bar ($p < 0.05$).

Table 2. Moisture content, swelling ratio, contact angle, and opacity of PHA-based films either incorporated or not with phloretin. Different small letters (a–d) indicate significant differences among the values reported in each bar ($p < 0.05$).

Phloretin (mg)	Moisture Content (%)	Swelling Ratio (%)	Contact Angle (θ)	Opacity (mm^{-1})
0	5.51 ± 0.24 [d]	3.88 ± 0.09 [d]	95.2 ± 1.6 [d]	31.06 ± 2.31 [a]
5	2.77 ± 0.12 [a]	1.90 ± 0.12 [a]	106.3 ± 0.7 [b]	33.88 ± 3.12 [ab]
7.5	3.42 ± 0.27 [b]	2.15 ± 0.10 [b]	110.9 ± 0.1 [a]	36.22 ± 2.45 [bc]
10	4.42 ± 0.19 [c]	3.56 ± 0.12 [c]	97.4 ± 1.2 [d]	39.09 ± 2.18 [cd]
20	4.68 ± 0.22 [c]	3.48 ± 0.07 [c]	100.7 ± 0.6 [c]	43.91 ± 2.56 [d]

2.2.3. FTIR Characterization

The ATR-FTIR spectra obtained for phloretin, PHA-based film, and PHA/phloretin film (the film loaded with 7.5 mg of phloretin was chosen as an example) are reported in Figure 2. In particular, the spectra of PHA and PHA/phloretin films show peaks between 2973–2962 cm^{-1} that can be assigned to the stretching vibration due to asymmetric CH_2 of the lateral monomeric chains and to the symmetrical methyl group of the polymer matrix. The sharp intense band at 1716 cm^{-1} is typical of ester carbonyl groups (C=O) of polymers. However, a broad but less intense shoulder peak at about 1744 is detectable. This can be attributed to the greater vibration energy of the amorphous ester carbonyl group domain.

Moreover, in the region 1500–1650 cm^{-1} of the PHA/phloretin sample, the characteristic stretching vibration of the ester carbonyl group of phloretin is clearly detectable. However, the spectrum revealed no real interaction among components but rather the dispersion of active compounds into the polymer matrix (Figure 2).

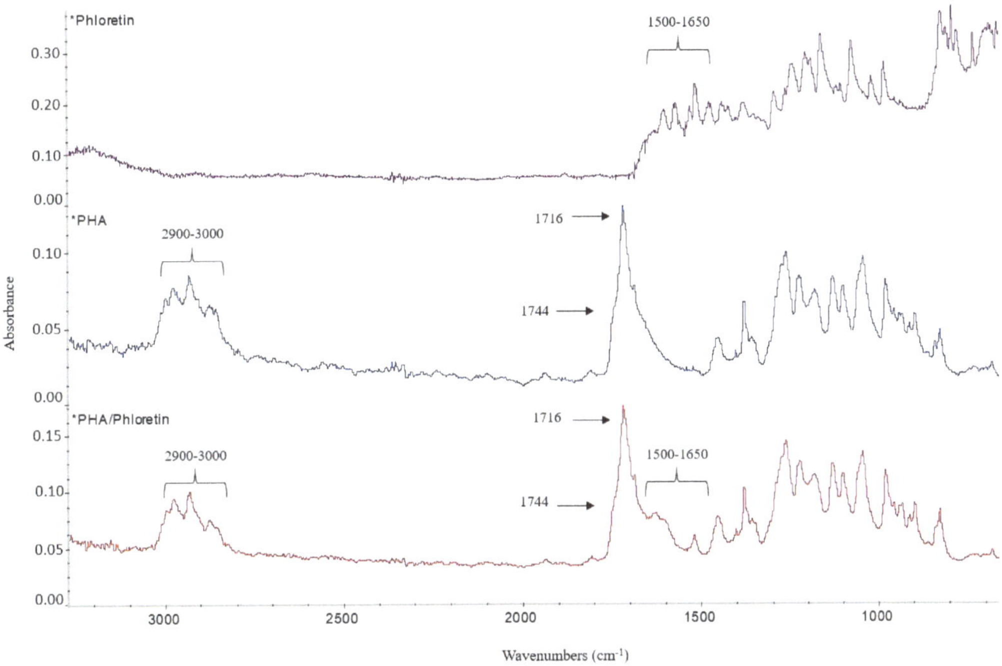

Figure 2. FTIR characterization of PHA-based films prepared in the absence and in the presence of 20 mg of phloretin. The phloretin spectrum is reported in the upper panel.

2.2.4. Raman Spectroscopy of PHA and PHA/Phloretin Film

Figure 3a compares the Raman response of PHA and PHA/phloretin films, which put in evidence a slight increase in fluorescence background for both excitation wavelengths, even if the fluorescence intensities are much higher at 638 nm (Figure 3b). Moreover, it is worth noting that PHA film shows a significant fluorescence emission even without phloretin when the excitation wavelength is set at 638 nm (black line in Figure 3b). However, the shape of the fluorescence emission is different when phloretin is also inside the film, pointing out its contribution to the total emission (red line in Figure 3b) even when the excitation line is matching only the tail of the absorption band [21].

In order to highlight the vibrational modes of the PHA/phloretin with respect to pure PHA, the fluorescence background was subtracted from the spectra and they were normalized to the vibrational mode of PHA at 1723 cm^{-1} (Figure 4a). Figure 4b is showing the vibrational Raman modes of phloretin obtained by subtracting the Raman contributions of PHA (black line in Figure 3a) from the Raman spectrum of PHA/phloretin (red line in Figure 4b), which is compatible with what reported from other studies [21,22]. In the high frequency region, it is clearly visible the aromatic ring (C–H) stretching vibration of phloretin at around 2930 and 3060 cm^{-1} or C–O stretching bands at around 1622 cm^{-1} [23]. On the other side, the bands at 737 and 1200 cm^{-1} are ascribed to the O-C-C and C-C-C bending modes, respectively [23]. Finally, in the range between 250 and 650 cm^{-1} we can detect other Raman peaks due to the torsional modes of phloretin [23].

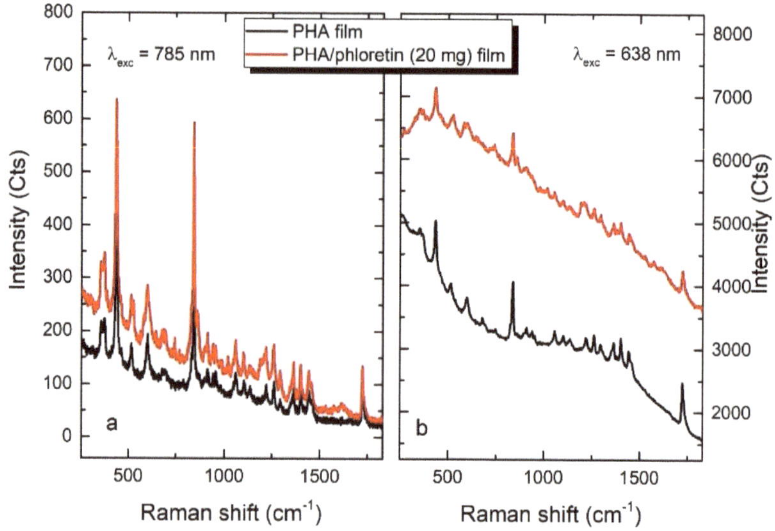

Figure 3. Raman spectra of PHA film (black lines) and PHA/phloretin (20 mg) film (red lines) taken at 785 (**a**) and 638 nm (**b**).

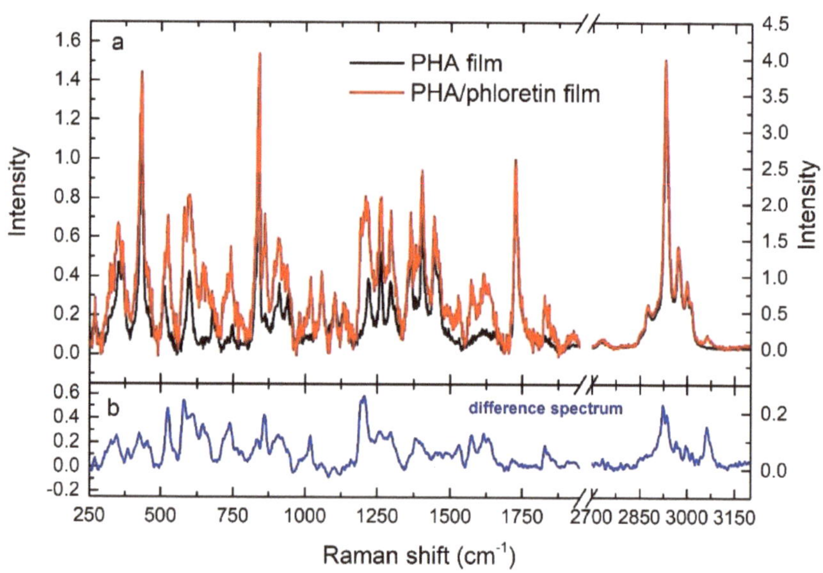

Figure 4. (**a**) Raman spectra at 638 nm of PHA film (black lines) and PHA/phloretin (20 mg) film (red lines) after background subtraction and normalization. (**b**) Phloretin spectrum obtained by subtracting the Raman contributions of PHA (black line panel (**a**)) from the Raman spectrum of PHA/phloretin (red line in panel (**a**)).

2.3. Release Assay

The analysis of the PHA/phloretin films upon incubation for different times with methanol, revealed the presence of the bioactive molecule released from the bioplastic. In fact, reverse-phase high-performance liquid chromatography with diode-array detection (RP-HPLC-DAD) analysis of the phloretin released from the material, detected at 278 nm, revealed the presence of a well-defined peak corresponding to the dihydrochalcone (Figure 5). The identification of the compound was also confirmed by the analysis of the same separation performed with the authentical standard and by spiking the starting material with the same compound. The results show that the release is independent of the concentration of phloretin present in the functionalized bioplastic and followed the same pattern. After approximately 24 h, the 80 $\pm$ 2% of phloretin was released, and in the following 24 h, a further 4–8% of the molecule was found in methanol. In the next 48, 72, and 96 h, the release did not change. The mechanical fragmentation led to a further release of phloretin of about 5–7%, reaching a total amount of phloretin incorporated in the film ranging between 92–95% of the starting concentration utilized for the preparation of the functionalized bioplastic. Using phosphate saline buffer (PBS), as release medium, no release was found.

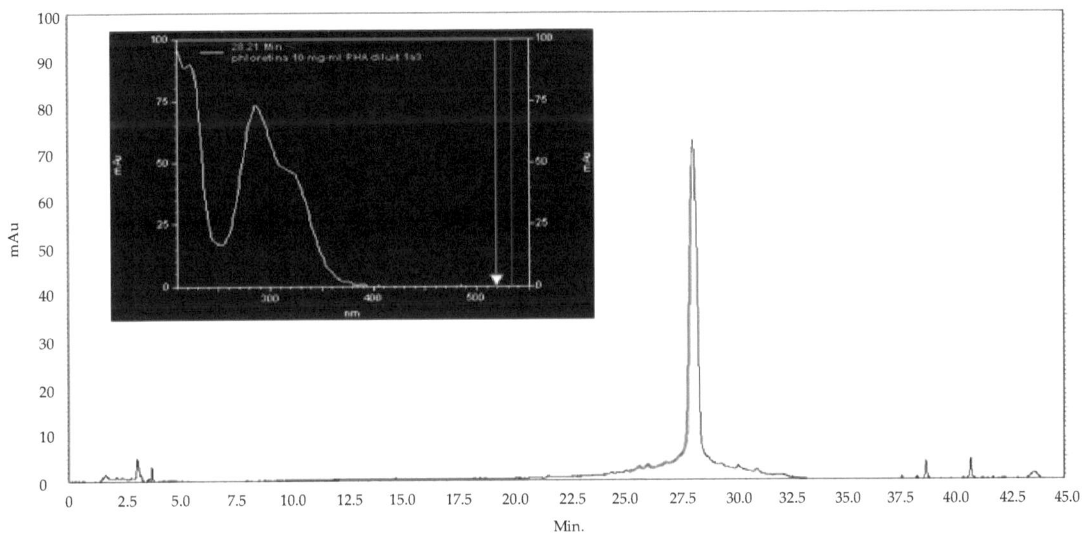

Figure 5. Representative RP-HPLC-DAD chromatogram obtained from the assay of the release of phloretin from PHA-films. In the inset, the UV spectrum of phloretin is reported.

Using PBS as a release medium, no release was reported.

2.4. Antioxidant Assay

In vitro biological assays (DPPH, TEAC, FRAP, ferrozine assay) were performed to evaluate the antioxidant activity of PHA films functionalized with different amounts of phloretin. Reactive oxygen species are dangerous molecules that can drastically change the quality of the food and its half-life, beyond being dangerous for living organisms that consume these matrices. Almost all tests showed that the free radical scavenging activity of PHA films was due exclusively to the presence of phloretin. The bioactive role of phloretin was first highlighted by using the DPPH assay (Figure 6A). It was observed that the free radical scavenging activity of PHA film was due exclusively to the presence of phloretin, considering that the film not functionalized did not show any activity. The reduction in the radical levels was found to be in a dose-dependent manner, reaching the maximum activity in the samples prepared with 10 mg of phloretin. Using the TEAC assay, it was observed that the films not functionalized did not show any activity (Figure 6B); instead, the films functionalized with 5 mg promoted about 80% of residual absorbance reduction and it increased to 90% when the amount of phloretin used for the functionalization was higher (7.5–10 and 20 mg). The same assumption was obtained with the FRAP assay, given that the ability of the films to reduce the ferric ion (Fe^{3+})–ligand complex to the intensely blue-colored ferrous (Fe^{2+}) complex was observed only with the PHA film functionalized with 7.5, 10, and 20 mg of phloretin and a certain dose-dependent activity was observed increasing the amount of phloretin. While no activity is reported for the PHA film without functionalization or with an amount of flavonoid of 5 mg (Figure 6C). The ferrozine assay showed the chelating capacity of all the PHA films, and this activity increased in a dose-dependent manner with the presence of the flavonoid (Figure 6D). The different antioxidant tests let us shed some light on the antioxidant potentials of the functionalized films and, in particular, for the presence of phloretin. This latter is a natural antioxidant derived from plant sources and it plays an important role by directly scavenging free radicals or increasing antioxidant defenses and decreasing, in general, the process of oxidation due to the presence of reactive oxygen species (ROS). The film functionalized with phloretin is able to perform antioxidant activity in both types of categories based on the chemical reaction for the scavenging of ROS: electron transfer (ET) reaction-based assays and hydrogen atom transfer (HAT) reaction-based assays. Thus, the functionalized films are able to scavenge radicals that require both the electron or hydrogen atom donating capacity due to the versatility of phloretin and of its pharmacophore (2,6-dihydroxyacetophenone). This is of particular interest because different kinds of ROS can be produced during food process transformation, conservation, and commercialization, decreasing its quality and shelf-life.

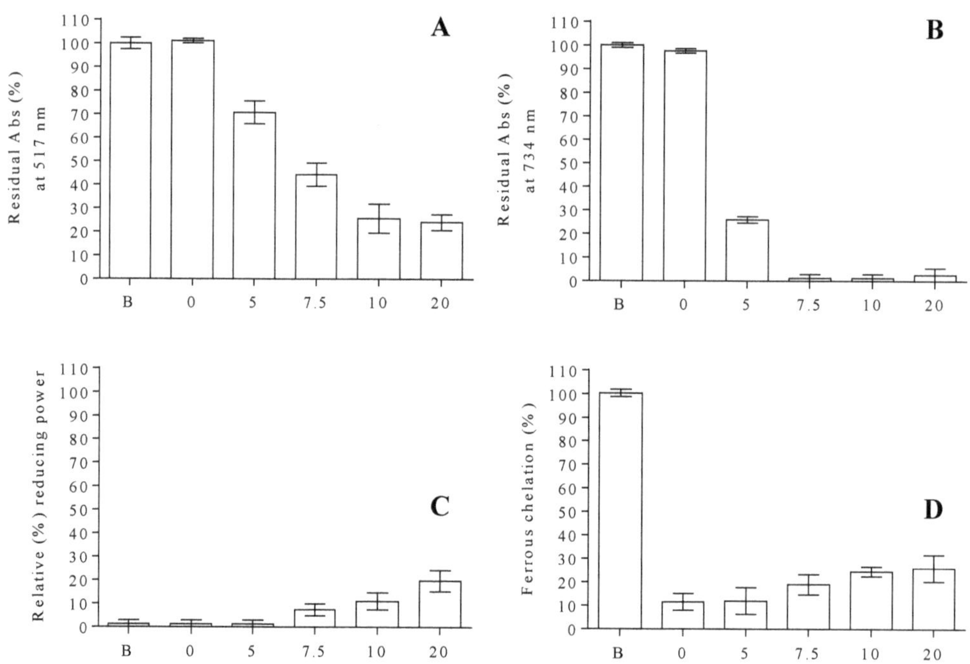

Figure 6. Evaluation of antioxidant activity of PHA films in the absence or in the presence of phloretin at different amounts (0, 5, 7.5, 10, and 20 mg) utilizing four antioxidant assays. (**A**) DPPH assay; (**B**) TEAC assay; (**C**) FRAP assay and (**D**) Ferrozine assay. B, below the histogram of each panel represents the sample blank performed in different assays.

2.5. Evaluation of Phloretin-PHA Antimicrobial Activity

The diffusion assay showed no inhibition zone for PHA films at all concentrations used. However, no growth was observed under all the films regardless of the presence of phloretin. The lack of inhibition halo was in accordance with the release test, which did not report phloretin release in the water solvent and explains the reason for the lack of antimicrobial properties around the film. The PHA films have been tested by incubation in 1 mL of the microbial liquid culture at the concentration of 5×10^5 CFU mL^{-1}. In line with our previous results on phloretin [7], the best activity was observed against *L. monocytogenes*, with a reduction of about 3-log compared with the control. Moreover, a dose-dependent effect was observed (Figure 7). It has been found that the antibacterial activity of phloretin may be attributable mainly to the ability to increase cell membrane permeability and aggregate nuclear acid materials [24]. No activity was reported for *E. coli*, but antagonism activity was observed using the films functionalized with 10 and 20 mg of phloretin, showing an increase in the microorganism growth. A negligible antimicrobial activity was, instead, detected for *S. enterica* serovar Typhimurium ATCC 13311. These results should be analyzed, taking into account that the phloretin was not released in the culture medium and the antimicrobial activity was limited to the microorganisms near the films. Moreover, the lack of activity against *E. coli* correlates with the MIC determination (Table 1), where no effect was detected at the maximum tested concentration. On the other hand, the high MIC and MBC values obtained with *S. enterica* serovar Typhimurium ATCC 13311 could explain the lack of activity of the functionalized films against these strains, given the phloretin concentration in the films did not reach the MIC values. The severity of *L. monocytogenes* disease indicates that safe handling of food is of paramount significance to ensure public health. Analyzing these data, it is important to mention that listeriosis was the fifth most reported zoonosis in the EU in 2021, with an increase in notification rate of 14% compared with 2020 in the EU as reported in "The European Union One Health 2021 Zoonoses Report" edited by EFSA and ECDC (available at https://efsa.onlinelibrary.wiley.com/doi/epdf/10.2903/j.efsa.2022.7666, accessed on 14 July 2023). *Listeria monocytogenes* contamination can be found in different foods (such as raw vegetables, cheeses, especially soft cheeses, meats, and smoked fish). The functionalized films can decrease the growth of *L. monocytogenes*, especially at the maximum tested concentrations, creating an interesting new active food packaging bioplastic. Taking into account that cooking at high temperatures kills the bacteria, but contamination of foods can happen also after production before it is packaged, the creation of active food packaging materials able to reduce the growth of *L. monocytogenes* can be an important step to reduce its proliferation. Moreover, even with the forming biofilm ability, new strategies to control the growth of *Listeria monocytogenes* are warranted. Regarding the biofilm formation, the phloretin-functionalized PHA films showed an inhibitory effect against *L. monocytogenes*. In particular, a biomass reduction compared to neat films of about 13%, 20%, and 33% was found with PHA films containing 7.5, 10, and 20 mg of phloretin, respectively. A negligible activity was instead detected against *E. coli* and *S. enterica* serovar Typhimurium ATCC 13311.

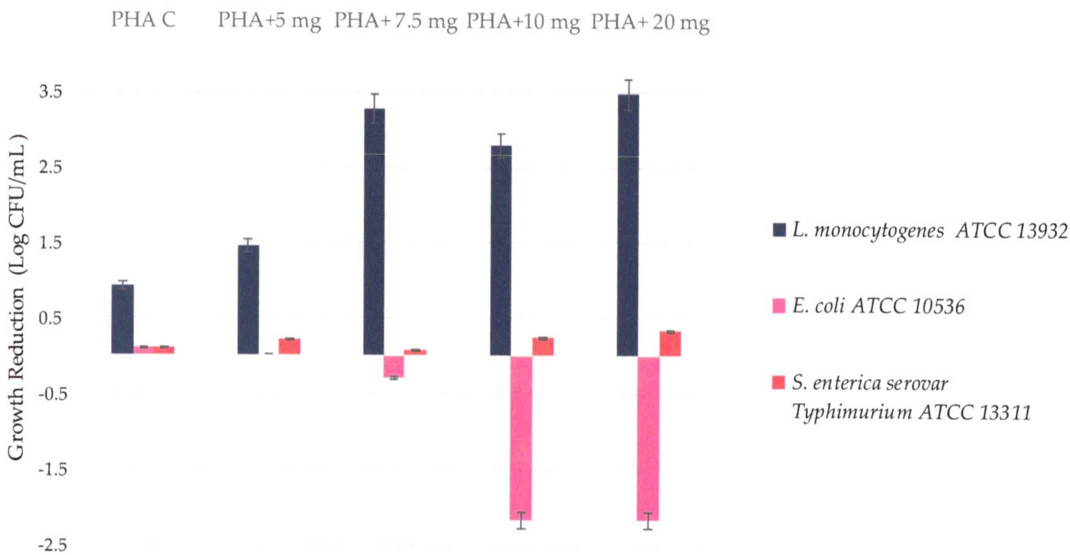

Figure 7. Growth reduction (i.e., decrease in log CFU/mL) of tested microorganisms. Initial inoculum of 5×10^5 CFU/mL; error bars represent standard deviation.

2.6. Food Fresh-Keeping Test

In order to study their fresh-keeping performance, the prepared films were used to pack fresh apple samples, as shown in Figure 8. These representative images of the apple samples were digitalized with a photo-camera and elaborated with the program Image J in function of their quality change due to browning reaction after different days of incubation (Figure 8a). It can be seen that apple samples exposed to air showed a browning phenomenon evident after 24 h, but marked after 48 and 72 h. The apple samples packed with a PHA bag functionalized with 20 mg of phloretin showed only limited color change after 24 and 48 h, but an obvious slight browning phenomenon can be observed after 72 h. In the presence of neat PHA, the samples packed show browning effects similar or superior to the samples not packed. To have a more accurate measure of the browning reaction we have monitored the changes in absorbance at 420 nm. As can be seen in the graph (Figure 8b) the changes in absorbance are almost completely superimposable with the data obtained by computer elaboration of the images. The changes in quality of the apple samples were also analyzed as a function of weight loss, change in antioxidant activity, and sugar content as a function of brix degrees. Figure 8c shows the weight loss of the apple samples after 24, 48, and 72 h. As can be seen, the PHA pack functionalized with 20 mg of phloretin showed a lower weight loss than those without a pack or with only a PHA bag for up to 48 h. After 72 h, the weight loss was the same for all the tested samples. A similar trend can be seen also in the monitoring of brix degrees changes with the samples packed in the functionalized PHA film showing the best performance (Figure 8d). The influences of the presence of phloretin in the functionalization of the PHA film are evident in the monitoring of antioxidant activity with the DPPH assay. As can be seen in the graph (Figure 8e), the apple samples packed in the PHA functionalized with phloretin maintained almost the same antioxidant activity also after 72 h of incubation, while in the other two samples, there is a decrease in this potentiality that is most marked in the samples not packed, that after 72 h lost about the half of its activity. The apple samples packed only with PHA film showed a decrease in this activity but are inferior to the ones not bagged and clearly superior to the ones fortified with phloretin. The analysis of the obtained results in the different tests shows that PHA-phloretin functionalized films may have food-preservation performance superior to the only film in agreement with the data characterization data obtained in the above section. In particular, the data obtained in the antioxidant determinations are in line with the one obtained in this section and are correlated with the ability of phloretin to scavenge radicals based on ET and HAT mechanism, preserving the endogenous antioxidant of the food and decreasing the browning reactions.

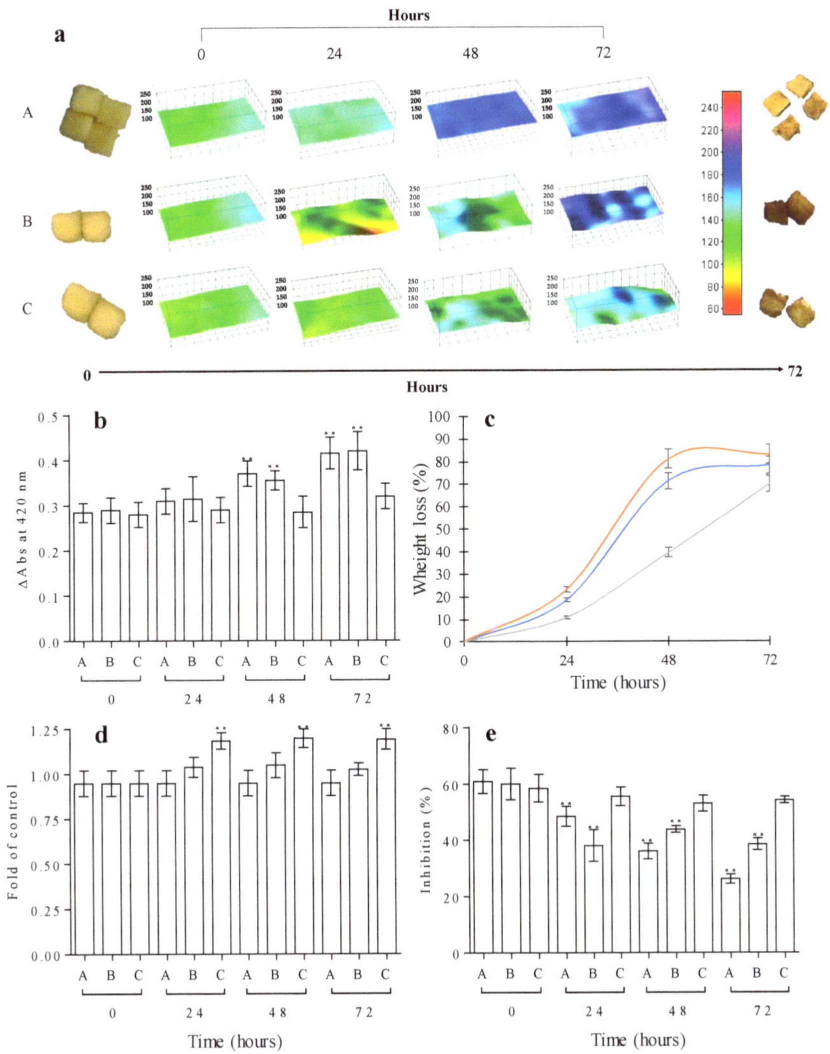

Figure 8. Analysis of the changes in apple samples packed or not packed after different intervals of time (0, 24, 48, and 72 h). (**a**) Representative images of the samples after incubation at different time intervals and their computer and graphic elaboration. (**b**) Changes in the browning of the samples monitored at 420 nm. (**c**) Weight loss after different times of incubation (blue line apple samples not packed; red line apple sample packed with PHA film; grey line apple samples bagged with PHA plus phloretin). (**d**) Changes in the brix grade of the samples. (**e**) Changes in the antioxidant potentials monitored by DPPH assay. The letters in the graph indicate: A, apple samples not packed; B, apple sample packed with PHA film; C, apple samples packed with PHA plus phloretin (20 mg). The ** indicates significant changes with respect to the control at $p > 0.05$.

3. Materials and Methods

3.1. Reagents and Standard Solutions

HPLC-grade acetonitrile and methanol were supplied by Sigma-Aldrich (St. Louis, MO, USA), dimethylformamide (DMF) by Carlo Erba (Milano, Italy). Phloretin (≥99%) was supplied by Sigma-Aldrich (St. Louis, MO, USA) and was used as standard. Mueller–Hinton broth and agar were supplied by Oxoid (Sigma, Milano, Italy).

3.2. Production and Characterization of Polyhydroxyalkanoates Based-Films

3.2.1. Polymer Production

For the production of poly(3-hydroxybutyrate) (PHB), microbial fermentation of *Cupriavidus necator* DSM 428 was carried out in a 5L BioFlo/CelliGen®115 (Eppendorf New Brunswick) following a two-step growth protocol as described in Mirpoor et al. [6]. Briefly, to induce polymer accumulation, the first step of growth in rich medium (TSB, Tryptic Soy Broth) for 24 h at 30 °C was followed by an additional 24 h in minimal medium MMCn (Budde, 2011) containing 20 g/L fructose. Fermentation parameters were set as follows: inoculum at 0.1 OD600, agitation rate 220 rpm, and the DO concentration maintained at 30% of air saturation. After 72 h the cells were harvested by centrifugation (6000 rpm, 20 min), lyophilized, and the polymer extracted according to Turco et al. [25]. Medium chain length PHA (mcl-PHA) was produced by microbial fermentation of *Pseudomonas resinovorans* NRL B-2649 in minimal medium E supplemented with 0.6% *v*/*v* oleic acid as C-source [6]. Fermentation parameters were set as follows: inoculum at 0.1 OD600, agitation rate 220 rpm, and the DO concentration maintained at 30% of air saturation. The cells were harvested after 48 h, and the polymer was extracted according to Turco et al. [25].

3.2.2. Film Preparation

PHA-based films were obtained by the solvent-casting method. PHA solutions were prepared by dissolving the polymers in chloroform at a concentration of 20 mg/mL. The binary blend was formulated at a 90:10 PHB/mcl:PHA ratio, for a total amount of 200 mg of polymers in a volume of 20 mL. Phloretin was dissolved in acetone and added to the polymer solutions to rich a final amount of 5, 7.5, 10, and 20 mg/film. The solutions were cast into glass Petri dishes (diameter of 10 cm), which were kept at room temperature until complete solvent evaporation occurred. Slow solvent evaporation was performed in a saturated chloroform atmosphere to avoid the formation of cracks and non-selective voids in the films and to guarantee their homogeneity.

3.2.3. Characterization of PHA-Based Films

- Mechanical properties

Mechanical properties of the films were determined by measuring tensile strength (TS), Young's modulus (YM), and elongation at break (EAB) according to ASTM standard method D882-12 (Testing & Materials–ASTM, 2012) using an Instron universal testing instrument (Model 5543 A, Instron Engineering Corp., Norwood, MA, USA). The film strips were cut into the dimension of 1 × 8 cm^2 and placed between the grips of the machine. The initial grip separation and cross-head speed were set to 40 mm and 5 mm/min, respectively.

- Moisture content, swelling ratio, and contact angle

Moisture content, swelling ratio, and contact angle of the films were measured according to the method described by Mirpoor et al. [26] with slight modification. The moisture content of the films (3 × 3 cm^2) was calculated by weighing the films before and after oven drying at 105 °C and dividing the weight loss due to oven drying by the film's initial weight. In order to measure the film swelling ratio, the initial weight (*Wi*) of the film samples (3 × 3 cm^2) was recorded and then immersed in 30 mL distilled water at 25 °C for

1 h. After that, films were removed and weighed after drying the surface water of the films with absorbent paper (*Ws*). The swelling ratio was calculated as follows:

$$Swelling\ ratio\ (\%) = (Ws - Wi) \times 100/Wi \quad (1)$$

The contact angle was measured by using a homemade contact angle goniometer. The film strip was placed on the horizontal stage and then 10 µL of distilled water was deposited on both sides of each film at different points. The image of the water drop was captured at the moment that the drop was in contact with the film surface. The mean value of the contact angle was measured with ImageJ software. Five measurements were reported as the average contact angle value.

- Opacity

To measure the film opacity, the absorbance of the four strips of each film (1 × 4 cm) was measured at a wavelength of 600 nm by using a UV–Vis spectrophotometer (SmartSpec 3000 Bio-Rad, Segrate, Milan, Italy) according to the method of Jahed, Khaledabad, Bari, and Almasi [27]. Then, the film opacity was calculated by dividing the obtained absorbance value by the film thickness (mm).

- Fourier transform infrared spectroscopy (FTIR-ATR)

FTIR analysis was performed in ATR (attenuated total reflection) modality. All the samples were analyzed at room temperature with an FTIR Nicolet 5700 spectrometer (Thermo Fisher Scientific, Waltham, MA, USA). Spectra were recorded as an average in a range of 4000–600 cm^{-1}, with a spectral resolution of 2 cm^{-1}. Spectra analysis was done with OMNIC 9 software (Thermo Fisher Scientific, Inc., Waltham, MA, USA).

- Raman spectroscopy of PHA and PHA/Phloretin film

Raman spectroscopy was carried out with an Xplora Plus microspectrometer (Horiba Scientific, Singapore) equipped with 638 and 785 nm diode lasers. The incident light was focused onto the sample surface using a 100× objective (laser power was set at 38 and 10.2 mW for 785 and 638 nm, respectively). The Raman signal was collected in a backscattering configuration through the same objective and dispersed by a 1200 L/mm diffraction grating onto a CCD detector (Sincerity–Horiba Scientific) using an integration time of 10 s.

3.3. *Antimicrobial Assays*

3.3.1. Microbial Strains and Culture Conditions

The Gram-positive bacteria used were *Enterococcus hirae* ATCC 10541, *Listeria monocytogenes* ATCC 13932, and 14 food isolates of *L. monocytogenes* belonging to serotypes 1/2a (4 strains), 1/2b (4 strains), 1/2c (3 strains) and 4b (3 strains). The Gram-negative bacteria used were *Escherichia coli* ATCC 10536, *Salmonella enterica* serovar Typhimurium ATCC 13311, and two clinical isolates of *S. enterica* serovar Typhimurium and *S. enterica*. All bacteria were grown in Mueller–Hinton broth (MHB, Oxoid, CM0405, Sigma, Italy) except for *L. monocytogenes* strains, which were grown on tTyptic Soy broth (TSB; CM0129, Oxoid, Sigma, Italy) at 37 °C (18–20 h).

3.3.2. Susceptibility Studies of Phloretin

The minimum inhibitory concentration (MIC) and the minimum bactericidal concentration (MBC) of phloretin were determined by the broth microdilution method according to CLSI for bacteria [28]. The MIC value was considered as the lowest concentration of phloretin giving a complete inhibition of visible bacterial growth after incubation for 24 h. The MBC value was defined as the lowest concentration of phloretin able to kill 99.9% of the inoculum after 24 h. The antimicrobial activity of PHA films was evaluated using both an adapted diffusion test and a broth microdilution assay for bacteria that grow aerobically, as recommended by the CLSI [28,29].

3.3.3. Disc Diffusion Assay of PHA Films

The PHA films were sterilized under a UV lamp before each assay. The diffusion test was performed as recommended by the CLSI 2018 [29]. Briefly, each film (1 cm^2), functionalized with 5, 7.5, 10, and 20 mg of phloretin, was deposed on plates of Mueller–Hinton Agar (MHA, Oxoid, CM0405, Sigma, Italy) or tryptic soy agar (TSA; CM0131, Oxoid, Sigma, Italy) previously swabbed with the standardized inoculum (5×10^8 CFU/mL). The susceptibility of bacteria to the functionalized films was measured as the clear area that appeared around and/or under the piece of film. Neat PHA films were included as controls.

3.3.4. Antibacterial Activity of PHA Films and Biofilm Biomass Measurement

Each PHA film (1 cm^2) functionalized with 5, 7.5, 10, and 20 mg of phloretin was incubated in a medium (1 mL) containing a standardized bacterial load at the concentration of 5×10^5 CFU/mL. The bacterial growth was evaluated after 24 h at 37 °C by plating serial dilutions. Plates were incubated at 37 °C for 24–48 h, and the number of colonies was reported as CFU mL^{-1}. Neat PHA films were included as controls. In addition, for the investigation of anti-biofilm properties, after 24 h-incubation the biofilm formed by the tested bacteria on the PHA films was evaluated by biomass measurement. The films were washed with PBS, dried, and stained with 0.1% safranin, as previously reported [30]. The stained biofilms were suspended in 30% acetic acid aqueous solution and the mean optical density at 492 nm (OD_{492}) was measured. The reduction percentage of biofilm was calculated using the following equation:

$$Biofilm\ Reduction\ (\%) = 100 - (OD_{492}\ PHA\ with\ phloretin)/(OD_{492}\ neat\ PHA) \times 100 \quad (2)$$

3.4. Identification of Phloretin Release

Prior to analysis, 1 mL of solvent (methanol or PBS) was added to each PHA film (1 cm^2) and incubated at room temperature. The solvent was then filtered through an Iso-Disc P-34, 3 mm diameter PTFE membrane, 0.45 µm pore size (Supelco, Bellefonte, PA, USA). The evaluation of the active molecule released from PHA films was performed by reverse-phase HPLC-DAD detection, using a Shimadzu HPLC system equipped with a UV–Vis photodiode array detector (DAD; Shimadzu, Kyoto, Japan) and a fluorescence detector (Hewlett Packard) according to Barreca et al. [31] at different intervals of time. Finally, the samples were mechanically homogenated with a mortal in the solvent utilized for release to obtain the maximum amount of phloretin present in the films. The column was a Discovery C18 (250 × 4.6 mm i.d., 5 µm) supplied by Supelco (Bellefonte, PA, USA), equipped with a 20 × 4.0 mm guard column, and held in a column oven set at 30 °C. The injection loop was 20 µL, and the flow rate was 1.0 mL/min. The mobile phase consisted of a linear gradient of acetonitrile/H_2O as follows: 5–20% (0–15 min), 20–30% (15–20 min), 30–100% (20–35 min), 100% (35–40 min), 100–5% (40–45 min), and 5% (45–55 min). UV spectra were recorded between λ 200 and 450 nm, and simultaneous detection by diode array was performed at λ 278 nm. Nitrogen was used as a sheath gas with a flow of 50 arbitrary units. Peak identity was confirmed by comparing the retention time and absorption spectra with the one of pure (≥99%) commercially available standard (concentration range 1–50 ug/mL). Each sample was tested three times and gave superimposable chromatograms.

3.5. Antioxidant Assays

2,2-Diphenyl-1-picrylhdrazyl (DPPH) and 2,2′-azino-bis (3-ethylbenzothiazoline-6-sulfonic acid) (ABTS) radical scavenging assays, ferric reducing power (FRAP), and the ferrozine assay were used to investigate the in vitro antioxidant efficacy of PHA films.

3.5.1. DPPH Assay

The free radical method was performed according to Barreca et al. [31], in short, by using the stable free radical 2,2-diphenyl-1-picrylhydrazyl (DPPH•). Briefly, each

1 cm^2 PHA film was mixed with 80 μM DPPH• in methanol in a final volume of 1 mL. The variations in absorbance at 517 nm were monitored over 30 min with a Varian Cary 50 UV–Vis spectrophotometer. The concentration of DPPH radical in the cuvette (1.0 cm path length) was chosen to give absorbance values less than 1.0. The inhibition (%) of radical scavenging activity was calculated by the following equation:

$$I\ (\%) = [(Ac - As)/Ac] \times 100 \quad (3)$$

where *Ac* is the absorbance of the control and *As* is the absorbance of the sample. All tests were run in triplicate and the results were expressed as means ± standard deviation (SD).

3.5.2. ABTS Radical Scavenging Assay

The 2,2′-azino-bis(3-ethylbenzothiazoline-6-sulphonic acid ABTS free radical-scavenging activity was carried out by a decolorization assay according to Smeriglio et al. [32]. The radical cation $ABTS^{\bullet +}$ was added to each 1 cm^2 film, and the absorbance changes at 734 nm were recorded in a spectrophotometer after 6 min.

3.5.3. Ferric Reducing Antioxidant Power (FRAP)

The FRAP assay was performed according to Barreca et al. [31]. The fresh working FRAP reagent was prepared daily by mixing 25 mL of acetate buffer (300 mM, pH 3.6), 2.5 mL of 2,4,6-Tris(2-pyridyl)-S-triazina (TPTZ) solution (10 in 40 mM HCl), and 2.5 mL of $FeCl_3$ (20 mM). The reagent was warmed up to 37 °C; and then 1500 μL were placed in a cuvette (1.0 cm path length) and the initial absorbance was read. Each 1 cm^2 PHA film was added to the cuvette and the absorbance was measured after 4 min at wavelength 593 nm with a Varian Cary 50 UV–Vis spectrophotometer. All tests were run in triplicate and the results were expressed as means ± standard deviation (SD).

3.5.4. Ferrozine Assay

The potential chelating activity of 1 cm^2 PHA films toward ferrous ions was analyzed according to Papalia et al. [33] with little modifications. EDTA (0.1 mM final concentration) was used as a reference compound. PHA films were added to a solution of 0.5 mM $FeSO_4$ (0.01 mL). After the addition of 5.0 mM ferrozine (0.4 mL) solution, the samples were shaken and left for 10 min at room temperature (RT). Finally, the absorbance at 562 nm of the solution was measured with a spectrophotometer. The inhibition (%) of ferrozine Fe^{2+} complex formation was found using the following equation:

$$\%\ \text{Inhibition} = [(Ac - As)/Ac] \times 100 \quad (4)$$

where *Ac* is the absorbance of the control and *As* is the absorbance of the samples where the phloretin should be released.

3.6. Analysis of Food Preservatives Properties

3.6.1. Apple Samples Design

We selected apples (*Malus domestica*, variety Pink Lady) that were consistent in size (medium fruit), color, and maturity and free from pests, diseases, and mechanical damage purchased from the local market. At room temperature, the apples were denied of the peel and cut in to pieces of 1.2 × 1.2 × 0.6 (L × W × H) cm with a professional cutter stainless steel with 1/2- and 1/4-inch blades, obtaining a weight of 1.5 ± 0.12 g for each apple sample. Each operation was performed under a biological fume hood to work under sterile conditions and all the following procedures were performed in the same conditions. Each sample of apple was packaged in a single-layer film, and the film was sealed and closed manually. In this study, two groups of self-made biodegradable plastic films were used to pack the apple samples, and filmless packaging was used as the control. The tests were carried out for several consecutive hours, maintaining the samples at a controlled room temperature of 25 °C, and every 0, 24, 48, and 72 h, three samples from each group were withdrawn and utilized for the following analysis.

3.6.2. Computer and Graphic Elaboration

The changes in the appearance of the apple samples were obtaining digitalizing the images of the samples, after the different intervals of time, by a professional digital camera. The digital images have been analyzed by ImageJ software (available at https://imagej.nih.gov/ij/download.html, accessed on 10 June 2023).

3.6.3. Browning Reaction and Determination of Brix Degree

The browning reaction was determined by monitoring the absorbance changes at 420 nm with UV–Vis spectrophotometry according to Xu et al. [34]. Each sample was diluted with distilled water in the ratio 1:1 (*w:v*), homogenated with a mortal pre-chilled, and centrifuged at 10,000 rpm, 4 °C for 10 min. The supernatant of each sample was utilized to monitor the changes at 420 nm. Distilled water was used as a blank control. The same samples were also utilized to analyze brix degree with a brix refractometer and changes in the antioxidant activity with the DPPH assay. For the DPPH assay, 10 µL of the samples has been withdrawn and tested as described in Section 3.5.1.

3.6.4. Weight Loss

The weight loss rates were calculated by the following equation:

$$Weight\ loss(\%) = \frac{m_o - m}{m_o} \times 100 \quad (5)$$

where m_o and m are the initial weight and weight after different intervals of time (0, 24, 48, 96 h) of apple samples [35]. The experiments were triply repeated to get the average value.

4. Conclusions

In this paper, the manufacture of films made of PHA as renewable sources for the production of novel bioplastics was reported. Such materials for the first time were further functionalized with different amounts of phloretin, a dihydrochalcone, belonging to the class of flavonoids found in different fruits and vegetables endowed with interesting biological properties, that significantly increased the antioxidant potential and antimicrobial properties of the produced films without drastically changing mechanical and hydrophilicity features. As a fact, the phloretin-grafted materials were also able to preserve apple samples' freshness envisaging their potential application as bio-based packaging systems to be applied for the extension of food shelf-life.

Author Contributions: Conceptualization, D.B., G.M., C.P., C.V.L.G. and T.G.; methodology, T.G., G.G., A.N., S.F.M., I.C., G.T.P., A.F. and P.G.G.; software, I.C., T.G. and S.F.M.; validation, T.G., C.V.L.G., D.B., C.P. and G.M.; formal analysis, T.G., G.G., S.F.M., I.C., G.T.P., A.F. and P.G.G.; investigation, T.G., D.B., S.F.M. and I.C.; resources, D.B.; data curation, T.G., S.F.M. and I.C.; writing—original draft preparation, T.G., S.F.M., I.C., A.N. and D.B.; writing—review and editing, D.B., G.M., C.P. and C.V.L.G.; supervision, C.P., D.B., G.M. and C.V.L.G.; funding acquisition, D.B. All authors have read and agreed to the published version of the manuscript.

Funding: Part of the activities were also carried out within MICS (Made in Italy—Circular and Sustainable) Extended Partnership and received funding from the European Union Next-GenerationEU (PIANO NAZIONALE DI RIPRESA E RESILIENZA (PNRR)—MISSIONE 4 COMPONENTE 2, INVESTIMENTO 1.3—D.D. 1551.11-10-2022, PE00000004).

Institutional Review Board Statement: Not applicable.

Informed Consent Statement: Not applicable.

Data Availability Statement: Not applicable.

Conflicts of Interest: The authors declare no conflict of interest.

References

1. Asgher, M.; Qamar, S.A.; Bilal, M.; Iqbal, H.M.N. Bio-based active food packaging materials: Sustainable alternative to conventional petrochemical-based packaging materials. *Food Res. Int.* **2020**, *137*, 109625. [CrossRef]
2. Dilkes-Hoffman, S.; Lane, J.L.; Grant, T.; Pratt, S.; Lant, P.A.; Laycock, B. Environmental impact of biodegradable food packaging when considering food waste. *J. Clean. Prod.* **2018**, *180*, 325–334. [CrossRef]
3. Wang, K.; Zhang, R. Production of Polyhydroxyalkanoates (PHA) by *Haloferax mediterranei* from food waste derived nutrients for biodegradable plastic applications. *J. Microbiol. Biotechnol.* **2021**, *31*, 338–347. [CrossRef] [PubMed]
4. Turco, R.; Santagata, G.; Corrado, I.; Pezzella, C.; Di Serio, M. In vivo and post- synthesis strategies to enhance the properties of PHB-based materials: A review. *Front. Bioeng. Biotechnol.* **2021**, *8*, 619266. [CrossRef]
5. Fernandez-Bunster, G.; Pavez, P. Novel Production Methods of Polyhydroxyalkanoates and Their Innovative Uses in Biomedicine and Industry. *Molecules* **2022**, *27*, 8351. [CrossRef] [PubMed]
6. Vijayamma, R.; Maria, H.J.; Thomas, S.; Shishatskaya, E.I.; Kiselev, E.G.; Nemtsev, I.V.; Sukhanova, A.A.; Volova, T.G. A study of the properties and efficacy of microparticles based on P(3HB) and P(3HB/3HV) loaded with herbicides. *J. Appl. Polym. Sci.* **2021**, *139*, 51756. [CrossRef]
7. Corrado, I.; Varriale, S.; Pezzella, C. Microbial processes for upcycling food wastes into sustainable bioplastics. In *Sustainable Food Science, A Comprehensive Approach*; Elsevier: Amsterdam, The Netherlands, 2023; Volume 4, pp. 51–74.
8. Mirpoor, S.F.; Corrado, I.; Di Girolamo, R.; Dal Poggetto, G.; Panzella, L.; Borselleca, E.; Pezzella, C.; Giosafatto, C.V.L. Manufacture of active multilayer films made of functionalized pectin coated by polyhydroxyalkanoates: A fully renewable approach to active food packaging. *Polymer* **2023**, *281*, 126136. [CrossRef]
9. Barreca, D.; Bellocco, E.; Laganà, G.; Ginestra, G.; Bisignano, C. Biochemical and antimicrobial activity of phloretin and its glycosilated derivatives present in apple and kumquat. *Food Chem.* **2014**, *160*, 292–297. [CrossRef] [PubMed]
10. Barreca, D.; Currò, M.; Bellocco, E.; Ficarra, S.; Laganà, G.; Tellone, E.; Giunta, M.L.; Visalli, G.; Caccamo, D.; Galtieri, A.; et al. Neuroprotective effects of phloretin and its glycosylated derivative on rotenone-induced toxicity in human SH-SY5Y neuronal-like cells. *Biofactors* **2017**, *43*, 549–557. [CrossRef]
11. Behzad, S.; Sureda, A.; Barreca, D.; Nabavi, S.F.; Rastrelli, L.; Nabavi, S.M. Health effects of Phloretin: From chemistry to medicine. *Phytochem. Rev.* **2017**, *16*, 527–533. [CrossRef]
12. Chen, J.; Zhang, H.; Hu, X.; Xu, M.; Su, Y.; Zhang, C.; Yue, Y.; Zhang, X.; Wang, X.; Cui, W.; et al. Phloretin exhibits potential food-drug interactions by inhibiting human UDP-glucuronosyltransferases in vitro. *Toxicol. In Vitro* **2022**, *84*, 105447. [CrossRef] [PubMed]
13. Habtemariam, S. The molecular pharmacology of phloretin: Anti-inflammatory mechanisms of action. *Biomedicines* **2023**, *11*, 143. [CrossRef] [PubMed]
14. Majdoub, Y.O.E.; Ginestra, G.; Mandalari, G.; Dugo, P.; Mondello, L.; Cacciola, F. The digestibility of *Hibiscus sabdariffa* L. polyphenols using an in vitro human digestion model and evaluation of their antimicrobial activity. *Nutrients* **2021**, *13*, 2360. [CrossRef]
15. Mandalari, G.; Bisignano, C.; D'arrigo, M.; Ginestra, G.; Arena, A.; Tomaino, A.; Wickham, M.S.J. Antimicrobial potential of polyphenols extracted from almond skins. *Lett. Appl. Microbiol.* **2010**, *51*, 83–89. [CrossRef] [PubMed]
16. Bisignano, C.; Filocamo, A.; Faulks, R.M.; Mandalari, G. *In vitro* antimicrobial activity of pistachio (*Pistacia vera* L.) polyphenols. *FEMS Microbiol. Lett.* **2013**, *341*, 62–67. [CrossRef]
17. Figueroa-Lopez, K.J.; Cabedo, L.; Lagaron, J.M.; Torres-Giner, S. Development of electrospun poly (3-hydroxybutyrate-co-3-hydroxyvalerate) monolayers containing eugenol and their application in multilayer antimicrobial food packaging. *Front. Nutr.* **2020**, *7*, 140. [CrossRef]
18. Rubini, K.; Boanini, E.; Menichetti, A.; Bonvicini, F.; Gentilomi, G.A.; Montalti, M.; Bigi, A. Quercetin loaded gelatin films with modulated release and tailored anti-oxidant, mechanical and swelling properties. *Food Hydrocoll.* **2020**, *109*, 106089. [CrossRef]
19. Sarıcaoglu, F.T.; Turhan, S. Physicochemical, antioxidant and antimicrobial properties of mechanically deboned chicken meat protein films enriched with various essential oils. *Food Packag. Shelf Life* **2020**, *25*, 100527. [CrossRef]
20. Mirpoor, S.F.; Giosafatto, C.V.L.; Di Girolamo, R.; Famiglietti, M.; Porta, R. Hemp (*Cannabis sativa*) seed oilcake as a promising by-product for developing protein-based films: Effect of transglutaminase-induced crosslinking. *Food Packag. Shelf Life* **2022**, *31*, 100779. [CrossRef]
21. Huang, S.; Xu, J.; Peng, Y.; Guo, M.; Cai, T. Facile tuning of the photoluminescence and dissolution properties of phloretin through cocrystallization. *Cryst. Growth Des.* **2019**, *19*, 6837–6844. [CrossRef]
22. Li, Y.; Xiang, H.; Xue, X.; Chen, Y.; He, Z.; Yu, Z.; Zhang, L.; Miao, X. Dual Antimelanogenic effect of nicotinamide-stabilized phloretin nanocrystals in larval zebrafish. *Pharmaceutics* **2022**, *14*, 1825. [CrossRef]
23. Govindammal, M.; Prasath, M.; Selvapandiyan, M. Spectroscopic (FT-IR, FT-Raman) investigations, quantum chemical calculations, ADMET and molecular docking studies of phloretin with B-RAF inhibitor. *Chem. Pap.* **2021**, *75*, 3771–3785. [CrossRef]
24. Zhao, P.; Zhang, Y.; Deng, H.; Meng, Y. Antibacterial mechanism of apple phloretin on physiological and morphological properties of *Listeria monocytogenes*. *Food Sci. Technol.* **2021**, *42*, 55120. [CrossRef]
25. Turco, R.; Corrado, I.; Zannini, D.; Gargiulo, L.; Di Serio, M.; Pezzella, C.; Santagata, G. Upgrading cardoon biomass into Polyhydroxybutyrate based blends: A holistic approach for the synthesis of biopolymers and additives. *Bioresour. Technol.* **2022**, *363*, 127954. [CrossRef] [PubMed]

26. Mirpoor, S.F.; Varriale, S.; Porta, R.; Naviglio, D.; Spennato, M.; Gardossi, L.; Giosafatto, C.V.L.; Pezzella, C. A biorefinery approach for the conversion of *Cynara cardunculus* biomass to active films. *Food Hydrocoll.* **2022**, *122*, 107099. [CrossRef]
27. Jahed, E.; Khaledabad, M.A.; Bari, M.R.; Almasi, H. Effect of cellulose and lignocellulose nanofibers on the properties of *Origanum vulgare* ssp. gracile essential oil-loaded chitosan films. *React. Funct. Polym.* **2017**, *117*, 70–80. [CrossRef]
28. *CLSI M100-S22;* Performance Standards for Antimicrobial Susceptibility Testing. CLSI: Wayne, PA, USA, 2012.
29. Clinical and Laboratory Standards Institute. *Performance Standards for Antimicrobial Disk Susceptibility Tests. M02 Standard,* 13th ed.; Clinical and Laboratory Standards Institute: Wayne, PA, USA, 2018.
30. Ferreri, L.; Consoli, G.M.L.; Clarizia, G.; Zampino, D.C.; Nostro, A.; Granata, G.; Ginestra, G.; Giuffrida, M.L.; Zimbone, F.; Bernardo, P. A novel material based on an antibacterial choline-calixarene nanoassembly embedded in thin films. *J. Mater. Sci.* **2022**, *57*, 20685–20701. [CrossRef]
31. Barreca, D.; Laganà, G.; Ficarra, S.; Tellone, E.; Leuzzi, U.; Galtieri, A.; Bellocco, E. Evaluation of the antioxidant and cytoprotective properties of the exotic fruit *Annona cherimola* Mill. (Annonaceae). *Food Res. Int.* **2011**, *44*, 2302–2310. [CrossRef]
32. Smeriglio, A.; Mandalari, G.; Bisignano, C.; Filocamo, A.; Barreca, D.; Bellocco, E.; Trombetta, D. Polyphenolic content and biological properties of Avola almond (*Prunus dulcis* Mill. D.A.Webb) skin and its industrial byproducts. *Ind. Crops Prod.* **2016**, *83*, 283–293. [CrossRef]
33. Papalia, T.; Barreca, D.; Panuccio, M.R. Assessment of antioxidant and cytoprotective potential of Jatropha (*Jatropha curcas*) grown in southern Italy. *Int. J. Mol. Sci.* **2017**, *18*, 660. [CrossRef]
34. Xu, Z.; Yang, Z.; Ji, J.; Mou, Y.; Chen, F.; Xiao, Z.; Liao, X.; Hu, X.; Ma, L. Polyphenol mediated non-enzymatic browning and its inhibition in apple juice. *Food Chem.* **2023**, *404*, 134504. [CrossRef] [PubMed]
35. Sun, Y.; Huang, Y.; Wang, X.-Y.; Wu, Z.-Y.; Weng, Y.-X. Kinetic analysis of PGA/PBAT plastic films for strawberry fruit preservation quality and enzyme activity. *J. Food Compos. Anal.* **2022**, *108*, 104439. [CrossRef]

International Journal of *Molecular Sciences*

MDPI

Article

Phoenix dactylifera (Ajwa Dates) Alleviate LPS-Induced Sickness Behaviour in Rats by Attenuating Proinflammatory Cytokines and Oxidative Stress in the Brain

Thippeswamy Boreddy Shivanandappa [1,*], Ghallab Alotaibi [2], Maheswari Chinnadhurai [3], Sudharshan Reddy Dachani [3], Mahmad Dabeer Ahmad [2] and Khalid Abdullah Aldaajanii [1]

1 Department of Biomedical Science, College of Pharmacy, Shaqra University, Al-Dawadmi Campus, Al-Dawadmi 11961, Saudi Arabia; khaledaldaajanii@gmail.com

2 Department of Pharmaceutical Sciences, College of Pharmacy, Shaqra University, Al-Dawadmi Campus, Al-Dawadmi 11961, Saudi Arabia; ghalotaibi@su.edu.sa (G.A.); mahmad@su.edu.sa (M.D.A.)

3 Department of Pharmacy Practice, College of Pharmacy, Shaqra University, Al-Dawadmi Campus, Al-Dawadmi 11961, Saudi Arabia; maki3kp@gmail.com (M.C.); suhasinraydaa@su.edu.sa (S.R.D.)

* Correspondence: drswamy@su.edu.sa or t_swamy@hotmail.com

Abstract: Traditional medicine claims that various components of the *Phoenix dactylifera* (date plant) can be used to treat memory loss, fever, inflammation, loss of consciousness, and nerve disorders. The present study aims to evaluate the effectiveness of *Phoenix dactylifera* fruit extracts (PDF) against rat sickness behaviour caused by lipopolysaccharide (LPS) by assessing behavioural and biochemical parameters. PDF was prepared by extracting dry fruits of *P. dactylifera* with a methanol:water (4:1, v/v) mixture. The PDF was evaluated for phenolic and flavonoid content and HPLC analysis of quercetin estimation. Adult Wistar rats were treated with LPS, PDF + LPS and dexamethasone + LPS. Water and food intake, behavioural tests such as locomotor activity, tail suspension and forced swim tests were conducted. Furthermore, alanine transaminase (ALT) and aspartate transaminase (AST) were estimated in plasma and malondialdehyde (MDA), reduced glutathione (GSH), nitrite, tumor necrosis factor-α (TNF-α) and interleukin-6 (IL-6), were estimated in the brain. PDF ameliorated LPS-induced sickness behaviour by reducing MDA, nitrite, IL-6, and TNF-α levels and improving GSH, behavioural alteration, water and food intake in the treated rats. In the plasma of the treated rats, PDF also decreased the levels of ALT and AST. The outcomes demonstrated the efficacy of PDF in reducing the sickness behaviour caused by LPS in rats. The authors believe that this study will provide the groundwork for future research to better understand the underlying mechanisms of action and therapeutic efficacy.

Keywords: Ajwa date fruits; sickness behaviour; antioxidants; quercetin; cytokines

Citation: Shivanandappa, T.B.; Alotaibi, G.; Chinnadhurai, M.; Dachani, S.R.; Ahmad, M.D.; Aldaajanii, K.A. *Phoenix dactylifera* (Ajwa Dates) Alleviate LPS-Induced Sickness Behaviour in Rats by Attenuating Proinflammatory Cytokines and Oxidative Stress in the Brain. *Int. J. Mol. Sci.* **2023**, *24*, 10413. https://doi.org/10.3390/ijms241310413

Academic Editors: Vassilis Athanasiadis and Theodoros G. Chatzimitakos

Received: 29 May 2023
Revised: 11 June 2023
Accepted: 13 June 2023
Published: 21 June 2023

1. Introduction

Dates (*Phoenix dactylifera* L., family—Arecaceae) are widely grown as a food and commercial crop in countries including India and Pakistan [1]. Ajwa dates are a unique variety of date fruit grown in Al-Madina Al-Munawwarah, Saudi Arabia and they have medicinal properties [1,2]. The carb-rich Ajwa date is a fantastic source of proteins, vitamins, high dietary fibre, minerals, and lipids [3]. They contain typical minerals such as iron, copper, zinc, calcium, potassium, cobalt, fluorine, sulphur, magnesium, manganese, phosphorus, selenium, and boron [2]. A variety of phytochemicals of Ajwa date include glycosides, polyphenols, flavonoids, and sterols. Ajwa dates have hepatoprotective [4], cardioprotective [5], nephroprotective [6], antioxidant [7,8], antihyperlipidemic [7], anti-inflammatory [8], antibacterial [2] and anti-cancer [1] properties.

In the Middle East, dates have been used as a staple meal for thousands of years [9]. Various religions give high importance to date fruits [10]. In Islam, to break the daylong fast during the holy month of Ramadan, dates are used [10]. Date palms are the best assets, according to the Prophet Muhammad, who also advised consuming dates and developing a preference for date palms [11]. Jews celebrate Palm Sunday and consider dates to be one of the seven holy fruits [11]. Many people in the Middle East believe that eating date fruits on an empty stomach can counteract the poisonous effects of any substances to which the individual may have been exposed [6]. In traditional medicine, the date plant is used to treat fever, inflammation, paralysis, nerve disorders, and memory loss [2,6]. All these overlap with the signs and symptoms of sickness behaviour [12]. The impact of dates on the CNS is extensively documented [6], on the other hand, there are no reports of its advantageous benefits on sickness behaviour.

A sophisticated and coordinated adaptive change brought on by tissue damage or acute infections is known as sickness behaviour [13–15]. The sickness behavioural pattern includes malaise, hyperalgesia, fever, lethargy, social retreat, inhibition, decreased locomotor activity, exploration, grooming, loss of libido, anhedonia, sleepiness, anorexia weight loss, disrupted concentration, and anxiety [14,16]. Even while sick behaviour is a normal immune response for faster recovery from an infection or damage, it nonetheless causes discomfort in the victims if it lasts for a longer period [12]. So, it is important to address sickness behaviour to reverse the patients' altered social, cognitive, and mental functioning.

The effect of medications on sickness behaviour is preclinically evaluated using a variety of animal models. Preclinical research uses lipopolysaccharide (LPS)-induced sickness behaviour in rodents the most frequently among others [12,17].

Based on the aforementioned information, the current investigation was carried out to determine the effectiveness of date fruit extracts against LPS-induced illness behaviour in rats by evaluating behavioural and biochemical markers.

2. Results

2.1. Standardisation of PDF

Total phenolic content in PDF was found to be 5.57 ± 0.31 mg/g of gallic acid equivalent weight (mean ± SEM, n = 3) and total flavonoid content was found to be 171.2 ± 1.15 mg/g of quercetin equivalent weight (mean ± SEM, n = 3).

The HPLC fingerprint profile of PDF was standardised with biomarker and it served as a standard for comparison in the subsequent preparation of PDF. HPLC chromatogram of the extract was found to contain constituents eluting between 2.0 min and 7.0 min. The presence of Quercetin in PDF at RT 6.171 min was confirmed by comparing its retention time and UV spectra with that of the standard Quercetin. The amount of Quercetin present in PDF was found to be 56.9 μg/mg. HPLC Chromatogram of PDF and quercetin is shown in Figure 1.

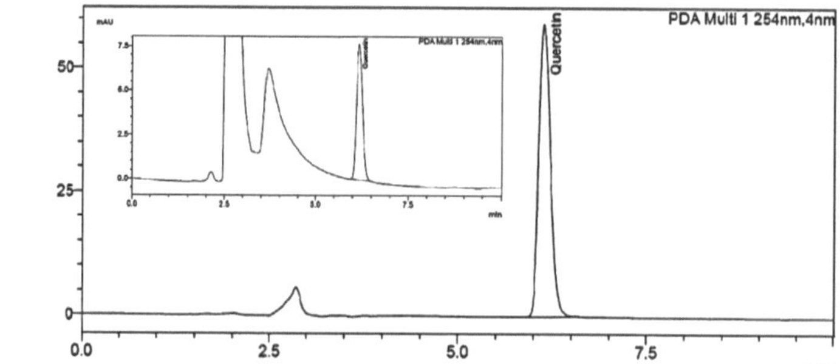

Figure 1. HPLC Chromatogram of PDF showing the presence of Quercetin (Inset), separated on a Waters xbridge ® RP C18 column (5 μm, 4.6 × 250 mm) using isocratic elution was performed using 70% methanol, 30% milli-q water and 0.1% trifluoro acetic acid at a total flow rate of 1.0 mL/min with a run time of 10 min. The chromatogram at 254 nm was analysed.

2.2. Effect of PDF on Water and Food Intake in LPS-Treated Rats

Treatment with LPS leads to reduced water and food intake in the rats by 63.77% and 55.66%, respectively, when compared to the vehicle control group. Administration of PDF 250 and PDF 500 mg/kg to LPS-challenged animals restored the water and food intake by 84.04% and 134.66%, respectively, when compared to the LPS control animals. Animals administered with PDF 500 mg/kg (15.3 ± 0.56 and 4.66 ± 0.33, respectively) consumed almost the same amount of water and food as animals treated with 1 mg/kg of dexamethasone (15.6 ± 0.59 and 4.83 ± 0.40, respectively). The effect of PDF on water and food intake in LPS-treated rats is shown in Table 1.

Table 1. Effect of PDF on water and food intake in LPS treated rats.

Parameters	Vehicle Control	LPS Control	PDF-250 mg/kg + LPS	PDF-500 mg/kg + LPS	Dex-1 mg/kg + LPS
Water intake (mL/6 h)	18.0 ± 0.60	6.52 ± 0.57	12.0 ± 0.46	15.3 ± 0.56	15.6 ± 0.59
Food intake (g/6 h)	6.00 ± 0.57	2.66 ± 0.16	4.50 ± 0.34	4.66 ± 0.33	4.83 ± 0.40

All the values are expressed in mean ± SEM. (n = 6). Where, LPS—lipopolysaccharide, PDF—*Phoenix dactylifera* fruit extracts and Dex—dexamethasone.

2.3. Effect of PDF on Behavioural Outcomes

2.3.1. Locomotor Activity Scores

Compared to the vehicle control group, the locomotor score in LPS control rats was drastically decreased ($p < 0.001$). Treatment with both doses of PDF improved ($p < 0.05$ and $p < 0.01$, respectively) the locomotor score in LPS-treated rats when compared to LPS control rats. The improvement in locomotor score in a higher dose of PDF was comparable with the scores of the standard drug dexamethasone ($p < 0.001$). The effect of PDF on locomotor score in LPS-treated rats is shown in Figure 2.

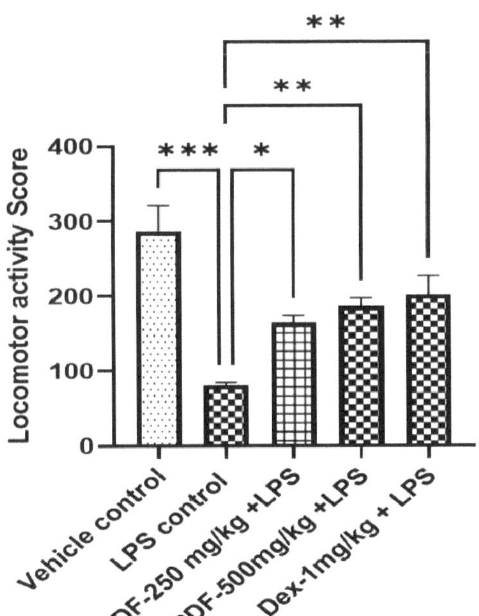

Figure 2. Effect of *Phoenix dactylifera* fruit extracts on locomotor activity score in actophotometer test of lipopolysaccharide treated rats. Numerical values are represented as mean ± SEM (n = 6), * $p < 0.05$, ** $p < 0.01$ and *** $p < 0.001$.

2.3.2. Tail Suspension Test

In the tail suspension test, compared to the vehicle control rats LPS control rats had significantly ($p < 0.001$) higher periods of immobility. Treatment of both the doses of PDF and the standard drug dexamethasone significantly reduced the immobility in LPS-treated rats when compared to the LPS-control rats. The effect of PDF on tail suspension test-induced immobility in LPS-treated rats is shown in Figure 3.

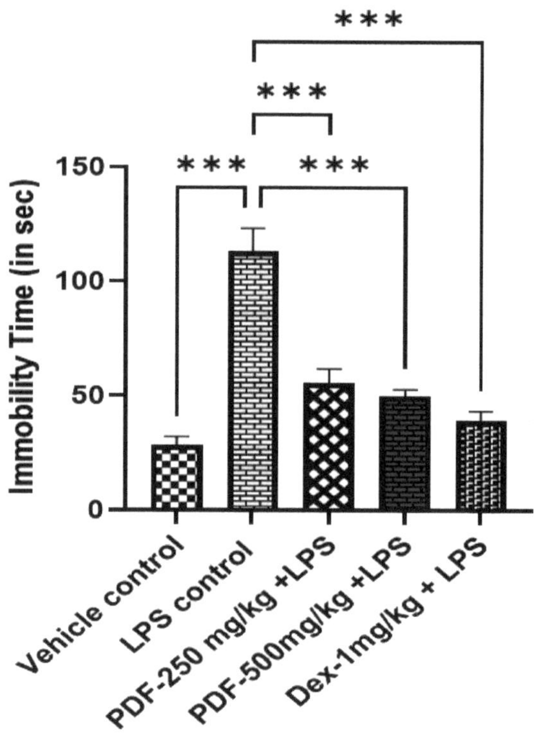

Figure 3. Effect of *Phoenix dactylifera* fruit extracts on tail suspension test induced-immobility in lipopolysaccharide treated rats. Numerical values are represented as mean ± SEM (n = 6), *** $p < 0.001$.

2.3.3. Despair Behaviour Test

The despair behaviour of rats was measured by the forced swim-induced immobility in the rats. Compared to the vehicle control rats the immobility time and the number of immobility states are significantly ($p < 0.001$) greater in LPS-control rats. Treatment with PDF at the dose of 250 mg/kg and 500 mg/kg drastically ($p < 0.01$ and $p < 0.001$, respectively) reduced the immobility time and number of immobility states in LPS-treated rats. The higher dose of PDF administration gives almost equivalent results to that of the standard control drug dexamethasone ($p < 0.001$). The results of the effect of PDF on forced swim test-induced immobility in LPS-treated rats are shown in Figure 4.

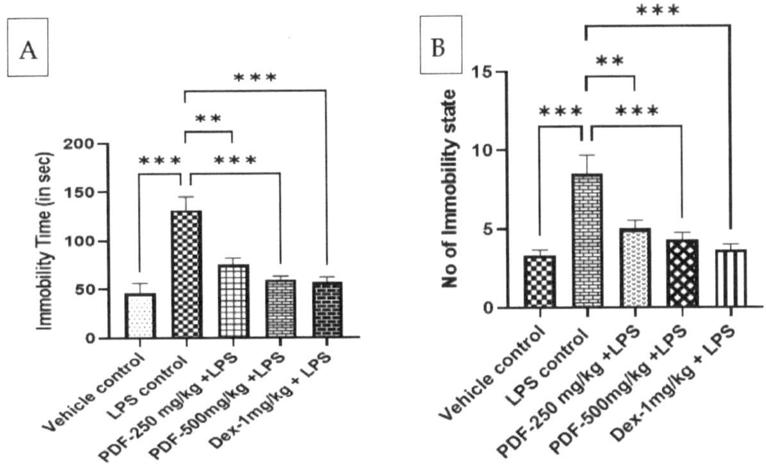

Figure 4. Effect of *Phoenix dactylifera* fruit extracts on forced swim test-induced immobility in lipopolysaccharide treated rats, (**A**) Immobility time and (**B**) Number of immobility state. Numerical values are represented as mean ± SEM (n = 6), ** $p < 0.01$ and *** $p < 0.001$.

2.4. Effect of PDF on ALT and AST Levels in Serum

The administration of LPS elevated the levels of liver marker enzyme ALT and AST in the serum when compared to the vehicle control rats and the values are found to be statistically significant with a p-value less than 0.001. The administration of PDF to LPS-treated rats significantly reduced the levels of ALT and AST when compared to the LPS control rats. The result of PDF-treated rats is comparable with the dexamethasone ($p < 0.001$) treated rats. The results of the effect of PDF on serum ALT and AST levels are shown in Figure 5.

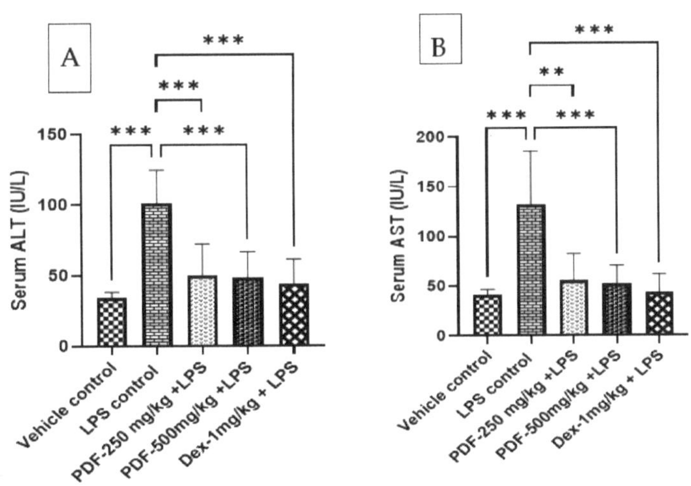

Figure 5. Effect of *Phoenix dactylifera* fruit extracts on serum (**A**) Alanine transaminase and (**B**) Aspartate transaminase levels in lipopolysaccharide treated rats. Numerical values are represented as mean ± SEM (n = 6), ** $p < 0.01$ and *** $p < 0.001$.

2.5. Tissue Biochemical Parameters

2.5.1. Effect of PDF on Oxidative and Nitrative Stress

The administration of LPS significantly elevated the levels of LPO ($p < 0.001$) and nitrite ($p < 0.001$) and reduced GSH ($p < 0.01$) in the rat brain tissue when compared to the vehicle control group. Treatment with both doses of PDF to LPS-treated rats restored the levels of LPO ($p < 0.001$), nitrite ($p < 0.001$) and GSH ($p < 0.001$) when compared to the LPS-control rats. These results are comparable with the dexamethasone ($p < 0.001$) treated rats. The effect of PDF on oxidative and nitrative stress in LPS-treated rats is shown in Figure 6.

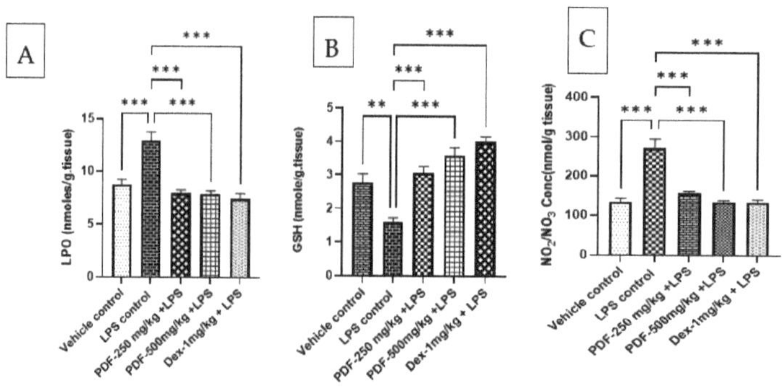

Figure 6. Effect of *Phoenix dactylifera* fruit extracts on oxidative and nitrative stress parameters in lipopolysaccharide treated rats, (**A**) Lipid peroxidation, (**B**) Reduced glutathione and (**C**) Nitric oxide metabolite. Numerical values are represented as mean ± SEM (n = 6), ** $p < 0.01$ and *** $p < 0.001$.

2.5.2. Effect of PDF on TNF-α and IL-6 Levels

The inflammatory markers TNF-α and IL-6 levels were significantly ($p < 0.001$) elevated in the brain tissue homogenate when compared to the vehicle control rats. The administration of PDF to LPS-treated rats reduced the levels of TNF-α and IL-6. The values of PDF at the dose of 250 mg/kg and 500 mg/kg found statistically significant TNF-α ($p < 0.001$ and $p < 0.01$) and IL-6 ($p < 0.05$ and $p < 0.05$) when compared to the LPS-control rats. Dexamethasone treatment also significantly ($p < 0.01$) reduced the levels of TNF-α and IL-6 when compared to the LPS control rats. The results of the effect of PDF on TNF-α and IL-6 levels are shown in Figure 7.

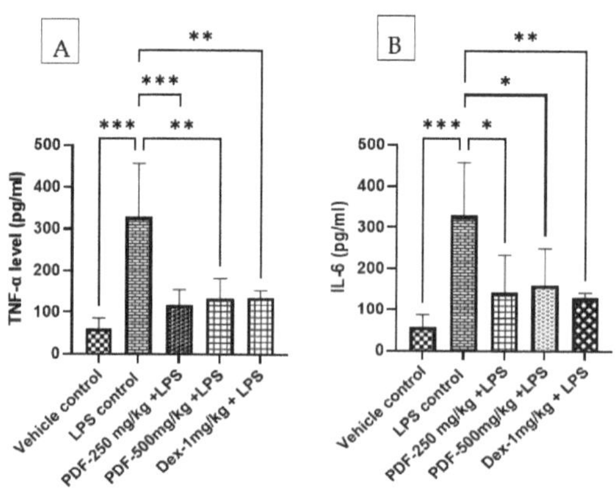

Figure 7. Effect of *Phoenix dactylifera* fruit extracts on (**A**) Tumor necrosis factor α and (**B**) Interleukin-6 in lipopolysaccharide treated rats. Numerical values are represented as mean ± SEM (n = 6), * $p < 0.05$, ** $p < 0.01$ and *** $p < 0.001$.

3. Discussion

In sickness behaviour, the soluble proteins secreted at the site of infection or injury make endocrine, autonomic, and behaviour changes in the victims [17–19]. The soluble proteins, interleukin (IL)-1, IL-6, and tumour necrosis factor-α (TNF-α) are essential pro-inflammatory cytokines that are released by activated immune cells such as macrophages and dendritic cells [9,17]. Controlling the immune system and coordinating cell-mediated immune responses depends heavily on the secreted soluble proteins [17].

Pro-inflammatory cytokines connect with the brain to alter behaviour in addition to controlling the localised inflammation [18,19]. Moreover, it stimulates the brain to release additional pro-inflammatory cytokines [18–20]. Mounting evidence suggests the involvement of pro-inflammatory cytokines IL-1, IL-6 and TNF-α in sickness behaviour [21,22]. Data from the current study convincingly confirm the aforementioned conclusions, showing that the brain tissue of LPS control mice has higher concentrations of IL-6 and TNF-α. These findings are consistent with the previously published reports on LPS-induced sickness behaviour in animals [22–24]. PDF attenuated IL-6 and TNF-α levels in LPS-treated rats. Moreover, a decrease in food and water intake is one of the primary signs of immune system activation, which is the result of raised cytokine levels [23]. These conclusions are supported by the findings of the current investigation. There was a drastic reduction in water and food intake in the LPS-treated rats compared to the vehicle control. PDF treatment restored the water and food intake in the LPS-treated rats. This may be due to reduced cytokine levels in the treated rats.

The brain creates significant quantities of peroxides and reactive oxygen species (ROS) as a result of a rapid inflammatory response to LPS [25,26]. The production of ROS is the primary cause of oxidative stress and is a significant risk factor for a variety of neuropsychiatric diseases [22]. In the present study, increased levels of MDA and nitrite and reduced levels of GSH in the LPS-treated animals support the above findings. Both doses of PDF reduced the restored GSH levels and attenuated the increased levels of MDA and nitrite in the LPS-treated rat brain. These results support the previously published research data on the antioxidant properties of Ajwa dates [7,8].

Reduced mobility and depressed behaviours are the characteristic features of LPS-induced illness behaviour [22]. Administration of LPS resulted in decreased locomotor activities, altered despair behaviour and increased immobility in tail suspension test in rats. Treatment with PDF at both the tested doses attenuated the behavioural alteration in the LPS-treated rats. These results support the protective effect of PDF against LPS-induced behavioural changes in animals. These results are in agreement with the recent study on the neuroprotective activities of Ajwa seed extract in toxin-induced neuronal insults in animals [27].

LPS is also known to cause hepatic injury in rodents, hepatocytes contain the liver enzymes AST and ALT, which are released from the hepatocytes when the liver is injured [28]. It was discovered that abnormalities in cerebral neurotransmission that lead to sickness behaviours are connected to inflammatory liver injury [29,30]. In line with the aforementioned findings, higher levels of ALT and AST were seen in the serum of LPS-treated rats in the current investigation. Nevertheless, giving PDF to LPS-treated rats reduced the levels of ALT and AST in the treated rat serum, showing that PDF had a protective effect against the hepatic damage that LPS had caused in the rats. This demonstrates how PDF protected the rats from LPS-induced sickness behaviour.

According to the behavioural and biochemical parameters of the current study findings, PDF protects rats from LPS-induced sickness behaviour. It was evident by its attenuation of behavioural changes, restoration of antioxidant levels, reduction of oxidative and nitrative stress, reduction of proinflammatory cytokines, restoration of water and food intake and attenuation of ALT and AST in LPS-treated animals. These findings are in agreement with research data published on Ajwa dates as neuroprotective and beneficial in memory improvement in neuro-compromised animals [27,31].

The standardisation of PDF revealed that it has a considerable amount of total phenolic and flavonoid content and quercetin. The positive effects of PDF on LPS-induced sickness behaviour in rats could be attributed to these active ingredients.

4. Material and Methods

4.1. Drug and Chemicals

Ajwa dates were purchased from the local market of Al Dawadmi, Kingdom of Saudi Arabia. Lipopolysaccharide (Product No-L2630, CAS No. 93572-42-0) and 5′-dithiobis-2-nitrobenzoic acid (Product No-D218200 CAS Number: 69-78-3) (Sigma–Aldrich, St. Louis, MI, USA), dexamethasone (Cadila Healthcare Ltd., Ahmedabad, India). Trichloroacetic acid and thiobarbituric acid were purchased from Loba Chemical Pvt. Ltd. in Mumbai, India. The remaining substances, including solvents, were all of the analytical grade.

4.2. Animals

Male adult Wistar rats (200–250 g) were purchased from King Abdul Aziz University's, Faculty of Pharmacy's Animal Centre, Jeddah, Saudi Arabia. The rats were housed on a 12-h light/dark cycle at 26 °C, and unlimited access to water and food. All animals were handled and cared for according to the guidelines of research on living organisms and its regulations and the rules governing ethics of scientific research at the University of Shaqra, Kingdom of Saudi Arabia. The Scientific Research Ethics Committee (ERC SU 20220071) of Shaqra University in Shaqra, Saudi Arabia, gave its approval to the study.

4.3. Extract Preparation

The fruits of Ajwa date were stripped of their pits, allowed to dry at room temperature, and then powdered in a stainless-steel blender. The powder of Ajwa date fruits was extracted with methanol and water (4:1, v/v) for 5 h using an orbital shaker. The obtained extract was filtered and centrifuged for 10 min at 4000× g. Then the supernatant was concentrated at 40 °C under reduced pressure for 3 h using a rotary evaporator. The crude extract of Ajwa date fruits (PDF) was stored in a freezer in a dark glass bottle until its use.

4.4. Standardisation of PDF

4.4.1. Total Phenolic Content and Flavonoid Content Determination

Total phenolics were determined in extract using the Folin-Ciocalteu method [32]. The amount of phenolic contents in each gram of extract was represented as mg gallic acid. The $AlCl_3$ method was used for the determination of the total flavonoid content in the extracts [33]. The amount of flavonoid in each gramme of extract was represented as mg quercetin.

4.4.2. Estimation of Quercetin Using HPLC

The HPLC chemo profiling of methanolic extract of Ajwa date was carried out using the Shimadzu HPLC system (Kyoto, Japan) equipped with dual pump LC-20AD binary system, photodiode array (PDA) detector SPD-M20A, RP C_{18} column (5 μm, 4.6 × 250 mm). Standard quercetin (10 mg) was taken in a 50 mL volumetric flask, dissolved, and diluted (standard stock). Further, 5.0 mL of the above solution was diluted to 10 mL with diluent. Exactly 10.54 mg of methanolic extract of Ajwa date was weighed, taken in a 50 mL volumetric flask (0.2108 mg/mL), dissolved and diluted to volume with diluent and filtered through a 0.45-micron syringe filter.

Isocratic elution was performed using 70% methanol, 30% milli-q water and 0.1% trifluoroacetic acid. The flow rate and the injection volume were set at 1.0 mL/min and 20 μL, respectively. The chromatograms at 254 nm were analysed and compared.

4.5. Experimental Grouping and Treatment

The male adult Wistar rats were weighed and randomised into five groups containing six animals each (n = 6). Group-1 (vehicle control) and Group-2 (LPS control) were orally administered with distilled water. Group-3 and Group-4 were orally treated with

PDF (250 and 500 mg/kg, respectively) and Group-5 was injected with dexamethasone (1 mg/kg, i.p.). All these treatments were given daily for 14 days.

On the 14th day 1 h after the above-mentioned treatment animals from Group-1 were intraperitoneally injected with normal saline and Groups 2–5 were challenged with 1 mg/kg of LPS (i.p.).

After 24 h of saline or LPS injection, animal food, and water consumption was monitored. The behavioural parameters such as locomotor activity, tail suspension test, and forced swim test, were conducted. Blood samples were collected by retroorbital plexus and used for the estimation of ALT and AST levels. Following a cervical dislocation, rats were killed and their complete brains were collected for biochemical analysis.

4.6. Behavioural Parameters

4.6.1. Locomotor Activity Scores

The locomotor activity scores of the animals were recorded as per the method described by Shaikh et al. (2016). To record the locomotor activity scores each animal was separately placed in the activity cage for 5 min and the score displayed on the box was recorded [23].

4.6.2. Tail Suspension Test

In the tail suspension test, individual rats were hung 50 cm from the ground for 6 min from the tip of their tails using adhesive tape. Immobility time was measured during the last 5 min by a video camera and recorded in seconds [34].

4.6.3. Despair Behaviour Test

The despair behaviour of the rats was measured by the forced swim test. Rats were individually placed into the open cylindrical transparent plastic container (diameter 30 cm, height 50 cm) filled with water (25–27 °C) to a depth of 30 cm. Rats were made to swim for 6 min, then were returned to their plastic cages. The duration of immobility during the test was videotaped as the absence of movement required to keep the head above the water or climb up against the wall during the last five min [35].

4.7. Estimation of ALT and AST Levels in Serum

The blood samples were centrifuged at 4500 rpm for 15 min, serum was separated and used for the estimation of ALT and AST by using commercially available diagnostic kits (Quimica Clinica Aplicada S.A. QCA, Tarragona, Spain).

4.8. Biochemical Estimations in Brain Tissue

4.8.1. Tissue Preparation

Brain tissues were dissected, and the hippocampus was separated on an ice-cold surface. Using a tissue homogenizer, a 10% homogenate was prepared in 0.1 M of ice-cold, pH 7.4 phosphate-buffered saline, and centrifuged at 10,000 rpm for 15 min at 4 °C [36,37]. The resulting supernatant was used to estimate biochemical parameters such as lipid peroxidation (LPO), reduced glutathione (GSH), nitrite, IL-6 and TNF-α.

4.8.2. Oxidative and Nitrative Stress Parameters

Lipid peroxide formation was analysed by measuring the thiobarbituric-acid reactive substances in the form of malondialdehyde (MDA) content in the homogenate. The MDA contents are expressed in nmoles/g tissue [38]. GSH levels were determined by Ellman (1959) method and expressed as mmol/mg of tissue [39]. The nitrite content in the supernatant was estimated using Griess reagent by recording the absorbance at 540 nm using a microplate reader (Biotek, Shoreline, WA, USA). The nitrite levels are expressed in nmol/g tissue.

4.8.3. Estimation of IL-6 and TNF-α Levels

TNF-α and IL-6 were estimated using specific ELISA kits (GENLISA, Rat IL-6 ELISA Cat: KB3068, Lot No RI60921; GENLISA, Rat TNF-α ELISA Cat: KB3145, Lot No RTA0122;

(Krishgen BioSystems, Mombai, India) as per the instructions of the kit manufacturer. The absorbance is recorded at 450 nm using a microplate reader (Biotek, USA).

4.9. Statistical Analyses

Graphpad Prism 8.0 (Graphpad Software Inc., Boston, MA, USA) was used for all Statistical analyses. All data were represented as mean ± standard error of the mean. One-way ANOVA followed by Tukey's test was used for analysis and $p < 0.05$ was considered statistically significant.

5. Conclusions

The results of the current study suggest that Ajwa dates have a protective effect against LPS-induced sickness behaviour in rats. Ajwa dates are thought to have health benefits because of their anti-inflammatory and antioxidant properties, and the presence of active ingredients such as quercetin. To clarify its molecular mechanism and validate its effects on human patients, more research is necessary.

Author Contributions: Conceptualisation, T.B.S. and G.A.; Methodology, T.B.S. and M.C.; Software, S.R.D.; Validation, T.B.S., M.D.A. and K.A.A.; Investigation, T.B.S.; Resources, G.A.; Data Curation, T.B.S.; Writing—Original Draft Preparation, G.A.; Writing—Review and Editing, T.B.S.; Visualisation, G.A.; Supervision, T.B.S.; Funding Acquisition, T.B.S. All authors have read and agreed to the published version of the manuscript.

Funding: The authors extend their appreciation to the Deputyship for Research and Innovation, Ministry of Education in Saudi Arabia, for funding this research work through project number: IFP2021-078.

Institutional Review Board Statement: All animals were handled and cared for according to the guidelines of research on living organisms and its regulations and the rules governing ethics of scientific research at the University of Shaqra, Kingdom of Saudi Arabia. The Scientific Research Ethics Committee (ERC SU 20220071) of Shaqra University in Shaqra, Saudi Arabia, gave its approval to the study.

Informed Consent Statement: Not applicable.

Data Availability Statement: All data generated or analyzed during this study are included in this published article.

Acknowledgments: Shivsharan B. Dhadde, Department of Pharmacology, Krishna Institute of Pharmacy, Krishna Vishwa Vidyapeeth (Deemed to be University), Karad, Maharashtra (India), is gratefully acknowledged for manuscript proofreading and editing.

Conflicts of Interest: The authors declare no conflict of interest.

References

1. Khan, F.; Khan, T.J.; Kalamegam, G.; Pushparaj, P.N.; Chaudhary, A.; Abuzenadah, A.; Kumosani, T.; Barbour, E.; Al-Qahtani, M. Anti-cancer effects of Ajwa dates (*Phoenix dactylifera* L.) in diethylnitrosamine induced hepatocellular carcinoma in Wistar rats. *BMC Complement. Altern. Med.* **2017**, *17*, 418. [CrossRef]
2. Alshwyeh, H.A. Phenolic profiling and antibacterial potential of Saudi Arabian native date palm (*Phoenix dactylifera*) cultivars. *Int. J. Food Prop.* **2020**, *23*, 627–638. [CrossRef]
3. Aljuhani, N.; Elkablawy, M.A.; Elbadawy, H.M.; Alahmadi, A.M.; Aloufi, A.M.; Farsi, S.H.; Alhubayshi, B.S.; Alhejaili, S.S.; Alhejaili, J.M.; Abdel-Halim, O.B. Protective effects of Ajwa date extract against tissue damage induced by acute diclofenac toxicity. *J. Taibah Univ. Med. Sci.* **2019**, *14*, 553–559. [CrossRef] [PubMed]
4. Bassem, Y.S.; Wael, M.E.; Abdulrahman, H.S.; Bassem Yousef, S.; Al-Moalim, M.A.B.L. Ajwa dates as a protective agent against liver toxicity in rat. *Eur. Sci. J. ESJ* **2014**, *10*, 358–368. [CrossRef]
5. Sabbah, M.; Alshubali, F.; Baothman, O.A.S.; Zamzami, M.A.; Shash, L.S.; Hassan, I.A. Cardioprotective Effect of Ajwa Date Aqueous Extract on Doxorubicin-Induced Toxicity in Rats. *Biomed. Pharmacol. J.* **2018**, *11*, 1521–1536. [CrossRef]
6. Pujari, R.R.; Vyawahare, N.S.; Kagathara, V.G. Evaluation of antioxidant and neuroprotective effect of date palm (*Phoenix dactylifera* L.) against bilateral common carotid artery occlusion in rats. *Indian J. Exp. Biol.* **2011**, *49*, 627–633.
7. Alqarni, M.M.M.; Osman, M.A.; Al-Tamimi, D.S.; Gassem, M.A.; Al-Khalifa, A.S.; Al-Juhaimi, F.; Mohamed Ahmed, I.A. Antioxidant and antihyperlipidemic effects of Ajwa date (*Phoenix dactylifera* L.) extracts in rats fed a cholesterol-rich diet. *J. Food Biochem.* **2019**, *43*, e12933. [CrossRef]

8. Zhang, C.-R.; Aldosari, S.A.; Vidyasagar, P.S.P.V.; Nair, K.M.; Nair, M.G. Antioxidant and Anti-inflammatory Assays Confirm Bioactive Compounds in Ajwa Date Fruit. *J. Agric. Food Chem.* **2013**, *61*, 5834–5840. [CrossRef]
9. Turner, M.D.; Nedjai, B.; Hurst, T.; Pennington, D.J. Cytokines and chemokines: At the crossroads of cell signalling and inflammatory disease. *Biochim. Biophys. Acta BBA Mol. Cell Res.* **2014**, *1843*, 2563–2582. [CrossRef]
10. Siddiqui, S.; Ahmad, R.; Khan, M.A.; Upadhyay, S.; Husain, I.; Srivastava, A.N. Cytostatic and Anti-tumor Potential of Ajwa Date Pulp against Human Hepatocellular Carcinoma HepG2 Cells. *Sci. Rep.* **2019**, *9*, 245. [CrossRef]
11. Al-Shahib, W.; Marshall, R.J. The fruit of the date palm: Its possible use as the best food for the future? *Int. J. Food Sci. Nutr.* **2003**, *54*, 247–259. [CrossRef] [PubMed]
12. Rahmani, A.H.; Aly, S.M.; Ali, H.; Babiker, A.Y.; Srikar, S.; Khan, A.A. Therapeutic effects of date fruits (*Phoenix dactylifera*) in the prevention of diseases via modulation of anti-inflammatory, anti-oxidant and anti-tumour activity. *Int. J. Clin. Exp. Med.* **2014**, *7*, 483–491.
13. Alotaibi, G.H.; Shivanandappa, T.B.; Chinnadhurai, M.; Reddy Dachani, S.; Dabeer Ahmad, M.; Abdullah Aldaajanii, K. Phytochemistry, Pharmacology and Molecular Mechanisms of Herbal Bioactive Compounds for Sickness Behaviour. *Metabolites* **2022**, *12*, 1215. [CrossRef] [PubMed]
14. Wilsterman, K.; Alonge, M.M.; Ernst, D.K.; Limber, C.; Treidel, L.A.; Bentley, G.E. Flexibility in an emergency life-history stage: Acute food deprivation prevents sickness behaviour but not the immune response. *Proc. R. Soc. B Biol. Sci.* **2020**, *287*, 20200842. [CrossRef] [PubMed]
15. Maes, M.; Berk, M.; Goehler, L.; Song, C.; Anderson, G.; Gałecki, P.; Leonard, B. Depression and sickness behavior are Janus-faced responses to shared inflammatory pathways. *BMC Med.* **2012**, *10*, 66. [CrossRef] [PubMed]
16. Cunningham, C.; Maclullich, A.M. At the extreme end of the psychoneuroimmunological spectrum: Delirium as a maladaptive sickness behaviour response. *Brain Behav. Immun.* **2013**, *28*, 1–13. [CrossRef]
17. Prather, A.A. Sickness Behavior. In *Encyclopedia of Behavioral Medicine*; Gellman, M.D., Turner, J.R., Eds.; Springer New York: New York, NY, USA, 2013; pp. 1786–1788.
18. Eisenberger, N.I.; Moieni, M.; Inagaki, T.K.; Muscatell, K.A.; Irwin, M.R. In Sickness and in Health: The Co-Regulation of Inflammation and Social Behavior. *Neuropsychopharmacology* **2017**, *42*, 242–253. [CrossRef]
19. McCusker, R.H.; Kelley, K.W. Immune-neural connections: How the immune system's response to infectious agents influences behavior. *J. Exp. Biol.* **2013**, *216*, 84–98. [CrossRef]
20. Dantzer, R.; BluthÉ, R.-M.; Castanon, N.; Kelley, K.W.; Konsman, J.-P.; Laye, S.; Lestage, J.; Parnet, P. Chapter 14—Cytokines, Sickness Behavior, and Depression. In *Psychoneuroimmunology*, 4th ed.; Ader, R., Ed.; Academic Press: Burlington, MA, USA, 2007; pp. 281–318.
21. Shattuck, E.C.; Muehlenbein, M.P. Human sickness behavior: Ultimate and proximate explanations. *Am. J. Phys. Anthropol.* **2015**, *157*, 1–18. [CrossRef] [PubMed]
22. Tastan, B.; Arioz, B.I.; Tufekci, K.U.; Tarakcioglu, E.; Gonul, C.P.; Genc, K.; Genc, S. Dimethyl Fumarate Alleviates NLRP3 Inflammasome Activation in Microglia and Sickness Behavior in LPS-Challenged Mice. *Front. Immunol.* **2021**, *12*, 737065. [CrossRef]
23. Shaikh, A.; Dhadde, S.B.; Durg, S.; Veerapur, V.P.; Badami, S.; Thippeswamy, B.S.; Patil, J.S. Effect of Embelin Against Lipopolysaccharide-induced Sickness Behaviour in Mice. *Phytother. Res. PTR* **2016**, *30*, 815–822. [CrossRef] [PubMed]
24. Kinra, M.; Ranadive, N.; Mudgal, J.; Zhang, Y.; Govindula, A.; Anoopkumar-Dukie, S.; Davey, A.K.; Grant, G.D.; Nampoothiri, M.; Arora, D. Putative involvement of sirtuin modulators in LPS-induced sickness behaviour in mice. *Metab. Brain Dis.* **2022**, *37*, 1969–1976. [CrossRef] [PubMed]
25. Sadraie, S.; Kiasalari, Z.; Razavian, M.; Azimi, S.; Sedighnejad, L.; Afshin-Majd, S.; Baluchnejadmojarad, T.; Roghani, M. Berberine ameliorates lipopolysaccharide-induced learning and memory deficit in the rat: Insights into underlying molecular mechanisms. *Metab. Brain Dis.* **2019**, *34*, 245–255. [CrossRef] [PubMed]
26. Wang, H.; Meng, G.L.; Zhang, C.T.; Wang, H.; Hu, M.; Long, Y.; Hong, H.; Tang, S.S. Mogrol attenuates lipopolysaccharide (LPS)-induced memory impairment and neuroinflammatory responses in mice. *J. Asian Nat. Prod. Res.* **2020**, *22*, 864–878. [CrossRef]
27. Mani, V.; Arfeen, M.; Dhaked, D.K.; Mohammed, H.A.; Amirthalingam, P.; Elsisi, H.A. Neuroprotective Effect of Methanolic Ajwa Seed Extract on Lipopolysaccharide-Induced Memory Dysfunction and Neuroinflammation: In Vivo, Molecular Docking and Dynamics Studies. *Plants* **2023**, *12*, 934. [CrossRef]
28. Dasgupta, A. Chapter 5–Liver Enzymes as Alcohol Biomarkers. In *Alcohol and Its Biomarkers*; Dasgupta, A., Ed.; Elsevier: San Diego, CA, USA, 2015; pp. 121–137.
29. D'Mello, C.; Le, T.; Swain, M.G. Cerebral microglia recruit monocytes into the brain in response to tumor necrosis factoralpha signaling during peripheral organ inflammation. *J. Neurosci. Off. J. Soc. Neurosci.* **2009**, *29*, 2089–2102. [CrossRef]
30. Chen, J.; Liu, S.; Wang, C.; Zhang, C.; Cai, H.; Zhang, M.; Si, L.; Zhang, S.; Xu, Y.; Zhu, J.; et al. Associations of Serum Liver Function Markers with Brain Structure, Function, and Perfusion in Healthy Young Adults. *Front. Neurol.* **2021**, *12*, 606094. [CrossRef]
31. Mani, V.; Arfeen, M.; Sajid, S.; Almogbel, Y. Aqueous Ajwa dates seeds extract improves memory impairment in type-2 diabetes mellitus rats by reducing blood glucose levels and enhancing brain cholinergic transmission. *Saudi J. Biol. Sci.* **2022**, *29*, 2738–2748. [CrossRef]

32. McDonald, S.; Prenzler, P.D.; Antolovich, M.; Robards, K. Phenolic content and antioxidant activity of olive extracts. *Food Chem.* **2001**, *73*, 73–84. [CrossRef]
33. Olajirea, A.A.; Azeez, L.A. Total antioxidant activity, phenolic, flavonoid and ascorbic acid contents of Nigerian vegetables. *Afr. J. Food Sci.* **2011**, *2*, 22–29.
34. Steru, L.; Chermat, R.; Thierry, B.; Simon, P. The tail suspension test: A new method for screening antidepressants in mice. *Psychopharmacology* **1985**, *85*, 367–370. [CrossRef]
35. Porsolt, R.D.; Bertin, A.; Jalfre, M. Behavioral despair in mice: A primary screening test for antidepressants. *Arch. Int. Pharmacodyn. Ther.* **1977**, *229*, 327–336.
36. Nagakannan, P.; Shivasharan, B.D.; Thippeswamy, B.S.; Veerapur, V.P. Restoration of brain antioxidant status by hydroalcoholic extract of Mimusops elengi flowers in rats treated with monosodium glutamate. *J. Environ. Pathol. Toxicol. Oncol. Off. Organ Int. Soc. Environ. Toxicol. Cancer* **2012**, *31*, 213–221. [CrossRef]
37. Shivasharan, B.D.; Nagakannan, P.; Thippeswamy, B.S.; Veerapur, V.P. Protective Effect of Calendula officinalis L. Flowers Against Monosodium Glutamate Induced Oxidative Stress and Excitotoxic Brain Damage in Rats. *Indian J. Clin. Biochem. IJCB* **2013**, *28*, 292–298. [CrossRef]
38. Ohkawa, H.; Ohishi, N.; Yagi, K. Assay for lipid peroxides in animal tissues by thiobarbituric acid reaction. *Anal. Biochem.* **1979**, *95*, 351–358. [CrossRef]
39. Ellman, G.L. Tissue sulfhydryl groups. *Arch. Biochem. Biophys.* **1959**, *82*, 70–77. [CrossRef]

MDPI AG
Grosspeteranlage 5
4052 Basel
Switzerland
Tel.: +41 61 683 77 34

International Journal of Molecular Sciences Editorial Office
E-mail: ijms@mdpi.com
www.mdpi.com/journal/ijms

www.ingramcontent.com/pod-product-compliance
Lightning Source LLC
Chambersburg PA
CBHW041356210926
51850CB00013B/3094
* 9 7 8 3 7 2 5 8 2 7 6 7 1 *